HUMBOLDT'S COSMOS

Alexander von Humboldt
and the Latin American Journey That
Changed the Way We See the World

GERARD HELFERICH

GOTHAM BOOKS

GOTHAM BOOKS
Published by Penguin Group (USA) Inc.
375 Hudson Street, New York, New York 10014, U.S.A.
Penguin Group (Canada), 10 Alcorn Avenue, Toronto, Ontario, Canada M4V 3B2
(a division of Pearson Penguin Canada Inc.); Penguin Books Ltd, 80 Strand,
London WC2R 0RL, England; Penguin Ireland, 25 St Stephen's Green, Dublin 2,
Ireland (a division of Penguin Books Ltd); Penguin Group (Australia),
250 Camberwell Road, Camberwell, Victoria 3124, Australia (a division of Pearson
Australia Group Pty Ltd); Penguin Books India Pvt Ltd, 11 Community Centre,
Panchsheel Park, New Delhi - 110 017, India; Penguin Group (NZ), Cnr Airborne
and Rosedale Roads, Albany, Auckland, New Zealand (a division of Pearson
New Zealand Ltd); Penguin Books (South Africa) (Pty) Ltd, 24 Sturdee Avenue,
Rosebank, Johannesburg 2196, South Africa

Penguin Books Ltd, Registered Offices: 80 Strand, London WC2R 0RL, England

Published by Gotham Books, a division of Penguin Group (USA) Inc.
Previously published as a Gotham Books hardcover edition.

First trade paperback printing, March 2005
1 3 5 7 9 8 6 4 2

Gotham Books and the skyscraper logo are trademarks of Penguin Group (USA) Inc.

THE LIBRARY OF CONGRESS HAS CATALOGED THE
GOTHAM BOOKS HARDCOVER EDITION AS FOLLOWS:
Helferich, Gerard.
Humboldt's cosmos : Alexander von Humboldt and the Latin American journey
that changed the way we see the world / Gerard Helferich.
p. cm.
ISBN 1-592-40052-3 (hardcover : alk. paper)   1-592-40106-6 (pbk.)
1. Humboldt, Alexander von, 1769–1859—Travel—Latin America.   2. Scientific
expeditions—Latin America—History—19th century.   I. Title.
Q143.H9 A6 2004
508.8—dc22   2004042518

Printed in the United States of America
Set in New Century Schoolbook with Modern MT Extended
Designed by Sabrina Bowers

Praise for *Humboldt's Cosmos*:

"Helferich's lush and engaging biographical adventure tale . . . successfully re-creates the New World when it was still very novel to European eyes."
—*Publishers Weekly*

"Alexander von Humboldt is surely one of the greatest but least remembered figures in scientific history. This immensely readable book . . . brings him wonderfully alive, whether testing electric eels in a jungle pool, counting bites from tropical insects, or discussing the alpine flora of Mount Chimborazo. *Humboldt's Cosmos* opens a fascinating window into the world of science and exploration two hundred years ago."
—Janet Browne, author of *Charles Darwin: Voyaging* and *Charles Darwin: The Power of Place*

"A concise appreciation of [Humboldt's] personality and scientific significance. . . . Sure to satisfy reader curiosity about the explorer who had so many places named for him."
—Gilbert Taylor, *Booklist*

"Gerard Helferich's lively portrait of the German naturalist and adventurer 200 years after his five-year, 6,000-mile exploration of South America should rekindle Alexander von Humboldt's reputation in the 21st century."
—Sharon Wootton, *The Olympian* (Olympia, WA)

"Colorful, accessible. . . . The description of his expedition on Venezuela's wild Orinoco River brings alive the sudden fierce winds and downpours, the water filled with piranhas, banks crowded with crocodiles, boas, vipers, and jaguars, and air thick with mosquitoes, biting flies, and gnats. . . . Readers will enjoy becoming acquainted with this remarkable man."
—Jean B. Crabbendam, *Sunrise*

"[A] heroic narrative. . . . Reading like an adventuresome diary, the book carries the reader along the Latin American route taken by Humboldt as he explored the Amazon basin. . . . This very humanitarian and inquisitive scientist was a generalist who should still serve as a model of the resolute scholar."
—Rita Hoots, *Library Journal*

"Well-written . . . celebrates anew contributions by scientific genius and consummate early-19th-century adventurer Humboldt. . . . Highly recommended."
—M. Evans, *CHOICE*

Gerard Helferich worked in book publishing for twenty-five years as an editor and publisher. He has traveled extensively through Latin America following the path of Humboldt's journey, and currently lives in San Miguel de Allende, Mexico, and Yazoo City, Mississippi.

With love to Teresa,
who first felt Humboldt's spirit

Everything is interrelated.

—ALEXANDER VON HUMBOLDT

CONTENTS

PREFACE

Humboldt's Ghost

THE PAN-AMERICAN HIGHWAY SOUTH OF QUITO IS A highway in name only. Indifferently paved, its two unmarked lanes are potholed in some places and awash with mud in others. But it is Ecuador's principal thoroughfare, and today the road is clogged with buses and trucks. The average speed is about thirty miles per hour, far less in some places.

We are tracing the route that Prussian scientist and explorer Alexander von Humboldt followed in 1802, but the countryside has been transformed in the intervening centuries. Strung out along the road now are a remarkable collection of structures in various stages of construction and decay. Mostly plain, cinder-block buildings with corrugated metal roofs, they include the expected gas stations (every third sign seems to advertise a *vulcanizadora*, or tire repair shop; apparently flats are a growth industry on the Panamericana). There are also places to sleep and to eat, but Motel 6 and Taco Bell they are not. How are the rooms, I wonder, at El Hotel Primitivo, hidden behind its raw cinder-block wall? And what is the specialty of the house at Café de la Vaca, a squat white building painted with exuberant black spots?

Gradually, the land becomes more rural, closer to what Humboldt would have seen. Commercial buildings give way to modest houses made of the rough local brick, with kitchen gardens sprouting behind them. Holsteins graze in the fields, and domestic pigs forage at will. Then the volcanoes appear on the horizon. It was Humboldt who named this region "the Avenue of the Volcanoes," and one can see why. The peaks come in quick, snowcapped succession—Pichincha, Pasochoa, Atacazo, Corazón, Illiniza, Yana-urcú, Rumiñahui, Cotopaxi. Even Chimborazo is visible, some fifty miles to the south.

At the town of Lasso, our bus turns onto a narrow lane. A few hundred yards farther on, we make another left and enter a set of stone-and-iron gates. Built in 1580, La Ciénega is one of the great historic haciendas of Ecuador, with a provenance including some of the country's most prominent families. The original land grant from the king of Spain stretched from Quito to Ambato, a distance of some fifty miles, but the vagaries of economics and politics have reduced the holdings to thirty acres, and instead of operating as a plantation, today the hacienda earns its keep as a *hostería*. But even in its reduced present, one can glimpse its glorious past, when the hacienda was the stopping place of presidents and kings—and Alexander von Humboldt, who, having already completed the first extensive scientific exploration of the Amazon Basin, was in the process of doing the same for the Andes.

Beyond the magnificent eucalyptus allée, we pass a faded picket fence and circle a grand fountain. The impressive stucco house has three stories, whitewashed walls, and thick stone columns flanking the door. We disembark from the bus and enter, suddenly feeling underdressed in our muddy hiking boots and dusty fleeces. Inside, a wide central hall extends through the house to a lovely patio with cobbled walkways, formal flower-beds, and another fountain. In the hallway, on a pedestal against the wall, is a bronze bust of Humboldt. Depicting him in his later years, it captures his high forehead, wide mouth, and prominent nose. He has the tousled hair of an adventurer and the penetrating gaze of a scientist.

To the right of the door is a reception desk. Andrés, our guide on this hiking trip, scoops a stack of room keys from the counter and fans them for the group. One of us will be lucky, he announces in his charmingly accented English. Because one of these keys opens the Humboldt Suite, the set of rooms where the great explorer stayed in 1802 while exploring nearby Cotopaxi. Preserved much as it was in the baron's time, the suite is the largest, finest accommodation in the hacienda. But it is a mixed blessing, Andrés warns, for the rooms are said to be haunted by the baron's ghost. Though not burdened by a belief in ghosts, aristocratic German ones or otherwise, I feel an uncanny certainty as I examine the keys. The first to choose, I pluck the key marked 7 from Andrés's hand. I'm not surprised when he tells me that I have picked the Humboldt Suite.

Congratulating ourselves, my wife, Teresa, and I rush up the broad staircase with visions of a king-size bed, crisp sheets, and a luxurious bath. But as we open the ancient door, we see that the suite is not the den of luxury we had imagined. The first room is a cavernous parlor with faded pink-and-white-striped wallpaper, heavy colonial furniture, and dusty draperies. Beyond is the barrel-vaulted bedroom, sheathed in somber paneling. And as we step into the unheated chamber we are greeted by a mustiness that seems to predate the hacienda itself. No wonder the room is thought to be the province of ancient spirits.

After dinner, as we lie in bed reading with the covers pulled up against the Andean chill, the wide, low door separating the bedroom from the sitting room suddenly swings open with a creak worthy of Vincent Price. Teresa and I look at each other and laugh. The hacienda is over four hundred years old, after all. Who would expect the doors to be plumb? A little while later, we're still reading when my hiking pole jumps from the wall where I had set it—doesn't slide down in a languorous arc, mind you, but seems to leap away from the plaster as though called to attention by some unheard voice. We laugh again, but now with a self-conscious edge. And when the time comes to go to sleep, jaded New Yorkers though we are, we feel an irrational reluctance to turn out the light. We lie in the dark for a time, straining for strange noises, then eventually drift off—only to be awakened in the wee hours by unexplained voices coming from the steep tile roof outside our window.

That day we hike the barren páramo around Cotopaxi, the volcano that Humboldt pronounced "unclimbable." The sky is cobalt, and the sun, magnified by the high altitude and the low latitude, seems perilously near. Jutting through a ring of clouds, impossibly huge, is the mountain's snow-draped cone. Buried in ash, strewn with huge blocks of obsidian, cut by rivers of mud— all evidence of its tortured geologic past—the terrain below the volcano is forbidding but irresistible. Even today, two centuries after Humboldt's journey, it is country that begs to be explored.

In the evening, as we sip the traditional *lazos* in the bar of the hacienda with Andrés and his brother Nelson, the talk turns to

Humboldt. I'm struck by how knowledgeable they are about him—his itinerary, his scientific contributions, his liberal politics, even the speculations about his sexuality. Throughout Latin America, everyone knows Alexander von Humboldt, they tell us. He is a pan-national hero, like Simón Bolívar, with streets, schools, hospitals, even babies, named in his honor. The obvious affection is impressive, considering Humboldt visited this hemisphere for only five years, two long centuries ago.

When Teresa and I confess the previous night's events in the Humboldt Suite, Andrés and Nelson betray no surprise. Neither do they share our facetiousness. Many other guests have reported strange happenings in the rooms, they tell us. Nelson himself spent one sleepless night there, troubled by a foreboding presence, and now avoids them. I feel my puckish skepticism begin to slip—and we still have another night in the rooms ahead of us.

That evening, we take sleeping pills to forestall any further apparitions.

TODAY, HUMBOLDT'S SPIRIT is felt far from La Ciénega, and even beyond Latin America. From 1799 to 1804, Humboldt and his traveling companion Aimé Bonpland accomplished what has been called "the scientific discovery of the New World," blazing a six-thousand-mile swath through what is now Venezuela, Colombia, Ecuador, Peru, Mexico, and Cuba. The expedition was even longer and more ambitious than Lewis and Clark's renowned trek across North America, which began the year that Humboldt and Bonpland ended theirs. And, whereas Lewis and Clark enjoyed the backing of the United States government and were accompanied by a thirty-man corps of discovery, Humboldt financed his expedition himself, and he and Bonpland traveled alone except for local guides and friends they met along the way.

Beyond its seminal role in the exploration of the Americas, the journey shaped scientific history. Humboldt lived in an age when the interior of every one of the world's continents save their own was terra incognita to European naturalists. And of all these vast landmasses waiting to be explored, none was more wild or exotic than the mountains and jungles of South America. The list of Humboldt's discoveries there—in anthropology, botany, geography, geology, geophysics, oceanography, physiology, and zoology—would fill a college catalog.

The first scientists to explore the Amazon Basin extensively, Humboldt and Bonpland collected some sixty thousand botanical specimens throughout Latin America—including more than three thousand species unknown in Europe—and made the first inventory of native American plants. They also greatly enhanced naturalists' knowledge of exotic New World creatures such as the monkey, alligator, and electric eel. By becoming the first to systematically study the effects of physical factors like altitude and geology on plant life, Humboldt gave birth to a new branch of science known as plant geography. He revolutionized geology by helping to resolve the controversy over how new landmasses are created and volcanoes are formed. He was instrumental in focusing scientists' attention on the need for accurate, systematic data collection, and his meticulous observations of the atmosphere and seas laid the cornerstones of climatology, meteorology, and oceanography. A pioneer in geomagnetism, he confirmed that the earth's magnetic field changes with latitude, located the planet's magnetic equator, and was the first to observe magnetic storms. He literally remade the maps of Latin America by fixing the latitude and longitude of hundreds of places (including Lima, Acapulco, and Havana) and by charting the courses of the Orinoco, Negro, and Casiquiare rivers. He even solved the riddle of how the new continent had come to be called America, and not Columbia. Keenly sympathetic toward Native American peoples, he introduced Europe to the glories of the Inca and Aztec cultures and suggested that American Indians had originally migrated from Asia. In his later years, Humboldt was an unstinting supporter of scientific talent and an important early advocate of international scientific collaboration. Today, more places are named after Humboldt than any other figure in history, including eleven towns in the United States and Canada, a mountain range in Antarctica, and even a sea on the far side of the moon.

During his lifetime, Humboldt was universally recognized as a genius. But his tremendous influence extended far beyond his unparalleled success as a data collector. As a great popularizer of science, Humboldt tirelessly promoted the appreciation of nature, from both a rational and an aesthetic viewpoint. In this he was no doubt animated by his own love of natural history. But he was also moved by humanitarian and political concerns.

Untouched by our twenty-first-century ambivalence toward "progress," he saw the advancement of science as a purely positive force that would benefit all mankind. Scientific knowledge was "the common property of all classes of society," as he wrote in the Introduction to *Cosmos*, an equalizing influence that would augment national prosperity and advance the republican ideals that he held dear.

Moreover, throughout his life Humboldt championed a particular way of viewing the natural world, one that sought to cut through the apparent dissimilarities among phenomena in order to lay bare the underlying unity of all nature. The advancement of this science, which Humboldt (who wrote primarily in French) called *la physique générale* and considered "one of the most beautiful fields of human knowledge," became the great quest of his life. And that quest began in earnest in the wilds of South America.

But Humboldt's scientific genius explains only part of his tremendous influence. By combining a love of travel and a flair for danger with his passion for discovery, Humboldt became the very prototype of the scientific adventurer. Lugging their instruments and boxes of specimens across the continent, he and Bonpland slogged through unmapped jungles and over some of the tallest mountains in the world. Tracing the course of the Orinoco in native canoes, they barely escaped treacherous cataracts. Attempting to climb the volcano Chimborazo, they reached a height of over nineteen thousand feet, setting an altitude record that would stand for nearly three decades and inspiring scores of mountaineers who followed, including the climbers who eventually conquered the Alps and the Himalayas.

Humboldt's expedition through Latin America was one of the great journeys of history, and it was his spirit of adventure as much as his love of science that made the young Prussian such a compelling figure among his contemporaries. His exploits, reported in the American and European newspapers (based on letters sent to friends and family en route) enthralled readers the same way that the adventures of Robert Scott, David Livingstone, and Charles Lindbergh would captivate future generations. At the conclusion of the journey, President Thomas Jefferson, who had just launched Lewis and Clark on their own journey of exploration, entertained Humboldt at the White House and Monticello.

On his return to Paris, Humboldt was welcomed as an international celebrity, drawing crowds to his lectures at the Institut National and to exhibits of his botanical specimens at the Jardin des Plantes. He was invited to Napoleon's coronation gala in the Tuileries. His books, especially *Aspects of Nature* and the monumental *Cosmos*, were snatched up by eager readers and translated into many languages.

Writers such as Honoré de Balzac, Victor Hugo, Lord Byron, Gustave Flaubert, and François-René de Chateaubriand all expressed their admiration. "One can truly say he has no equal in information and lively knowledge," wrote his friend Goethe. "Whatever one touches he is everywhere at home and overwhelms one with intellectual treasures." Emerson was even more laudatory: "Humboldt was one of those wonders of the world, like Aristotle, like Julius Caesar, like the Admirable Crichton [Scots scholar James Crichton, renowned for his intellectual acumen], who appear from time to time as if to show us the possibilities of the human mind, the force and range of the faculties,—a universal man."

Humboldt's books inspired American artists Frederick Church and George Catlin to journey to South America to paint. Latin American intellectuals and revolutionaries acknowledged his inspirational role in the eventual liberation of the Spanish colonies, and Simón Bolívar stated, "Alexander von Humboldt is the true discoverer of South America." It has been said that in the first half of the nineteenth century, Humboldt's fame throughout Europe was second only to that of Napoleon himself. During those years, in the judgment of paleontologist and author Stephen Jay Gould, "Humboldt may well have been the world's most famous and influential intellectual."

Humboldt was well loved in the United States as well. When he died in 1859, the event was reported in all the New York newspapers. His obituary in the *Times* ran more than a full column. The *Tribune* wrote, "His fame belonged not only to Europe, but to the world." The *Herald* was even more effusive, splashing the obituary in the center of page one and lauding Humboldt as "one of the greatest men of this age or of any other." A decade

later, on the centenary of his birth, *The New York Times* devoted
the *entire front page* (plus a continuation) to a description of the
myriad festivities commemorating his legacy. Across the country,
Humboldt's hundredth birthday was a cause for celebration, as
speakers, citing his humanitarianism as well as his scientific
perspicacity, hailed Humboldt as a citizen of the world and a benefactor of all mankind.

But in the vast army of those who felt Humboldt's impact,
perhaps one stands out above the others. He was a young,
dreamy British naturalist who was so moved by Humboldt's accounts of his journey that he committed whole passages to memory
and longed to make a similar voyage one day. When he was offered a post aboard a ship of scientific discovery in 1831, the
young man quickly accepted, packing in his seabag his copy of
Humboldt's *Personal Narrative*. The ship was the *Beagle*, the
young man Charles Darwin. Throughout his own epic voyage,
Humboldt's text was his constant companion and guide. In *The
Voyage of the Beagle*, Darwin cited Humboldt no fewer than seventeen times. After his arrival in Brazil, he wrote, "I formerly admired Humboldt, now I almost adore him; he alone gives any
notion of the feelings which are raised in the mind on entering
the Tropics."

Darwin was indebted to Humboldt for more than just the itch
to travel. The German also inspired him to devote his life to science. In his autobiography, Darwin wrote, "During my last year
at Cambridge, I read with care and profound interest Humboldt's
Personal Narrative. This work and Sir J. Herschel's *Introduction
to the Study of Natural Philosophy* stirred up in me a burning
zeal to add even the most humble contribution to the noble structure of Natural Science. No one or a dozen other books influenced
me nearly so much as these two. I copied out from Humboldt long
passages on Tenerriffe, [*sic*] and read them aloud. . . ."

The Briton also made use of some of Humboldt's myriad data
in constructing his revolutionary theory of natural selection. In
marshaling his evidence in *On the Origin of Species*, Darwin
makes two arguments from nature. The first is that the (admit-
tedly incomplete) fossil record shows that older species have, over
huge expanses of time, been supplanted by newer species. This
conclusion stems from his reading in geology, including Charles
Lyell's monumental *Principles of Geology*, which he had also

packed in his seabag (and which itself borrowed from Humboldt's work).

Darwin's second argument draws more directly from Humboldt, who had previously shown that differences in climate alone were not enough to explain the diversity of species that were seen from place to place. After all, most continents had hot areas, cold areas, and areas in between. So why was the kangaroo found only in Australia and not in the deserts of North America? Why did apes live in Africa but not in the rain forests of South America? Darwin's solution was migration followed by long isolation: A species wandered into new territory, was cut off from its brethren, and over a vast period, breeding only among itself, evolved into what was eventually recognized as a unique species. Besides providing the conceptual starting point for this part of the argument, Humboldt, by discovering thousands of plant species unique to South America, also provided copious evidence of exactly the process that Darwin was describing. In the midst of this discussion, in fact, Darwin cited "the illustrious Humboldt" for his contribution.

But it wasn't just Humboldt's inspirational example and his powers of observation that had such an effect on Darwin. It was also his worldview. Like Humboldt, Darwin was a synthesizer, one of those iconic figures who propel science forward through their compulsion to create order (*cosmos*) out of the apparent disarray (*chaos*) of natural phenomena. The result in Darwin's case was *On the Origin of Species*, which, as biologist and writer Steve Jones points out, single-handedly propelled the science of biology from a collection of disparate facts into a "system of knowledge." Though the crucial insight of natural selection was Darwin's, the synthesizing impulse behind it owed a debt to Humboldt. In fact, considering the profound influence he exerted on the young Briton, it's arguable that without Humboldt there would have been no Darwin. Or, as Darwin himself put it: "I shall never forget that my whole course of life is due to having read and re-read as a youth [Humboldt's] *Personal Narrative*."

Yet, luminary that he was during his lifetime and immediately beyond, Humboldt's celebrity has been eclipsed over the past century and a half by his scientific successors. Instead of his all-encompassing *physique générale*, science today is the province of ever more narrowly focused specialists. Even the

xx   *Preface: Humboldt's Ghost*

Humboldt Current, the cold upwelling along the Pacific Coast of South America that he studied, is now apt to be called the Peru Current. Although many North Americans have a vague sense of Humboldt's name and a hazy recollection that he had something to do with Latin America and perhaps the Avenue of the Volcanoes, most would be hard pressed to give particulars—as I had been before visiting La Ciénega.

WHAT KIND OF MAN would travel thousands of miles to strike into a wilderness where no European had ever ventured before? How does a person develop such an eclectic, obsessive curiosity, ranging from the distribution of the dragon tree, to the origin of basalt, to the grammatical structure of ancient Indian languages? Was Humboldt driven by hubris and self-centeredness (as his brother, Wilhelm, and sister-in-law, Caroline, suspected), or by something more altruistic and heroic? Where did he find the incredible resilience to sustain such a journey for five years, and to accomplish so much during that period? And why was he showered with adulation during his lifetime and immediately after, yet all but forgotten in recent decades?

Over the course of my research I have come to see Humboldt as a unique commingling of the Enlightenment and the Romantic Era, of intellect and feeling, of contemplation and action. I have also come to see him as a surprisingly modern figure, with quirks and enthusiasms and concerns similar to ours. There is a great deal to admire in him—his intelligence and curiosity, certainly, and his reverence for nature and respect for cultures different from his own. Above all, there is his courage—not just the physical courage required on the journey but the even greater courage needed to leave home in the first place. The courage to give up a responsible profession and comfortable circumstances in order to chase one's dream, the courage to believe that it's never too late to reinvent oneself, to discard doubt and convention and to pursue one's true calling—the courage to become the hero of one's own life. It was perhaps his individuality and audacity, as much as his scientific prowess and his enlightened politics, that underlay Humboldt's tremendous popularity. Devouring his *Personal Narrative*, teachers, shopkeepers, clerks could also leave their everyday life behind and ship out for South

America, if only for a while. And perhaps they would return from that vicarious journey a bit changed—a little more adventurous, a little more trusting of their own internal compass. Along with Humboldt, I, too, felt "spurred on by an uncertain longing for what is distant and unknown, for whatever excited my fantasy: danger at sea, the desire for adventures, to be transported from a boring daily life to a marvelous world."

In the two centuries since Humboldt's visit, the former Spanish colonies have become independent nations. Slavery has been abolished. The population of metropolitan Mexico City, which Humboldt called "the City of Palaces," is approaching twenty million—and palaces aren't necessarily the first impression of the visitor. Yet many things remain unchanged. Snow-cloaked Chimborazo still towers above the Andean Highlands, and the coastal fog still shrouds Lima from May through October. Though Spanish rule is gone, the sad legacy of colonial misgovernment and malfeasance lingers. Areas of crushing poverty remain, especially among the Indians. Mexico City is still the largest metropolis in the New World, just as it was in Humboldt's day. Much colonial architecture has been preserved there and elsewhere, and Humboldt would instantly recognize today's zócalos in the Mexican capital, Lima, Quito, and many other cities. Mexico's La Valenciana silver mine, which once produced a fifth of the world's supply of the precious metal, is still in operation, and the fabulous eighteenth-century church built from its wealth, boasting three huge gold altars, still attests to the riches that came out of the earth on the backs of the native people. Humboldt's world may have passed, but tantalizing traces remain. And even today his spirit is very much alive throughout Latin America—in the respect for cultural heritage, in the striving for modernity, in the spirit of social progress.

The fact is that Humboldt helped to shape the world as we know it, and his influence is still felt around the globe, even where his name is not widely recalled. The product of a rich intellectual tradition stemming back to the ancient Greeks and encompassing such disparate titans of the Enlightenment as Francis Bacon, Isaac Newton, René Descartes, and Immanuel Kant, Humboldt passed that tradition to his own successors in science, including Charles Darwin, Albert Einstein, Max Planck, and

Edwin Hubble. Humboldt's *physique générale* is a link in the conceptual chain comprising such touchstone theories as evolution, relativity, quantum mechanics, and the Big Bang.

Even amid today's rampant scientific specialization, Humboldt's search for "the unity of nature" not only survives but thrives, still yielding some of the most provocative areas of contemporary investigation, such as superstrings, complexity, emergence, and the elusive "theory of everything." In his brilliant, ambitious book *Consilience*, E. O. Wilson takes the synthesizing impulse to the ultimate, arguing that all tangible phenomena, including human behavior and culture, are ultimately reducible to the laws of physics. "I have argued that there is only one class of explanation," he says. "It traverses the scales of space, time, and complexity to unite the disparate facts of the disciplines by consilience, the perception of a seamless web of cause and effect." Though Humboldt's approach to science differed from Wilson's in fundamental ways, the peripatetic Prussian undoubtedly would have applauded this latest effort to discover the ultimate "unity of nature."

HUMBOLDT'S COSMOS

ONE

Tegel

BORN IN 1769, ALEXANDER VON HUMBOLDT GREW UP IN one of the most exciting periods of maritime exploration that Europe had ever seen. Though nearly 250 years had passed since Ferdinand Magellan had set off on mankind's first voyage around the world, by the middle of the eighteenth century the Pacific Ocean was still a vast unknown. But with the end of the Seven Years' War in 1763, France and Great Britain determined to carry their political, scientific, and commercial rivalry to the farthest corners of the earth. By 1800, they had dispatched half a dozen voyages of circumnavigation, in what would become known as "the Second Great Age of Discovery."

Like their predecessors in the fifteenth and sixteenth centuries (the so-called Great Age of Exploration), these new expeditions were launched primarily in the hope of financial gain from expanding overseas trade. But they were also products of the decidedly rational, secular tenor of the time, and they differed from the earlier voyages in two important respects. First, unmotivated by religious zeal, they did not seek to make converts to any faith. And second, they were undertaken partly out of scientific curiosity, in order to advance discoveries in natural history. Thus, the captains of this era took on board not only sailors and marines, but astronomers and naturalists as well.

The preeminent French explorer of the eighteenth century was the charming, brilliant Louis-Antoine de Bougainville, who, having fought with Montcalm in Canada during the Seven Years' War, became the first of his countrymen to circle the earth. Sailing from Europe in 1766 with the speedy *Boudeuse* and *L'Étoile*,

Bougainville called on Tahiti, which, following its recent discovery by Samuel Wallis, had already assumed its place as the prototypical Pacific island paradise. The party stopped at other islands, including Samoa, the New Hebrides, and the Solomons, skirted the Great Barrier Reef, then, barely surviving ferocious storms in the Coral Sea, returned to France three years later with a trove of natural history specimens, including the first marsupials ever seen in Europe. Bougainville was welcomed as a national hero, and his book, *Voyage autour du monde,* proved a huge best-seller.

But the greatest navigator of the era was the Englishman James Cook. Having abandoned a promising career commanding coal ships in the North Sea in order to join the Royal Navy at the advanced age of twenty-seven, Cook was known for his good nature, his solicitude toward his men, and his consummate skill at navigation. He specialized in charting coastlines, and his masterful survey of the St. Lawrence River was credited with allowing the British to take Quebec in 1759, during the Seven Years' War.

When Cook departed on his first voyage, his express purpose was to sail to Tahiti to observe a rare transit of Venus across the sun, which would allow a more precise calculation of the earth's distance from our star and provide other astronomical data useful to navigators. However, Cook was also charged with a more politically sensitive, strategic mission—to resolve the two most pressing issues in Pacific exploration. First, he was to determine whether the Great Southern Continent, posited by Ptolemy fifteen hundred years before, actually existed. Second, he was to ascertain whether New Zealand was a cape jutting out from New Holland (Australia) or an island in its own right.

Choosing the *Endeavour,* one of the stout but sluggish coal ships he knew well, Cook departed England in August 1768. After observing the transit of Venus from Tahiti, he sailed somewhat beyond 40 degrees south latitude, then, finding no Southern Continent, spent the next six months surveying New Zealand, which he proved to be made up of islands, and charting part of the eastern coast of Australia. He also penetrated the Great Barrier Reef, on which he very nearly lost his ship when it ran aground on the treacherous shoals. With the *Endeavour* repaired, Cook sailed through the Torres Strait, confirming that New Guinea was also not joined to Australia, then continued back to England.

Cook's first circumnavigation has been called the most successful exploration of the Pacific ever conducted, but there was still a great deal to be learned about the world's largest ocean. From 1772 to 1775, Cook made a second voyage in two more refitted coal ships, the *Resolution* and the *Adventure*, during which he became the first person to cross the Antarctic Circle. Making this circumnavigation at the lowest latitude ever attempted, he also disproved once and for all the existence of the Great Southern Continent, at least in any region habitable by humans. (Antarctica wouldn't be discovered until 1820, by the Britons William Smith and James Bransfield and the American Nathaniel Palmer.) On this voyage, Cook also discovered South Georgia and New Caledonia, among other islands of the South Pacific.

On his third and final voyage, Cook left England in 1776, again in command of the *Resolution* and the *Adventure*. This time he was charged with sailing up the Pacific Coast of North America in search of the supposed Northwest Passage; though he failed to find the elusive waterway, he did make one of his greatest discoveries, the Sandwich Islands (Hawaii), where in 1779 he was killed by indigenous people in a contretemps over a stolen boat. The *Resolution* and *Adventure* returned to England in October 1780.

Cook had explored the Pacific more thoroughly than anyone before him, and his discoveries were unparalleled. In fact, with their improved maps and tremendous trove of natural history specimens, Cook's voyages were far more successful from a scientific perspective than from a commercial one. Accompanied on his first expedition by famed naturalist Joseph Banks, as well as botanist Daniel Solander and two botanical illustrators, Cook returned to Europe with myriad new species of exotic plants and animals. On his second voyage, he had with him the renowned German naturalist Johann Forster and Forster's son Georg, as well as Swedish botanist Anders Sparrman and two astronomers. Forster's resulting book, *Observations Made During a Voyage Round the World*, was a masterful blend of natural history, anthropology, and travelogue and became an international bestseller, further spurring Europe's fascination with the exotic lands on the other side of the world.

Despite the myriad other achievements, Cook's greatest scientific breakthrough was against scurvy, the wasting disease that had been the bane of sailors for centuries; on his first and second voyages, he experimented with citrus fruits, sauerkraut, fresh vegetables, even grass, and, incredibly, didn't lose a single man to the illness, which is caused by a deficiency of vitamin C. It was actually for his conquest of scurvy, not his geographic discoveries, that Cook received the Royal Society's highest award, the Copley Medal, in 1776. Meanwhile, his naturalist, Joseph Banks, became a world-famous advocate of the sciences, serving as president of the Royal Society until his death in 1820.

BORN IN THE YEAR that Bougainville completed his circumnavigation, the young Alexander von Humboldt was thrilled by these great voyages of scientific discovery. Though he found himself in a landlocked country known for its mountains and mines—or perhaps because of that accident of birth—he had a passion for the sea from the youngest age, and he longed to sail to far-off places seldom visited by Europeans. As a boy, he would spend rainy afternoons poring over travel books and maps, retracing the routes of Bougainville and especially Cook, and letting his imagination wander over the exotic place names of the Southern Hemisphere. Later, after he became famous, Humboldt would admit that he had seen himself as a land-roving Captain Cook, bent on laying bare the interiors of the continents in the way Cook had mapped the Pacific.

There were not many other happy memories from those early years. Whereas his older brother, Wilhelm (later a famous linguist, educator, and diplomat, and principal founder of Germany's Humboldt University), was bright and outgoing, Alexander was quiet, moody, sickly. His tutors pronounced him slow.

Their father, Major Alexander Georg von Humboldt, came from an old Prussian family. Although family legend had it that their forebears had been noblemen, Alexander was never formally granted a title, and the "Baron" later affixed to his name was a simple courtesy; Wilhelm's children would be the first Humboldts officially elevated to the nobility, in 1875. The major, an adjutant during the Seven Years' War to the duke of Brunswick, who was also Alexander's godfather, was later appointed court chamberlain to the brother of Frederick the Great.

Thanks to their father's position, Alexander and Wilhelm made the acquaintance of scholars, courtiers, even royalty. Ac-

cording to one story, when the young Alexander was introduced
to Frederick the Great, the king asked him if he'd like to conquer
the world like his namesake, Alexander the Great of Macedonia.
Humboldt's prescient reply was: "Yes, Sire, but with my head."
The major hated court routine, however, and whenever he could
get away he would take Wilhelm and Alexander to roam their estate, Tegel, two hours by coach from Berlin, where they would inspect the mulberry trees, dig in the sandpits, and swim in the
cool, green lake. Those long days, bristling with the wonders of
nature and burnished with the companionship of his father and
brother, were the sweetest moments of Alexander's boyhood.

The major had married well. Alexander and Wilhelm's mother
was Maria Elisabeth von Hollwege, daughter of the Huguenot
Colomb family and a wealthy widow. The town house in Berlin
was a legacy from her mother, and Tegel was a bequest from her
first husband, as was another estate, Ringenwalde, on the banks
of the Oder. People often complimented her on her good looks and
her education. But while she was cool and reserved, her husband
was warm and outgoing. In some ways, it wasn't an ideal match.

In January 1780, when Alexander was ten and Wilhelm twelve,
came the crushing blow of their father's death. With him, it
seemed, all the warmth went out of the Humboldt household.
Burdened with a difficult son from her previous marriage, Frau
Humboldt determined to take a firm hand with Wilhelm and
Alexander. The boys' care was turned over to their tutors, who
drilled them in German, French, the classics, and history. Their
second tutor, Karl Sigismund Kunth, became a lifelong friend
and advisor, but even this friendship was no substitute for their
father's affection, and their mother wasn't able to close the emotional distance with her sons. It was a particularly painful time
for Alexander, who'd always been the more sensitive of the boys.
It was excruciating, he later wrote a friend, "to be living among
people who loved me and showed me kindness, but with whom I
had not the slightest sympathy, where I was subjected to a thousand restraints and much self-imposed solitude, and where I was
often placed in circumstances that obliged me to maintain a close
reserve and to make continual self-sacrifices." He and Wilhelm
clung to each other during this miserable time, and the brothers
became inseparable.

But even the gregarious Wilhelm was unable to break through

Humboldt's self-portrait, from 1814.
From *Life of Alexander von Humboldt* by Karl Bruhns (editor), 1872.

his brother's loneliness. "Happy he will never be," Wilhelm later wrote, "and never tranquil, because I cannot believe that any real attachment will ever steal his heart. . . . A veil hung over our innermost feelings which neither of us dared to lift." For his part, Alexander seemed to embrace his solitude. "Isolation has much in its favor," he later confided. "One learns thereby to search inwardly to gain self-respect without being dependent on the opinions of others."

Even more than to his older brother, Alexander turned for solace to the forests and fields he'd tramped so often with their father. "Nature can be so soothing to the tormented mind," he found, "—a blue sky, the glittering surface of lake water, the green foliage of trees, may be your solace. In such company, it is even possible to forget the reality of one's personal existence." Besides this transcendence, he experienced a sense of freedom in the woods, an ability to be himself, that he never felt under the sharp gaze of his mother and the tutors. It was during this period that he first sensed the great restlessness that would drive him forever after, and he spent many hours wandering the estate, exercising his considerable talent for sketching. He also collected rock, plant, and insect specimens, which he painstakingly classified, labeled, and displayed in his room in such profusion that he was mockingly known in the household as "the little apothecary."

Alexander grew into a good-looking young man of medium height and slight build. His eyes were blue and keen, his mouth wide and expressive. His nose was prominent and his chin strong, and his brown hair tumbled over his forehead, partially hiding the scars from a childhood bout with smallpox. By the age of sixteen, he had renounced his boyhood ambition of becoming a soldier like his father, and had determined to become a man of science. As with everything he did, he threw himself into these pursuits with a determined energy. Around this time, he carted his plant and rock specimens to the home of the renowned German botanist Karl Ludwig Willdenow, for the great man's approval and explication. Though only a few years older than Alexander, Willdenow was already renowned for his landmark book *Flora of Berlin*. It was from Willdenow that Alexander first heard the word *botany*.

Alexander also called on the famous etcher Daniel Chodowiecki, director of the Academy of Arts, to show him his sketches. Chodowiecki was sufficiently impressed to take him on as a student, providing important technical training for a budding naturalist in that prephotographic age. Alexander attended physics lectures by Marcus Herz, and even cajoled his mother into installing a lightning rod at Tegel, the second in all of Germany (after the University of Göttingen).

Despite her son's obvious passion for science, his mother was determined that Alexander would study finance and prepare for a career in government. When he turned seventeen, he and Wilhelm were sent to the Academia Viadrina in Frankfurt an der Oder, near the Polish border. Alexander took courses in law, medicine, and philosophy, none of which interested him. The school was chosen for its proximity more than its excellence, and he wrote to friends complaining of the inferior teaching: "The goddess of Knowledge has no temple here." Wilhelm, ever the student, immersed himself in his course work, and the two brothers, once inseparable, grew apart.

Alexander turned elsewhere for companionship. His closest

friend at Frankfurt was a theology student named Wilhelm Gabriel Wegener, and over the next few years, whenever they were separated, Alexander wrote Wegener passionate letters. "Since that February 13," went one note, "as we swore eternal brotherly love to each other, I feel that none of my other acquaintances can give me what you have for me. . . . My fervent love and sincere friendship for you are as imperishable as the soul that gives them birth." Another proclaimed: "When I measure the longing with which I wait for news from you, I am certain that no friend could love another more than I love you. When I recall all the signs of your friendship, I feel tormented in the thought that I don't love you as much as your sweet impressionable soul, your attachment for me, deserve."

These were not typical expressions of male friendship, even by the standards of eighteenth-century Europe. Indeed, Alexander himself was disturbed by such feelings, and he discovered that demanding intellectual work helped to scour them from his mind, at least temporarily. As he wrote Wegener, "Serious themes, and especially the study of nature, become barriers against sensuality." The following year, Alexander returned to Berlin, and the two boys gradually fell out of touch.

Several years later, Humboldt formed another intense attachment, this time to a young army officer named Reinhard von Haeften. In one letter to "my dearly beloved Reinhard," Humboldt professed, "I know that I only live through you, my good precious Reinhard, and that I can only be happy in your presence." Later, when von Haeften married, Humboldt wrote him in despair over the cooling of their friendship:

Two years have passed since we met, since your fate became mine. . . . I felt better in your company, and from that moment I was tied to you as by iron chains. Even if you must refuse me, treat me coldly with disdain, I should still want to be with you. . . . Never would I cease to remain attached to you, and I can thank heaven that I was granted before my death the grand experience of knowing how much two human beings can mean to each other. With each day my love and attachment for you increase. For two years I have known no other bliss on earth but your gaiety, your company, and the slightest

expression of your contentment. My love for you is not just
friendship, or brotherly love—it is veneration, childlike
gratefulness, and devotion to your will is my most
exalted law.

For Humboldt, the attachments to Wegener and von Haeften
were the beginning of a lifelong pattern of infatuation with and
subsequent estrangement from other men. And thus was Wilhelm's prediction, that no permanent attachment would ever
steal his brother's heart, confirmed—though perhaps not in exactly the way he'd meant.

Science would become Alexander's lifetime passion instead.
Burdened with an insatiable curiosity—about history, art, and
language, as well as physics, geology, and botany—he was always
working, never able to sit still. *Restless* was the word that family,
friends, and acquaintances applied to him time and again. As he
himself wrote, "I am anxious, agitated, unable to enjoy anything
I have finished, and never content except when undertaking
something new and doing three things at the same time." With
the condescension of an older brother, Wilhelm put it more pith-
ily: "Alexander maintains a horror of the single fact. He tries to
take in everything."

Alexander didn't always wear his learning lightly, and his
bustling preoccupation was sometimes seen as self-absorption.
In 1797, two years before Humboldt left for Latin America, the
great German poet and dramatist Johann Schiller wrote of his
young friend, "I am afraid that despite all his talents and restless activity he will never contribute much that is important for
science. There is a little too much vanity in all his doings. . . ."
And Wilhelm had written to his wife, Caroline, "[Alexander's] is a
lovable nature, yet he maintains very peculiar ideas for which I
really don't care. . . . He is a busybody, full of enterprise that to
others necessarily must seem like vanity. He peddles the wares
of his knowledge with much ado, as if he desperately needed either to dazzle people or to beg for their sympathy. . . . All this
wanting to impress others really stems from a desire to impress
oneself."

In spring of the monumental year 1789, Alexander was en-
rolled as a law student at Göttingen, which he entered after a

year in Berlin brushing up on his Greek and studying industrial processes as part of his training in economics. As the premier university in Germany, Göttingen was also the center of German scientific scholarship. Here, free from the backwater of Frankfurt an der Oder and immersed in the intellectual mainstream at last, Alexander found ample opportunity to pursue his scientific interests in addition to his set courses. It was also here that he was exposed to the writings of Kant, whose ideas had taken hold at the university. In fact, in later life Humboldt credited his time at Göttingen as the most influential period of his education. At the university he threw himself into his studies and did so well that he was accepted into the prestigious Philosophical Society, through which he met some of Europe's greatest naturalists. There were also lectures on anthropology, anatomy, art, and mythology. But even in this stimulating new environment, he wasn't able to escape completely the fits of melancholy and ill health he'd suffered throughout his youth.

The son-in-law of one of Alexander's professors at Göttingen happened to be Georg Forster, who as a teenager had accompanied his father on Cook's second voyage around the world. The younger Forster was now a best-selling author in his own right, and his books, such as the 1777 work *A Voyage Round the World*, were, like his father's writing, a highly readable mixture of scientific observation and travelogue. (In fact, it has been noted that Humboldt's *Personal Narrative* owes a serious stylistic debt to Forster's work.) Forster was now planning a trip across Europe to London, to search for a publisher for his next project and to try to straighten out some difficulties in his deceased father's affairs. For Alexander, who managed to invite himself along, his first journey across Europe would be a turning point in his life. In fact, in many ways, it was that journey that had set him on course for his great trek through Latin America.

In the spring of 1790, Forster and Alexander started down the Rhine, stopping at Cologne to admire the cathedral. From there they continued to Belgium, then, despite the revolution raging in France, went on to Dunkerque, where Alexander had his first, intoxicating glimpse of the sea. Hiking across Flanders, they narrowly skirted the French army, then arrived in Lille just in time to witness the rebellion against the monarchy. In London, they toured Westminster Abbey and the Houses of Parliament and paid a call on Sir Joseph Banks, the brilliant naturalist of Cook's first voyage, then continued across the island to Lancashire. During the journey, Alexander impressed Forster with his powers of observation and his endless drive. The great man was so taken with Humboldt's comments on Cologne Cathedral, in fact, that he later included them in his book *Sketches of the Lower Rhine*.

But the long journey also took its toll on Alexander's already tenuous health. Brimming with a nervous energy, determined to inspect every cavern and mine, tour every manufactory and monument, he exhausted himself. Forster, in whom he confided, attributed the young man's frailty to an "overexertion of the mind" caused by "certain people in Berlin" who were forcing him into a career he didn't want. Whatever the cause of Alexander's chronic illnesses, they certainly didn't bode well for a future of scientific exploration.

In July, Alexander and Forster recrossed the Channel and arrived in Paris, which was seething with revolutionary zeal. Forster longed to stay and throw himself into the battle for the rights of man. Alexander was also moved by the suffering of the people but abhorred the violence of the mob. Besides, there was no question of his remaining in France. In the autumn, his mother expected him to matriculate in the Hamburg School of Commerce.

The traveling companions returned to Mainz, on the Rhine, and said their farewells. Later, in 1792, the French occupied the city, and Forster became a member of the provisional government. But in 1795, while he was in Paris on official business, Mainz was recaptured by Prussian and Austrian troops, and Forster was denounced in Germany as a traitor. Unable to return home, he died in Paris that year, deeply disillusioned.

Alexander never saw the charming, imprudent Forster again, but he still felt indebted to his mentor. For during their months together, he had had the first, hazy glimpse of his future life. For one thing, he was deeply impressed by his direct experience of the French Revolution, which in those days before the descent into the Terror was greeted as proof of human progress and perfectibility; for the rest of his life, Humboldt would remain a staunch republican and an outspoken advocate for the oppressed and neglected.

In addition, as he listened night after night to the circumnavigator's adventures, Alexander was seized with a passion for everything having to do with the sea, and he burned with the Romantic's ambition to visit tropical lands. It was during this journey with Forster, Humboldt wrote in the *Personal Narrative*, that he determined to undertake a voyage of discovery to the New Continent—the journey that would be the pivotal event of his life and the basis of his international celebrity. Yet his mother had other ideas, and it was an ambition he doubted he could ever realize during her lifetime. Reluctantly, he returned to Berlin and prepared to take up his studies in business and political economy.

However, in those few weeks before leaving for Hamburg, it appears that Alexander also embarked on a very different course, intended to subvert his mother's plans for him. Seemingly without her knowledge, he applied for admission to the famous Bergakademie in the old silver-mining town of Freiberg, in Saxony. Founded in 1765 to provide practical training in assaying, engineering, and mineral science, the Mining Academy had developed into one of the premier schools of mineralogy in Europe. As such, it was an ideal training ground for a would-be scientific explorer.

Alexander made his application directly to the academy's director, Abraham Gottlob Werner, the premier geologist of the time and one of the principal founders of the science. Developing a whole new system of mineral identification based on color, hardness, texture, smell, and taste, Werner had provided a practical means of identifying minerals in the field and had brought order to a previously chaotic study. It was Werner's system that Humboldt would employ in Latin America and throughout his publications.

Werner had also developed a major theory of the earth's history. Before, although a good deal was known about minerals and rock formations, there was no generally accepted theory of the earth's processes. Most literate Europeans, in fact, still subscribed to the view that the earth was only six thousand years old, based on highly speculative calculations made from biblical sources. Although erosion had long been recognized as the process by which landforms were broken down, it was Werner, drawing on the earlier, theologically inspired Flood theory of John Woodward, who had put forward the leading explanation of how they had been built up. Known as neptunism, this Bible-compatible theory held that all minerals had precipitated out of a vast primordial ocean that had once covered the entire planet. As the ocean had receded, the landforms had been exposed. Since the earth had no molten core under this view, volcanoes were said to be caused by fires smoldering in underground coal beds, and lava was simply thought to be sedimentary rock that had been melted by these fires, then ejected.

For decades, neptunism had been geology's reigning paradigm. But in recent years, Werner's ideas had been challenged by the controversial Scots geologist James Hutton. Drawing on earlier ideas of Anton Moro, the amateur Hutton was advancing a radical new vision of the earth as a dynamic, self-regulating system that must have been at work far longer than the biblical span of six thousand years, and that, in Hutton's famous phrase, showed "no vestige of a beginning,—no prospect of an end." This theory, called vulcanism or plutonism, suggested that landforms were actually created by molten rock oozing from deep within the planet. Its fundamental principle, uniformitarianism, held, with startling simplicity, that the same geologic processes that formed the earth were still at work and still observable. As a Deist, Hutton believed that the universe functioned according to its own unchanging laws, without additional intervention by a supreme being. Landforms hadn't been sedimented out of a primordial ocean in a one-time act of creation but had been thrust up, and were still being thrust up, as a result of heat deep within the earth's core—in the same process seen in volcanoes. Although some rocks were undoubtedly formed by sedimentation, even there the earth's internal heat provided the energy to fuse the sediment into stone.

At the time Humboldt applied to the Mining Academy, the neptunist/vulcanist debate was still raging, and he was leaning toward the Werner camp. He apparently wasn't swayed by the religious argument for the neptunist theory, but Werner enjoyed tremendous prestige and personal charisma, and it was hard to refute such authority. Humboldt had previously published a paper defending the neptunist point of view, and, hoping to make a good impression, he now enclosed a copy along with his application to the Bergakademie.

But, always a reluctant correspondent and now overburdened with teaching and administrative duties, Werner by that point

had stopped opening his mail. Not to be put off, Humboldt wrote
again, but still there was no answer. Approaching the end of his
studies at Hamburg, Humboldt was growing increasingly anxious over his future. So in spring 1791, he went over Werner's
head and applied directly to Herr Heinitz, minister of industry
and mines and founder of the Mining Academy. Would it be possible, he wanted to know, to attend the academy for an abbreviated
period, then take up a position in the ministry? Heinitz was impressed with the application, and his acceptance came at the end
of May. Frau Humboldt approved the plan, with its promise of a
civil service position, not realizing that the Mining Academy was
only a step toward Humboldt's ultimate, unspoken goal of scientific exploration.

Humboldt arrived at Freiberg in June. Determined to make
the most of his limited time there, he immersed himself in his
studies, including geology, surveying, and mathematics, with a
dedication that bordered on obsession—six hours in the morning
for practical study underground, lectures all afternoon, and either plant collecting in the woods or studying in his room all
evening. Under this grueling schedule, he fell behind in his correspondence, neglected friends and family, even begged off Wilhelm's wedding. He'd never been so busy in his life, and his
less-than-robust health suffered as a consequence.

Wilhelm and his new wife, Caroline, grew concerned about
him. They also found him self-centered and overly ambitious.
Caroline confessed that at times she feared for her brother-in-
law's sanity. It was a concern Humboldt shared. "I work so hard,"
he wrote, "yet for all my efforts I realized how much needs to be
done. . . . I am frightfully busy, so that I cannot spare myself.
There is a drive in me that, at times, makes me feel as though I
am losing my mind." Now that he'd finally begun to train in
earnest for his life's work, Humboldt determined that family ties,
friendship, nothing, would deter him. And despite—or perhaps
because of—the overwhelming workload, he was very happy at
Freiberg.

Nine months after matriculation, at the age of twenty-three,
Humboldt completed his studies at the Mining Academy. A week
later he was appointed, at his own request, inspector of mines for
Bayreuth and the Fichtel Mountains, a geologically rich but badly
neglected area where a number of different ores were mined,

ranging from coal to gold. Seizing this opportunity to put his geology training into practice, Humboldt became an outstanding mine inspector, winning the confidence of the men, dramatically increasing production, inventing safety equipment such as a new type of lamp and a gas mask, and even financing out of his own salary the country's first training program for miners. He also had the opportunity to travel widely through Prussia and neighboring states on official business, which provided important field experience for a would-be scientific explorer.

In addition to his work for the Ministry of Mines, Humboldt somehow also found time to continue his investigations in the life sciences. One of his papers on plant distribution, "Freiberg Flora," won him a gold medal from the elector of Saxony and recognition from scholars across Europe. A cornerstone of the science known today as plant geography, the paper exhorted naturalists to give up their obsession with nomenclature and categorization (an obsession Humboldt had shared as a boy) and instead to study plants in all their complexity and diversity as they appear in nature. "Observations of individual parts of trees or grass is by no means to be considered plant geography," he'd written in the paper; "rather plant geography traces the connections and relations by which all plants are bound together among themselves." It was to discuss his own studies of plants that Goethe (himself a former inspector of mines), sought out Humboldt, and the two would remain close friends until the older man's death in 1832.

Around this time, inspired by Italian physician Luigi Galvani's research on "animal electricity," Humboldt began extensive experiments on nerve conduction, testing the effects of electrical impulses on himself and on animals. When this work was finally published—in two volumes detailing four thousand separate experiments—it created a sensation among European scientists. Published in 1797, the research still stands out as an important early effort in the field.

In January 1794 Humboldt was promoted to counselor of mines, and in February 1795 he was offered the position of director of mines in Silesia. But he had long ago determined that his ambitions lay elsewhere. "I am considering a complete change in my mode of life," he wrote in declining the promotion, "and I intend to withdraw from any official position with the state." He'd never planned to make a career with the ministry, he'd explained, only to get practical experience in geology in order to prepare for his true calling in scientific exploration. "As I have a deep conviction that such an expedition is highly important for increasing our knowledge of geology and physical science," he continued, "I am exceedingly eager to devote my energies immediately to this end." Yet the government was loath to lose its rising star in Industry and Mines, and Minister of State Hardenburg intervened personally to persuade Humboldt to accept the post of counselor at the Upper Court of Mines, with the proviso that he be permitted time off for travel. On those terms, Humboldt agreed to put off his resignation until February 1797.

It's not clear how Humboldt planned to finance his explorations or to reconcile his mother to his abrupt abandonment of a promising career. But in the event, he didn't have to. Frau Humboldt had been suffering horribly from breast cancer, and in the autumn of 1796 Alexander received word that she'd taken a turn for the worse. He rushed home to Berlin and was at her side when she died that November, at age fifty-five. He didn't feel stricken at her passing, he wrote later, but composed, since her agony had finally come to an end. In fact, he seems to have taken his mother's death with a sangfroid bordering on coldness. "My heart could not have been much pained by this event," he explained, "for we were always strangers to each other." His unhappy childhood memories didn't vanish with his mother's death, and for months afterward she appeared to him in dreams.

Under the terms of their mother's will, Wilhelm inherited Tegel, and Alexander inherited Ringenwalde. The brothers suddenly found themselves wealthy men, and Alexander's annual income was now six times his previous salary. Finally out from under his mother's financial and emotional shadow, he was free at last to plan his future without regard for anyone else's expectations. He completed his work with the ministry, as agreed, then traveled with Wilhelm and Caroline and their two small children first to Dresden and then to Vienna, with the idea of spending the winter in Italy. But when the outbreak of war made that impossible, Wilhelm and his family went on to Paris, while Alexander took off for the Tyrolean Alps with a friend from Freiberg, the brilliant but eccentric geologist Leopold von Buch.

"I was so pleased to see him," Humboldt wrote. "He is brilliant and an excellent observer but his manners—as if he came from the moon. . . . I have taken him with me to see people but it did not turn out well. Usually he puts on his glasses and goes and studies the cracks in the stove tiles on which he is dead keen, or he slinks round the walls like a hedgehog and looks at the skirting board. On his own he is an interesting and charming person— a treasure of knowledge from which I profit." Von Buch would later propose one of the most important scientific hypotheses of the nineteenth century, the theory of "craters of elevation," which suggested that mountains were formed when trapped gasses and molten rock caused the earth's crust to swell like a blister; volcanoes supposedly resulted when the blister popped. Adopted by both Humboldt and Darwin, the theory would become the main precept of volcano science for the next four decades.

That winter in the Tyrol, Humboldt took advantage of the opportunity to build his constitution and to practice his astronomical, topographical, and meteorological measuring skills, all in preparation for a career in scientific exploration. His and von Buch's measurements also resulted in many corrections to the maps of the region, including the exact latitude of Salzburg. Finally, in spring 1798, he arrived in Paris.

Paris was reclaiming its place as the cultural and intellectual capital of the world, and thanks to Caroline's salons and to his own growing reputation as a naturalist, Alexander had the opportunity to meet many prominent scientists of the day, including the zoologist Georges Cuvier, the chemist Claude-Louis Berthollet, and the astronomer Pierre-Simon Laplace (who taught Humboldt his new method of finding altitude by measuring changes in barometric pressure, a technique he would use countless times on his travels). He was invited to lecture at the Institut de France on nitrogen gas and on the applications of chemistry to agriculture, and, in light of his research on nerve conductivity, his opinion was solicited on the subject of galvanism. He lived with Wilhelm and Caroline during this period, but, utterly immune to Caroline's efforts to domesticate him, he barely stopped in for meals between visits to the observatory and the Jardin des Plantes.

Paris was an exciting, revelatory interlude, but he never forgot the reason he'd gone there—to find a voyage of scientific discovery. The search had consumed nearly a year. First the aged,

Aimé Bonpland.
Detail of a painting by E. Ender.
Courtesy Akademie der Wissenchaften, Berlin.

eccentric Lord Bristol offered him a place on his voyage to Egypt. Though Humboldt hadn't contemplated a journey to the Near East, he accepted, and planned to extend his travels afterward through Syria and Palestine. But the expedition was canceled after Napoleon's invasion of Egypt, and Bristol, an Englishman, was arrested for alleged intrigues against the French.

Then Humboldt's boyhood hero, the great Louis-Antoine de Bougainville, invited Humboldt to join his new circumnavigation via South America, the South Pacific, Africa, and perhaps even the South Pole. Humboldt quickly accepted, but the Directoire first replaced the seventy-one-year-old Bougainville with the younger but less impressive Nicolas Baudin, then postponed

the expedition for a year on account of the war—not to mention
the lack of three hundred thousand livres. After that, Humboldt
nearly despaired of escaping lovely, civilized Paris.

But at least one good thing had come of all that waiting in the
French capital: He had met Aimé Bonpland. Four years younger
than Humboldt, Bonpland was the son of a surgeon from La
Rochelle. Having trained as a physician, he was now in the city
to study botany and zoology, and he had also been chosen for
the Baudin expedition. Tall, with dark good looks, Bonpland was
clever and outgoing and shared Humboldt's progressive views on
science and politics.

After the postponement of Baudin's voyage, Humboldt and Bonpland resolved to find another expedition together. They traveled to
Marseilles, where they were to join a voyage to Algiers led by the
Swedish consul, a Mr. Skiöldebrand. But after waiting in vain for
two months in the French port, they discovered that their ship,
the frigate *Jaramas*, had been caught in a storm off the coast of
Portugal and had been forced into Cádiz to refit. Rather than
wait in Marseilles till spring for the ship to be repaired, Humboldt and Bonpland arranged passage on a small vessel bound
for Tunis.

But they canceled their passage at the last minute, after discovering that they and their gear would be forced to share the
great cabin with livestock. Not that they minded the discomfort,
but they feared for the safety of their precious instruments.
Later they learned that the setback had actually been a near escape: About the same time, the North African tribes had launched
a rebellion against their new French masters and were imprisoning everyone disembarking from a French port; if Humboldt and
Bonpland had followed through on their plans, they would have
journeyed no farther than a Tunisian prison.

Finally, the pair traveled to Spain, hoping to find passage to
Africa the following spring, when, they believed, the political climate would have improved. (Never one to waste an opportunity,
during the six-week journey to Madrid, Humboldt took detailed
barometric measurements—at night, to avoid the jeers of hostile
townfolk—which showed for the first time that the interior of
Spain is a large plateau instead of a valley, as had been previously believed.)

In Madrid, Humboldt hit on an even more ambitious scheme.

When the Prussian envoy showed no interest in his idea, Humboldt paid a call on Baron Philippe von Forell, Saxony's minister to the Spanish court, whose brother was an acquaintance of Humboldt in Germany. Keenly interested in the sciences, Forell was aware of Humboldt's recent publications on nerve conductivity, and now he was favorably impressed by the young man's energy and intelligence. He arranged for Humboldt to be presented at court, where King Charles IV himself endorsed the young Prussian's proposition. It was left to the Spanish foreign minister, Mariano Luis de Urquijo, to manage the details. Fantastic though it seemed, Humboldt was charged by the Crown with making the first extensive scientific exploration of Spanish America.

At the time, Spain's vast New World holdings sprawled across North America from San Francisco to St. Louis and southward to the southern tip of Argentina, taking in California, Texas, Florida, Mexico, Cuba, the Caribbean, and all of Central and South America except for Brazil (which belonged to Portugal) and neighboring Guiana (which between 1781 and 1803 changed hands among Holland, France, and Great Britain before ending up with the latter). There was no pretense that this vast empire existed for any purpose but to enrich the mother country. Spain sought to control all trade from its American colonies, and though slavery of the Indians had been abolished by this time, slaves were still imported from Africa to work the plantations. In some places, Native Americans still lived in virtual bondage thanks to an onerous system of peonage that kept them in perpetual debt to the hacienda owners. Although the slaves and Indians were treated with barbaric cruelty, they generally lacked the power to threaten the established order. However, the Creoles, the ninety-five percent of the white population who were born in the New World of supposedly pure Spanish descent, were growing increasingly restive under their colonial masters.

Over the next few months, as he and Bonpland prepared for the expedition, Humboldt lay in bed many a night, unable to sleep, mulling the incredible opportunity that had been given him. He understood that Spain's principal interest was hardly the advancement of the sciences. The Spanish economy had become dependent on New World gold and silver, and with Humboldt's experience in mineralogy, Madrid was hoping he'd unearth

new riches in their American colonies. And as long as he wasn't a
spy for Britain, what did the Crown have to lose? After all, Humboldt
was financing the expedition, including Bonpland's share,
out of his inheritance.

Their royal passports gave Humboldt and Bonpland (who was
listed as Humboldt's secretary) passage on all His Majesty's vessels
and total freedom of movement in the Spanish colonies,
and authorized him "to freely use his physical and geodesical instruments
. . . [to] make astronomical observations, measure the
height of mountains, collect whatever grew on the ground, and
carry out any task that might advance the Sciences." It also
called on colonial officials to assist him in any way they could. It
was an extraordinary document.

Over the past three centuries, the Spanish government had
allowed six scientific missions into their extensive New World
colonies. But never had a traveler been given greater concessions,
and never had the Spanish government placed more
confidence in a foreigner. In fact, with few exceptions, Spain's
policy—based on a strong tradition of xenophobia, the demands
of state security, the need to maintain the colonial trading monopoly,
and a desire to protect the purity of the Catholic faith—
had for hundreds of years been to bar non-Spaniards from their
New World colonies altogether.

Not only was Humboldt a foreigner, he was a Protestant. Over
centuries of warfare against the Muslim Moors, Spanish Catholicism
had taken on a vehemence not seen elsewhere in Europe,
giving rise to such outrages as the Inquisition. But by the end of
the eighteenth century, such attitudes seemed increasingly irrelevant,
as Europe and even America began to forge a new
international community of philosophers, scholars, and scientists.
In these changing times, a more liberal attitude could prevail,
making it possible to grant Humboldt an unrestricted
passport to Spain's New World colonies.

The last time a non-Spanish scientist had been granted
entrée to Spain's American territories was more than sixty
years before, when Charles-Marie de la Condamine went to
Quito in 1735 to measure the shape of the earth. Some years
before, the great British physicist Isaac Newton had suggested
that the earth was oblate, that is, slightly flattened at the poles
and slightly bulging toward the equator. He argued that, since

the planet spins faster at the equator than at the poles (just as a wheel spins faster at the circumference than at the hub), the extra speed must produce greater centrifugal force, which should be sufficient to create a measurable distortion. However, the Italian-born French astronomer Giovanni Cassini, using a different logic, countered that the planet was instead a perfect sphere. With two prominent scientists from rival powers involved, the issue soon mushroomed from a question of scientific principle to a matter of national prestige. There were important practical considerations as well—if the earth were oblate, the astronomical tables that mariners used for navigation would have to be recomputed to take the deformation into account.

The only way to resolve the issue was to measure identical arcs of longitude in two different parts of the globe, then compare them. If the arc closer to the equator were longer, Newton's hypothesis would be confirmed; if the opposite were true, Cassini would be vindicated. Eager to defend their country's honor, Paris dispatched two expeditions. One traveled far north to the snows of French-controlled Lapland, while the other (by special permission of Madrid, which had been indebted to France for help in the War of the Spanish Succession) journeyed far south to the mountains outside Quito (present-day Ecuador). For seven arduous years, La Condamine and his team made painstaking astronomical sightings and hammered reference marks into the ground. But hampered by local resistance and the rugged terrain, the South American party fell behind schedule.

Eventually, La Condamine received a letter from the Académie des Sciences that the issue had been settled without him: After the Lapland expedition's return, their data had been compared to measurements already made right in France. Much to the annoyance of the French, Newton had been proved correct: The earth was indeed oblate. After years of painstaking work so far from home, La Condamine was naturally dispirited to hear that the question had been resolved with data collected in his own country. But he persisted, knowing that his measurements, being made closer to the equator, would yield a more precise result than those taken farther north. Ultimately, it was determined that a degree of arc measured 110,600 meters in Ecuador, versus 111,900 in Lapland.

His assignment finally completed, La Condamine decided to return home by sailing east down the Amazon River, whose full length had been previously traveled by only three other explorers, none in the past century and never by a scientist. On his river journey, from July to September 1743, he made cursory studies of the geography, plant life, and native peoples of the vast region. On his return, he also introduced to Europe the antifever drug quinine, the nerve poison curare, and rubber (which he named *latex* from the Spanish *leche*, in reference to its milky appearance). Ultimately, in fact, it was his two-month sail down the river more than his seven years of onerous geodesic work that elevated La Condamine to international celebrity. Though he never returned to South America, for many years his name was synonymous with the Amazon, and until Humboldt's journey nearly seventy years later, the Frenchman was recognized as the individual who had done the most to excite popular interest in that part of the world.

HUMBOLDT HARBORED GREAT HOPES for this expedition to the New World. It was a daring, some would say vain, notion, but it was one that had been in his mind for years, long before his arrival in Paris. On the eve of his departure from the Spanish port of La Coruña, Humboldt sat in his cabin aboard the frigate *Pizarro* and spelled out his hopes for the journey: "In a few hours we sail round Cape Finisterre. I shall collect plants and fossils and make astronomic observations. But that's not the main purpose of my expedition—I shall try to find out how the forces of nature interact upon one another and how the geographic environment influences plant and animal life. In other words, I must find out about the unity of nature."

For Humboldt, "the unity of nature" meant the interrelation of all the physical sciences—such as the conjunction between biology, meteorology, and geology that determined where specific plants grew—which the scientist unraveled by discovering patterns in myriad, painstakingly collected data. This ambition to view nature as a whole wasn't unique to Humboldt, though. It was a quest that historians believe had begun with the ancient

Greek philosopher Thales of Miletus, in the sixth century B.C. Recognized as the founder of Greek geometry, Thales also taught that all matter is ultimately composed of water. Though dead wrong, the theory was still an intellectual turning point, since it marked the first time anyone had tried to explain natural phenomena without appeal to religious dogma. It was also the first time that anyone had tried to explain the whole, divergent physical world in one grand unifying principle. Thales, now recognized as the first natural philosopher, had tried to penetrate the obvious differences in things—between a rock and a person, say— and to see the not-so-obvious similarities.

Two hundred years later, Aristotle had been the first to use extensive observation to try to tease the truth from the natural world. But the next great contribution along these lines had come nearly two millennia later, from Francis Bacon. Lord chancellor under James I as well as a philosopher, Bacon had defined the scientific method, thereby laying the conceptual groundwork of modern science. And science, in E. O. Wilson's phrase, "was the engine of the Enlightenment," that great eighteenth-century movement to replace dogma and tradition with reason and observation and to supplant tyranny and slavery with humanism and social progress. It had been during these rousing years that René Descartes had suggested that all knowledge could be expressed mathematically. And it hadn't been long afterward that Isaac Newton had used a few simple formulas to explain the functioning of much of the physical world. Newton's achievement had been a masterstroke of unification that would change the world forever.

However, despite its transcendent successes, by the beginning of the nineteenth century, faith in the old paradigm was fading. A new worldview was rising in rebellion against the cool rationalism of the Enlightenment, suggesting that there is a higher truth that man can know, not through reason, but only through the emotions—the worldview of Romanticism. The approach to science in Germany had always been less mechanistic, more spiritual, and more speculative than in England and France, and by the late eighteenth century a school of thought had arisen there called *Naturphilosophie*. More influential in biology than in the physical sciences, the nature philosophers suggested that every

animate and inanimate object was infused with the eternal
World Spirit, the driving force behind the development of the
universe. The school's principal advocate was Johann Wolfgang
von Goethe, who in addition to being one of the great poets of
all time, was also an avid naturalist, and its leading philosopher was Friedrich Schelling, who suggested that there existed
an underlying unity of all things that man could never discern
through logic alone.

This was the world in which Humboldt had come of age; in
the waning days of the Enlightenment and the dawn of Romanticism, and along with figures such as Goethe, he formed a
bridge between the rational and intuitive modes of understanding the universe. In his belief in rigorous quantification, Humboldt was a child of the Enlightenment. All too often, he
believed, the nature philosophers (among whom he had once
counted himself) spun imprecise observations and half-baked
generalities into elaborate "scientific" theories that threatened
to topple at the slightest nudge of logic. Instead, he envisioned a
more stringent methodology in which painstaking observations
were cemented, brick by scrupulous brick, into an enduring
foundation.

Toward this end, Humboldt gathered perhaps the most sophisticated armamentarium of scientific instruments ever before
assembled. Each of the forty-two instruments, nestled in its own
velvet-lined box, was the most accurate and most portable of its
kind yet devised. There were thermometers for measuring the
temperature of air and water, barometers for fixing elevation
above sea level, quadrants and sextants for determining geographic position (including a sextant small enough to fit in a
pocket), telescopes, microscopes, a balance scale, chronometers,
compasses, a rain gauge, substances for performing chemical
assays, electric batteries, electrometers (for measuring electric
current), a Leyden jar (a glass vessel capable of storing static
electricity), theodolites (surveyors' instruments for measuring
vertical and horizontal angles), hygrometers (for measuring atmospheric moisture), a dip needle (for measuring variations in
the orientation of the earth's magnetic field), and eudiometers
(for measuring the amount of oxygen in the air). Everything
would be measured, using the finest instruments and most

sophisticated techniques available, for such data were the basis of all scientific understanding. This exacting methodology, in fact, would become known as "Humboldtian science."

But despite his unyielding empiricism, Humboldt was also touched by the new Romantic spirit of the age. He was not content simply to measure and catalog nature. Combining meticulous observation with inspired description, scientific rigor with almost childlike wonder, he had an abiding passion for the transcendental beauty around him. *Grandeur* and *marvel* are words he used often to describe natural phenomena. For what was scientific understanding without aesthetic appreciation? What was the good of knowing that the earth's atmosphere was seventy-eight percent nitrogen, if one couldn't be moved by the beauty of a cloudless summer sky? What was the use of measuring the acceleration of falling water if one couldn't be awed by a raging cataract? Even Humboldt's countryman Immanuel Kant, the leading philosopher of the Enlightenment, had argued that reason alone, because it was limited by input from the senses, could never yield a complete understanding of reality. Aesthetic appreciation must complement pure reason, if one were ever to grasp the true nature of the world.

"Nature herself is sublimely eloquent," Humboldt wrote. "The stars as they sparkle in the firmament fill us with delight and ecstasy, and yet they all move in orbit marked out with mathematical precision." To truly understand nature, one must feel the ecstasy as well as grasp the mathematics.

Kant's ideas influenced Humboldt in other ways as well. In his lectures on physical geography, Kant had taken the great Swedish naturalist Carl Linnaeus to task for his narrow, categorizing view of botany. Instead of trying to pigeonhole the natural world into prescribed classifications, Kant had argued, scientists should work to discover the underlying scientific principles at work, since only those general tenets could fully explain the myriad natural phenomena. Thus Kant had extended the unifying tradition of Thales, Newton, Descartes, et al. But besides arguing for the unity of knowledge, Descartes had also introduced the idea of reductionism—dividing the world into smaller units that can be studied separately—which has fueled the phenomenal growth of Western science in the centuries since. Humboldt agreed with Kant that a different approach to science was needed, one that could account for the harmony of nature that lay beneath the apparent diversity of the physical world. The scientific community, despite its prodigious discoveries, seemed to have forgotten the Greek vision of nature as an integrated whole. Content to collect and label rocks, they never thought to ask how those specimens were related to the surrounding types of soil, or what influence they exerted on the local flora. "Rather than discover new, isolated facts I preferred linking already known ones together," Humboldt later wrote. Science could only advance "by bringing together all the phenomena and creations which the earth has to offer. In this great sequence of cause and effect, nothing can be considered in isolation." It was in this underlying connectedness that the genuine mysteries of nature would be found.

This was this deeper truth that Humboldt planned to lay bare—a new paradigm from a New World. For only through travel, despite its accompanying risks, could a naturalist make the diverse observations necessary to advance science beyond dogma and conjecture. Although nature operated as a cohesive system, the world was also organized into distinct regions whose unique character was the result of all the interlocking forces at work in that particular place. To uncover the unity of nature, one must study the various regions of the world, comparing and contrasting the natural processes at work in each.

The scientist, in other words, must become an explorer. And the New World, with its lofty mountains, volcanoes, and inexhaustible variety of plant life, offered "ample fields for the labors of the naturalist. On no other part of the globe is he called upon more powerfully by nature to raise himself to general ideas and the cause of phenomena and their mutual connection." Humboldt planned to investigate Cuba, then to explore Spain's vast holdings in North America. In the laboratory provided by the unspoiled New Continent, he hoped to discover how nature's forces act upon one another and how the geographic environment works on animals and plants. It was toward this end that all his studies— of plants and minerals, physics and astronomy, history, art, mythology—his travels across Europe, even his experiments on human nerves and muscles, had been ineluctably building.

And now, at last, he was embarked on a journey that would allow him to test these principles in the field. Aboard the *Pizarro*, Humboldt had the realization, at once heady and daunting, that the purpose of his life was about to be fulfilled.

Assuming, that is, they could slip past the British blockade.

Tenerife

THE ANCIENT CITY OF LA CORUÑA, ON THE NORTHWEST tip of Spain, rested on a hammer-shaped peninsula in the Atlantic Ocean. From the battlements, one could take in the sweep of town and water below. Propped south of the hammer's head, like a nail waiting to be pounded in, was a dot of an island holding the sixteenth-century fortress known as Castle San Antonio. A little more than a mile to the northwest, on the tip of the hammer's claw, was the port's other principal monument, the lighthouse called the Tower of Hercules, after La Coruña's apocryphal founder. Ninety-two feet high, with square stone walls and an incongruous baroque roof, the tower had been rebuilt in the eighteenth century on a foundation laid by the Roman emperor Trajan some sixteen hundred years before. To the west sprawled La Pescadería, the fishermen's quarter. And beyond that, on the hammer's handle, stretched a long arc of waterfront enclosing the city's raison d'être—the harbor.

One of Spain's most ancient seaports, La Coruña had been occupied by the Celts, Phoenicians, Romans, Moors, and Portuguese before the Spanish. It was from La Coruña that Prince Philip of Spain had set sail in 1544 for his wedding with Mary Tudor of England. It was also from here, in 1588, that the Spanish Armada had departed for Britain, bent on a less subtle brand of diplomacy. The following year, Francis Drake, whose name was still cursed in these parts, had sailed into the harbor and sacked the town in retribution for that ill-fated attack. More recently, the English had scored a naval victory against the French there in 1747. And it was from La Coruña, on June 5, 1799, that twenty-nine-year-old Alexander von Humboldt embarked on his monumental journey of discovery to Latin America.

The *Pizarro* had been loaded and ready to sail for two days, but the weather had refused to cooperate. Then on the evening of June 4, the wind finally shifted to the northeast. By morning the fog was thick enough to obscure Castle San Antonio across the harbor, but the *Pizarro*'s Captain Cagigal consulted with the master of the *Alcudia*, a packet ship also delayed in port by the British blockade, and it was decided. The ships would sail that afternoon.

The *Pizarro* nearly didn't make it out of the harbor. Weighing anchor at two o'clock, the ship made eight short tacks in the confines of the port. But three of the maneuvers proved useless in the contrary wind, and the vessel stalled under the battlements. The crew and passengers watched with mounting terror as the current drove them helplessly toward shore. Only when it was practically on the rocks did the *Pizarro* finally gain some headway and sail out of danger. It was not an auspicious beginning to a long ocean voyage.

Soon after, the ship passed under Castle San Antonio, and Humboldt couldn't help thinking of the poor Marqués de Malaspina incarcerated there. Admiral Alessandro Malaspina had been the last man Spain had entrusted with making an exploration of their New World territory, a 1789 expedition in search of the Northwest Passage. Having failed in that mission, Malaspina had been arrested for his unconventional political beliefs and had spent the past decade imprisoned at La Coruña without benefit of trial. As he sailed by the castle where his most recent predecessor was being held, the lesson wasn't lost on Humboldt. "On the point of leaving Europe to visit the countries which this illustrious traveler had visited with so much advantage," he wrote, "I could have wished to have fixed my thoughts on some object less affecting."

By six-thirty the ship had rounded the point and passed the Tower of Hercules, whose weak coal fire barely penetrated the afternoon fog. As night deepened, the wind picked up and they began to make better speed, though the seas had risen as well. Leaving the harbor, they tacked to the northwest to skirt the British squadron that had been reported offshore—a necessary precaution because the *Pizarro* was a Spanish ship.

Though Spain had originally allied herself with England and the other monarchies as France sought to export its revolution to the rest of Europe, it had made peace with Paris in 1795, along with Holland, Tuscany, and Humboldt's own Prussia. Accordingly, if the *Pizarro* were captured, she would be conveyed to Spain's neighbor and Britain's ally, Portugal. Though Humboldt and the other civilian passengers would be released, they would lose their passage to the New World. After the myriad frustrations he and Bonpland had endured to get even this far, the thought of starting all over again was more than Humboldt could bear.

Though he'd dreamed of this moment since childhood, the prospect of leaving Europe for the first time filled Humboldt with an unexpected solemnity. "Separated from the objects of our dearest affections, and entering in some sort on a new state of existence, we are forced to fall back on our own thoughts," he wrote, "and we feel within ourselves a dreariness we have never known before." At nine o'clock on that first evening out of port, the *Pizarro* passed the town of Sisarga. There was a light burning in a fishing hut there, and, knowing it was the last object they would see in Europe, Humboldt and Bonpland sat on deck and watched the lamp slip from sight. "So many memories are awoken in our imagination by a dot of light in a dark night, flickering above the rough waves, signaling our homeland!" Humboldt found.

The next day, the *Pizarro* passed Cape Finisterre, southwest of La Coruña. On June 8, at sunset, they spied the English squadron scudding southeast along the coast and abruptly changed course. Named in honor of the Spanish conqueror of Peru, the *Pizarro* has been variously described as a corvette or a light frigate. The difference is semantic—corvettes and frigates were similar classes of fighting ship, with three square-rigged masts and with their guns carried on a single deck. The corvette was smaller and faster, with only twenty cannon or so, versus the thirty or forty that would be shipped on a full-size frigate.

Whether a small frigate or a corvette, the *Pizarro* was built for speed more than for firepower. This made her an ideal vessel for the long passage to Spain's New World colonies, but it meant that en route she would have to rely on her agility to evade enemy warships, especially England's huge ships-of-the-line, which carried sixty to a hundred cannon stacked in three deadly tiers.

In the interest of stealth, there would be no lamps permitted in the great cabin for the remainder of the voyage, standard procedure on Spanish ships at sea during those perilous times.

Though he appreciated the need for such a precaution, the ever restless Humboldt despaired at all those dark, empty evenings without his books or his work. "In the torrid zone, where twilight lasts a few minutes, our operations ceased almost at six in the evening," he complained. "This state of things was so much the more vexatious to me as from the nature of my constitution I never was subject to seasickness, and feel an extreme ardor for studying during the whole time I am at sea."

Humboldt and Bonpland's days aboard the *Pizarro* quickly settled into a scientific routine. One can fairly see them standing on deck, hunched over their instruments and notebooks, while the crew scurries around them and the other passengers watch with frank curiosity or outright mistrust. Taking advantage of the calm seas between Madeira and the coast of Africa, the pair unpacked their dip needle. Rotating freely on its axis, the dip needle acted as a kind of vertical compass, measuring the orientation of the earth's magnetic field. Humboldt and Bonpland also took water-temperature readings during this part of the voyage and confirmed Benjamin Franklin and Jonathan Williams's counterintuitive observation that Atlantic waters are actually cooler over shoals than over deeper regions, due to the upwelling of cold water from the ocean floor.

By now the current was tugging the ship southeast toward Gibraltar and the Canary Islands. The *Pizarro* passed Cape St. Vincent, the southwest tip of Portugal, where in the fifteenth century Prince Henry the Navigator had seen off Europe's first great voyages of discovery. Born in Oporto in 1394, the third son of John I, Henry had led Portugal's army to victory over the Moors at Ceuta in North Africa—marking the beginning of Portugal's overseas expansion—then, at the age of twenty-one, had retired to Sagres, on Cape St. Vincent. With the intertwined goals of eliminating Arab middlemen from the spice trade, promulgating Christianity, and advancing geographical knowledge, the young, reclusive prince had assembled an international team of scholars and engineers and had established Europe's foremost observatory and school of navigation. Working furiously, never marrying, Henry had directed his experts in improving nautical instruments, devising more accurate charts, and, by adapting the design of the Arab dhow to the requirements of Atlantic sailing, developing the Portuguese caravel, which, with its big hull and lateen rig, had the range, speed, maneuverability, and cost-efficiency to become the ideal ship of discovery.

It had taken Henry's crews nineteen years to pass the psychological barrier of Cape Bojador, the jut of the Sahara beyond which lay the "Sea of Darkness," in which white men supposedly turned black, and by the time Henry died in 1460, his caravels had ventured only a third of the way down the African coast. Yet the prince's innovations and single-mindedness had generated a momentum that survived him. By 1488, Bartolomeu Dias had managed to round the Cape of Good Hope, and just ten years later, Vasco da Gama had sailed all the way to India. In the interim, Christopher Columbus had persuaded Portugal's rival Spain to attempt a western route to the spice lands of the East, and, commanding caravels, had "discovered" a whole New World. Without both Henry and Columbus, Humboldt knew, he and Bonpland would never be embarked on their own journey of exploration. But at a distance of 240 nautical miles they couldn't make out Cape St. Vincent to the east, only some sea swallows and a few dolphins following the *Pizarro*.

Over the next several days, the mysteries of the open ocean continued to reveal themselves. On the eleventh, the ship encountered a huge school of jellyfish. "The vessel was almost becalmed," Humboldt wrote, "but the mollusca were borne towards the southeast with a rapidity four times greater than the current. Their passage lasted nearly three-quarters of an hour. We then perceived but a few scattered individuals, following the crowd at a distance as if tired with their journey."

In the evenings they "were never weary of admiring the beauty of the nights; nothing can be compared to the transparency and serenity of an African sky." Many years later, Humboldt would recall those nights in a wistful frame of mind: "All who possess an ordinary degree of mental activity, and delight to create to themselves an inner world of thought, must be penetrated with the sublime image of the infinite when gazing around them on the vast and boundless sea, when involuntarily the glance is attracted to the distant horizon, where air and water blend together, and the stars continually rise and set before the eyes of the mariner. This contemplation of the eternal play of the elements is clouded, like every human joy, by a touch of sadness and of longing."

ON JUNE 16, at five o'clock in the afternoon, the lookout sighted the Canaries, off the coast of Africa. In ancient times, when they were believed to be the western limit of the earth, the Canaries had been known as the Fortunatae Insulae, "Fortunate Islands." Rediscovered by the Portuguese in 1341, they had been granted to Spain by papal bull three years later. Now they were the usual first port of call for Spanish vessels en route to the New World. The captain had orders to lay over so that Humboldt and Bonpland could climb the volcano on Tenerife. But a delay longer than four or five days would be impossible, he reminded them, on account of the English blockade.

First to come into view was the foreboding, lava-encrusted island of Lanzarote. That night, Humboldt and Bonpland sat on the afterdeck, admiring how the moon, just a few degrees above the horizon, shone silver on the volcano's ash-covered flanks. The night was beautifully serene and cool, and the sea was sparkling with phosphorescence. Then, after midnight, great black clouds obscured the moon, and they could spy fishermen's torches onshore. The two sat up into the wee hours: Their first glimpse of these islands off the African coast had likely left them too excited for sleep.

The next morning was foggy, and the ship picked its way through the islands of Alegrana and Montaña Clara, taking soundings the entire way. The small islands north of Lanzarote reminded Humboldt of the Rhine near Bonn, which he had visited with Georg Forster nearly a decade before.

On June 17, the *Pizarro* passed the western shore of the island of Lanzarote, which was black, parched, and devoid of vegetation. Rising above this otherworldly landscape was the island's volcano, Timanfaya. With a rounded, not entirely conical top, the volcano seemed quiet enough that day, though it had erupted violently some seven decades before, destroying nine villages with lava and earthquakes.

As the ship made its way down the western side of the island, the captain thought he saw a castle on the coast. Hoisting the Spanish flag in salute, he sent a boat to inquire about English ships in the area. But as the launch approached shore, the land-

ing party discovered that the castle was just a tall rock and that
the land wasn't a continuation of Lanzarote at all but part of the
small neighboring island of Graciosa. Although Graciosa had a
large bay, its volcanic earth was barren, and the island was uninhabited. Despite the desolation, Humboldt was moved by the
rugged and wild beauty of the place. In the background, beyond
the bleak shoreline, the cultivated fields on Lanzarote provided a
less somber, more domestic perspective at sunset. "In the narrow
pass between two hills, crowned with scattered tufts of trees,
marks of cultivation were visible. The last rays of the sun gilded
the grain for the sickle. Even the desert is animated wherever we
can discover a trace of the industry of man."

The wind dropped, and rather than return the way they had
come, the captain decided to maneuver out of the bay by tacking
between the nearby island of Clara and a protruding, twenty-
foot-high mass of lava known as West Rock but marked on the
old charts "Hell." Around midnight, an infernal current gripped
the *Pizarro* and propelled it directly toward the crag. In the light
wind, the frigate no longer obeyed the helm, and for the second
time in a fortnight, the ship seemed destined for disaster. All
night the crew worked the sails, struggling to hold the vessel off
the outcropping. Finally, toward morning, the wind freshened
and the ship was able to negotiate the channel, past the rock
named for the place where unlucky sailors were consigned by
these treacherous waters.

The next day was hazy, but that evening the company sighted
Grand Canary, the island that was the granary of the archipelago. The wind picked up, and by the morning of the nineteenth
the fog was so thick they had to drop anchor. When the mist
lifted, the first rays of sunlight caught the distant top of el Pico
del Teide, Tenerife's volcano, thought in ancient times to be the
highest mountain in the world. (By 1799, Chimborazo in the Andes was believed to hold that honor.) Humboldt and Bonpland
rushed to the bow for a glimpse of the peak they had come to
challenge. The 12,200-foot active volcano was snowless at this
time of year, they saw, and terminated in a piton, or sugarloaf,
with a small, truncated cone. The flanks were formed by black
lava, with lush vegetation in some places, barren rock in others.

They didn't have long to admire the view. As the mist dissipated, it also exposed four English warships very near the

Pizarro's stern. Apparently the enemies had passed unnoticed in the fog. But, luckily, the *Pizarro* was now under the guns of the Spanish fort, and the British squadron dared approach no closer. The corvette sailed on.

Horatio Nelson, the great British admiral, had attacked that same fort just two years before. In July 1797, Nelson came to Tenerife in search of a Spanish treasure ship rumored to be anchored here. Late on the night of July 24, the British attempted an amphibious assault on Santa Clara, but the landing party was greeted by withering cannon fire from the fort. As Nelson stepped from his longboat onto shore, he was shot through the right elbow. His stepson Josiah Nisbet ferried him back to his flagship, the *Theseus*, where the limb was amputated just beneath the shoulder. Although some of Nelson's men did manage to hold part of the town for a time, the assault was a disaster, with 153 English killed and 105 wounded, and Nelson limped back to England certain that his career was over. But only a year later he would become a national hero after his brilliant rout of the French at Abukir Bay, also known as the Battle of the Nile.

From the deck of the *Pizarro*, Humboldt took in the coastal town of Santa Clara. The village was neat, with dazzlingly white, windowless houses set on a narrow beach under a wall of black rock. But it also was breathlessly hot and devoid of vegetation, and Humboldt pronounced it gloomy.

As he disembarked, Humboldt noticed that, after two weeks of fresh sea air, the once familiar smells of animals and vegetation and even the earth itself came as an overpowering assault to the senses. Still, he and Bonpland were elated to be taking their first steps on non-European soil. As Colonel Armiagra, the commander of the local infantry regiment, showed them around his garden, they gaped at bananas, papaws, colorful poincianas, and other plants they had previously seen growing only in hothouses.

Before dawn on June 20, the two men set out with guides and mules for the port city of La Orotava, on the western flank of the volcano. Wearing his customary open-necked shirt, loose striped trousers, short jacket, high black hat, and tall boots with the tops turned over, Humboldt followed the guide over a tortuous, upward-leading road with a stream running along one side. Not long after, they passed through the town of San Cristóbal, which clung to a basalt ridge forming a broad girdle around the Peak of

Tenerife. Stopping to examine rocks along the way—to the growing impatience of their guides—they eventually came to lovely La Laguna, capital of Tenerife.

Built on a small plain and protected by a forested hill, the town of nine thousand people was surrounded by gardens and by windmills for grinding grain. There were also small chapels, called *ermitas* in Spanish, that were erected on little hills and encircled by evergreens. The town, with its very old, solidly built houses, was pleasantly cool, and Humboldt felt that from its streets Tenerife seemed a happy, peaceful place. "No abode appeared to me more fitted to dissipate melancholy, and restore peace to the perturbed mind, than that of Tenerife or Madeira," he wrote. "These advantages are the effect not of the beauty of the site and the purity of the air alone: the moral feeling is no longer harrowed up by the sight of slavery, the presence of which is so revolting in the West Indies, and in every other place to which European colonists have conveyed what they call their civilization and their industry."

As the climbing party left La Laguna, the landscape was filled with a profusion of exotic plant species, and Humboldt delighted in them all—date and coconut palms, orange trees, vines, ferns, myrtles, cypresses. Cactus and agaves formed hedges to mark the property boundaries. The Canaries deserved their ancient name of Fortunate Islands, he wrote: "I own that I have never beheld a prospect more varied, more attractive, more harmonious in the distribution of the masses of verdure and of rocks, than the western coast of Tenerife." Yet, he reminded himself, although the indigenous people weren't slaves, they were nevertheless suffering in an artificial poverty, with the land concentrated in the hands of a few wealthy planters. In addition, the volcano that had created all this beauty could easily destroy it again. "Happy the country, where man has no distrust of the soil on which he lives!" Humboldt concluded.

They passed through the town of Juan de la Rambla, whose rising hills were cultivated like a garden, followed by the pleasant hamlets of Victoria and Matanza. Its name meaning "Slaughter" in Spanish, Matanza was the site where the native people had temporarily turned back the Spanish conquest in the fifteenth century. On the outskirts of La Orotava, the party stopped at the botanical garden, where they met Monsieur LeGros, the

French vice-consul. Many years before, LeGros had left France with Captain Baudin, bound for the West Indies. En route they had been shipwrecked at Tenerife, and declining to be rescued, M. LeGros had been living contentedly on the island ever since. He had climbed el Pico del Teide many times, and he now agreed to show them the way.

The next morning, June 21, the party set out for the peak, whose top was shrouded in mist. Besides M. LeGros and the porters, they were accompanied by M. Lalande (secretary to the French consul), and an English gardener from the botanical garden. Passing through the steep, seemingly deserted streets of La Orotava, Humboldt found the houses solidly built but thought the place a dreary town, like Santa Clara. The foliage was spectacular, though, and they walked for a time atop an aqueduct covered with fine ferns. In the gardens, they found it odd to see fruit trees from northern Europe mixed with tropical species such as orange, pomegranate, and date. They also stopped to admire an ancient, renowned specimen of dragon tree, sixty feet tall and fifty feet around the trunk. A species of slow-growing tropical evergreen, the dragon tree had many branches divided like candelabra, with tufts of leaves sprouting from the end of each. Yet this extraordinary specimen also presented a mystery, the first of many riddles of plant distribution that Humboldt would encounter over the course of his journey. The tree is native to the West Indies, and never grows wild in Africa. So how did it come to be cultivated on the Canaries and neighboring islands? (This particular specimen, which still grows today, was the same one that Darwin later longed to see on the voyage of the *Beagle*. But, although the ship called at Tenerife, a quarantine prevented him from going ashore.)

Leaving Orotava, a narrow, stony path led through a lovely forest of chestnut trees. At a place called Monteverde, amid a mixture of brambles, laurels, and heaths, the party filled their canteens at a spring and admired the magnificent view of the sea. Next came a region of ferns, whose roots, M. LeGros explained, the natives ground to powder and mixed with barley meal into a conglomeration called *gofio*, which they ate for sustenance. Humboldt was disturbed that the practice was necessary in a country of such natural abundance.

Next they entered a wood of junipers and firs that had been

severely damaged by a hurricane. There was a narrow pass between two basalt hills, then the great plain known as Spartium,
after the species of broom plants growing there. After this, the
lush vegetation of the lower elevations dropped away. For two
and a half hours they crossed the Llano del Retama, a vast, hot,
desolate sea of sand punctuated by the tufts of the *retama*, a
nine-foot-tall flowering shrub, with only a few goats and rabbits
to break the solitude. The plain was littered with blocks of obsidian ejected from the volcano, and the pumice dust, kicked up as
they walked, was suffocating.

After the Llano del Retama the men passed though a series of
narrow ravines, then, at about eight thousand feet, came to the
Estancia de los Ingleses ("English Halt"). Consisting of two inclined boulders offering some protection from the wind, the Estancia was the traditional place for climbers (who were generally
Englishmen) to rest for the night before going on to the summit.
As darkness came, the peak above them was covered with clouds,
and there was a strong northerly wind that exposed the moon on
and off, giving spectacular, intermittent views of the volcano. The
temperature dropped to just above freezing, and having no tents
or cloaks, the men tried to rig a windscreen by tying cloths together, but the makeshift mess blew too close to the fire and
burned. The wind drove the woodsmoke toward them, making
it hard to breathe. Despite the uncomfortable night, Humboldt
was exhilarated. "We had never spent a night on a point so elevated. . . . The view of the volcano threw a majestic character
over the nocturnal scenery. Sometimes the peak was entirely hidden from our eyes by the fog, at other times it broke upon us
in terrific proximity; and, like an enormous pyramid, threw its
shadow over the clouds rolling beneath our feet."

At three A.M., the party packed their gear and started for the
summit, lighting the way with torches. Ascending the steep northeast face, they came after two hours to a small plateau known as
Alta Vista, where in the winter workers called *neveros* (from the
Spanish word for "snow") collected ice and snow to sell in the villages. Above this plain was the bleak Malpays ("Badlands"), covered with fragments of lava and totally devoid of vegetation. Just
below the winter snow line was a cavern forming a kind of natural icehouse. As they left this ice-filled cave, the sun broke the
horizon. During the night, they saw, a layer of fleecy clouds had

collected beneath them, concealing the sea and shimmering in the thin light like a field of snow, with the other islands jutting through like rocks in a snow-covered pasture.

Pausing here, the climbers saw a strange optical effect. Seven or eight degrees above the horizon, luminous points appeared in the sky, first traveling vertically, then horizontally, like rockets thrown into the air. At first they thought the volcano on Lanzarote was erupting, but after about eight minutes, when the fireworks abruptly stopped, they concluded that the points of light were stars, distorted by the mingling of layers of air of different temperatures and densities near where the sun was about to rise.

The ascent across the Malpays was steep, and the footing was treacherous, with blocks of lava constantly rolling out from under their feet. The porters, who had tried to persuade the party to turn back at the English Halt the night before, became more phlegmatic as the party advanced. They insisted on sitting down to rest every ten minutes, and when they thought they weren't being watched, they threw away the specimens of obsidian and pumice they had been charged with carrying. Eventually it became evident that none of them had ever been to the summit of the volcano.

The ash-covered piton was the steepest part of the climb, and would have been virtually impassible except for an old current of lava, which allowed uncertain handholds. When covered in snow, the piton was even more treacherous. As they passed, M. LeGros pointed to the spot where Captain Baudin, making this same climb, had nearly been killed in the winter of 1797, when he'd lost his footing and rolled partway down the mountain before fetching up on a heap of lava.

They reached the summit at eight A.M. At the top of the piton there was scarcely room to sit, and a circular wall of lava surrounded the crater like a parapet, preventing a view of the volcano's interior. There was an exquisite vista of the other islands, though, spread out before them like a map. The wind was so strong out of the west that they could scarcely stand, and it roared through the crevices in the rock. The temperature was just above freezing, and the party suffered in their lightweight clothing.

On the eastern side there was a breach in the lava wall, and the party began their descent into the crater. Humboldt saw that the caldera was elliptical, about three hundred feet by two hundred and only some one hundred feet deep. To the southwest, where the rim was lower, he noticed an enormous mass of cooled lava perched atop the crater. Also in that direction a large opening in the rim gave a spectacular view of the sea beyond. In the bottom of the crater sat great blocks of lava, but it was clear that the volcano had not erupted through the caldera for many years. However, the volcano was emphatically active. Just one year before, lava streams had broken out on the sides of the mountain, there had been occasional subterranean noises like the firing of cannon, and stones the size of houses had been hurled four thousand feet into the air.

Picking their way through the broken lava, the party passed crevices in the crater where vapor escaped and they heard a curious buzzing. At these vents, they could feel the heat from deep within the earth, and Humboldt measured the temperature of the ground at up to 108 degrees Fahrenheit. He collected some air in a corked vial in order to test the oxygen content later on. The walls of the caldera were bleached snow-white by sulfuric acid, and it was impossible to sit for long, lest the climbers' clothes be corroded. When Humboldt wrapped some sulfur crystals in paper, thinking to take them back to the ship, the wrapper, and even parts of his journal, were quickly eaten away. There were no insects on the summit, he noted, but he did find some dead bees inside the caldera, apparently carried up the mountain by air currents.

Climbing back to the piton, the party stopped to admire again the amazing beauty laid out before them. The sky was a stunning cobalt, and arrayed on the island below, in pleasing contrast, were the steep, barren flanks of the volcano, desolate plains, forests, vineyards, gardens, and finally the towns along the coast. In the thin, clear air, they could even make out the masts of the ships anchored off La Orotava. What an incredible place was Tenerife, brimming with tropical plants, vines, fruit trees, even roses. Every road was lined with camellias—and the people actually fattened pigs on apricots. Though he'd scarcely left Europe, Humboldt was tempted to let the *Pizarro* sail on without him. He would be quite happy, he thought, to settle in the Fortunate Islands, like M. LeGros, the castaway who had never gone home.

From this vantage point, it was obvious to Humboldt how the island's vegetation grew in five distinct zones, according to elevation, soil, and availability of water—grasses on top of the peak; then the tall flowering shrubs known as retama; then, in descending order, pines and heaths; verdant forests of laurel, oak, chestnut, myrtle, and other trees, with thick growths of ferns at their trunks; and finally, the cultivated land along the coast, graced with vines, grain, fruit trees, olives, and tropical species such as date palms, figs, and banana trees. Humboldt made a sketch of the various bands. This relationship between altitude and plant life was exactly the sort of observation he'd hoped to make on this journey, just one clue to the coherent, interlocking whole of nature. If, when he was barely out of Europe, such secrets were already presenting themselves, he could only imagine what discoveries waited in the untrammeled wilds of America. Humboldt was shaken out of his reverie by the cold westerly wind, which drove the party to seek shelter behind the piton. Their hands and faces were frozen, and they reluctantly began their descent.

During the long climb down, Humboldt had ample opportunity to consider the irony that the visit to El Teide, his first experience of an active volcano, had posed more questions than it had answered. Are volcanoes constructed entirely of lava, he wondered, or do they sit on a base of nonvolcanic rock? Do volcanoes begin as domes of softened rock pushed up by expanding subterranean gases, even before the crater forms and the lava begins to flow? What causes the underground fire that produces a volcano, and why is it sometimes explosive and other times subdued? Is the fire near the surface, or at some immense depth?

There was maddeningly little data to resolve any of these issues. But whatever the ultimate answers, Humboldt was convinced that they would be the same everywhere on earth. In the past, researchers had focused too narrowly on the specific geological formation in view. Distracted by surface differences— Was this volcano domed or conical? Was it isolated or surrounded by other mountains?—these well-meaning investigations had emphasized superficial contrasts at the expense of underlying

truths. Since geological phenomena were subject to regular laws,
just like the laws of biology, "the ties which unite these phenomena . . . are discovered only when we have acquired the habit
of viewing the globe as a great whole; and when we consider in
the same point of view the composition of rocks, the causes which
alter them, and the productions of the soil, in the most distant
regions." Such as the New World.

Altogether, it had been twenty-one hours of walking from
La Orotava to the peak and back. Humboldt was disappointed
to have so little time to explore the island, but the *Pizarro*'s master, Captain Cagigal, was eager to resume their journey. Yet
when they returned to the ship, they discovered that their departure had been postponed a couple of days, to the twenty-fourth
or twenty-fifth, because the English squadron had been sighted
again outside the harbor. Humboldt was frustrated—had they
known, they could have spent longer on El Teide or perhaps even
gone on to Tenerife's smaller volcano, Chahorra. Instead, he and
Bonpland spent the extra time exploring the area around La
Orotava.

Not content to study only the geological and botanical features of the Canaries, Humboldt—as he would throughout his
journey—also turned his attention to the islands' human occupants. What had become of the Guanches, he wondered, the indigenous people of the islands, "whose mummies alone, buried in
caverns, have escaped destruction"? The answer was brutally
simple: slavery, especially as practiced by Portugal and Humboldt's nominal patron, Spain. "The Christian religion," he went
on, "which in its origin was so highly favorable to the liberty of
mankind, served afterwards as a pretext to the cupidity of the
Europeans" who felt no compunction at shipping off to the slave
market in Seville any unbaptized person they could lay hands
on. Those who escaped capture had gradually intermarried with
the Spanish colonists, until the Guanches, as a people, had ceased
to exist.

Yet, despite his obvious sympathy for the Indians, Humboldt
was adamant that they not be idealized for their supposed physical prowess and gentle character. Their physical stature had
been exaggerated, he observed, judging from the remaining
mummies. And prior to the European conquest, the Guanches

labored under a violent, feudal society, far from the "perpetual fe-licity" that jaded Europeans might ascribe to a distant race sup-posed to be living as "noble savages."

Humboldt's compassion and probity thus rose to the surface at his first encounter with non-European peoples. It is a theme he would sound time and again over the next five years, and in-deed, throughout his life. In his culminating work, *Cosmos,* he would famously write: "While we maintain the unity of the hu-man species, we at the same time repel the depressing assump-tion of superior and inferior races of men. There are nations more susceptible of cultivation, more highly civilized, more enno-bled by mental cultivation than others, but none in themselves nobler than others. All are in like degree designed for freedom. . . . "

Though he came to the New World by the grace of the Spanish king, Humboldt clearly felt no obligation to support the monar-chy's colonial practices in his writings. He had been well aware of Spain's American policies before ever setting out for Madrid. Yet, having exhausted all leads in France and desperate to begin a voyage of discovery, Humboldt hadn't hesitated to accept Spain's patronage, even if it meant concealing his own views—which, had they become known, would certainly have prevented him from receiving his royal passport. So, even as the Crown was in a sense hoping to manipulate Humboldt, to use his discoveries to tighten its financial and political stranglehold in the New World, Humboldt was secretly withholding his allegiance as well. (The irony that his ship was named after one of the most brutal con-quistadors of all couldn't have been lost on Humboldt.) Though his journey was first and foremost one of scientific discovery, once in South America the young Prussian would have no compunc-tion about exposing—loudly and often—practices and conditions that he found reprehensible, however embarrassing those disclo-sures would prove to his patrons.

UNDER WAY ONCE MORE, the *Pizarro* cut the Tropic of Cancer on June 27. A few days later, the ship approached the area marked on the chart as "Bank of Maal-strom" and toward night-fall changed course to avoid the region. The original Maelstrom, or *Moskenstraumen* in Norwegian, was an infamous channel of treacherous currents and unforgiving winds located in the Lofton Sea, off northern Norway. Humboldt strenuously doubted that such a thing existed in these calm tropical waters, but knowing that sailors were a superstitious lot, he kept his thoughts to himself. In any event, a northwest current kept them from altering their course as much as the captain had planned.

The ship passed 150 nautical miles west of the Cape Verde Islands, which had become a possession of Portugal after Alvise da Mosto, sailing for Henry the Navigator, had claimed them in 1456. Some land birds appeared, driven from shore by a storm, and followed the boat for several days before flying off. North of the Cape Verdes, the ship entered the great mass of floating seaweed called the Sargasso Sea, which Columbus had reported on his first voyage. He had compared the "sea" to an extensive meadow, and the metaphor is apt. A huge, elliptical mass of free-floating brown seaweed (also called "sea holly" because of the berrylike air sacs that hold it afloat and the sawtooth edges on its leaves), the Sargasso Sea was an eerie, oceanic desert of light wind, weak current, warm water, and little life. At latitude 17 degrees, 42 minutes, and longitude 34 degrees, 21 minutes, they came across a ghostly sight—a wrecked and abandoned vessel, mostly submerged and covered with floating seaweed. Since it seemed unlikely that the ship could have sunk in these tranquil waters, Humboldt speculated that it had foundered in the rough North Atlantic and drifted south with the currents. Whatever its source, it was a chilling sight to men in midocean.

As the *Pizarro* sailed on, Humboldt and Bonpland continued their painstaking measurements—of latitude and longitude, air and water temperature, humidity, barometric pressure, the purity of the air and its electric charge, the salinity of the ocean, even the color of sky and sea. On very calm days, they were able to use their dip needle to measure the angle of the earth's magnetic field. All their observations were duly recorded in notebooks for later analysis.

Humboldt and Bonpland passed the evenings on deck, never tiring of the majestic tropical sky and the new constellations that revealed themselves night by night, replacing the familiar constellations they had known since childhood. They saw the Ship and the Clouds of Magellan drift into view, followed, on the night of July 4–5, by the fabled Southern Cross. Consisting of five stars, the not-quite-symmetrical Cross has been used for centuries to tell time in the Southern Hemisphere, since the constellation grows nearly perpendicular as it passes the meridian each night. Watching in silence, Humboldt found himself in a pensive mood as one of the dreams of his youth was realized. In the solitude of the sea, he hailed the Cross as a long-lost friend—yet the constellation was also a poignant reminder of their distance from Europe. He wrote, "Nothing awakens in the traveler a livelier remembrance of the immense distance by which he is separated from his country than the aspect of an unknown firmament."

Overall, the ship had an easy passage to America. It traversed nine hundred leagues (about 2,700 nautical miles) in the next twenty days, following the same route that Columbus and nearly all his successors had taken—almost due south to the Tropic of Cancer, then west to pick up the trade winds, which in the beginning blew from the east-northeast, then due east, propelling them steadily to the New Continent. Since Europeans had been plying these waters for three centuries, the sailors had a good idea of what to expect. En route, they enjoyed the calm seas the Spanish called el Golfo de las Damas ("the Ladies' Sea"), and as they sped on before the wind, the sailors barely had to touch the sails.

As they approached the West Indies, the islands encircling the Caribbean, they encountered the dark skies and *las brisas pardas* ("the dark winds"). Here frequent storms would threaten, and brooding, well-defined clouds would appear in the east. The breeze would die, there would be lightning (but no audible thunder), and a few large raindrops would fall; then the wind would rise, the squall would pass harmlessly, and the crew would unfurl the topsails again. The storms were like nothing Humboldt had ever experienced, but they were common in the tropics in June and July. The constancy of the daily temperatures also made an impression on him: During the day his thermometer fluctuated between about seventy-three and seventy-five degrees, and at night between about seventy-one and seventy-three. "Nothing," he suggested, "could equal the beauty and mildness of the climate of the equinoctial region on the ocean."

Then, as the *Pizarro* neared the Antilles, on the northeast edge of the Caribbean, an event occurred that shattered the peaceful voyage: A typhus epidemic broke out on board.

Common aboard ships of the time (and wherever people were packed together in unsanitary conditions) typhus is caused by a bacterium transmitted to humans by louse feces. An infected louse bites its host, and the host scratches the bite and breaks the skin, allowing the bacterium to enter the bloodstream. After an incubation period of one to two weeks, the first symptoms appear—headache, fever, body aches, nausea, prostration, and a characteristic rash. If the victim survives, the disease runs its course in another two weeks, after which the fever suddenly breaks and a long convalescence begins. If the fever fails to subside, delirium and coma set in, followed by death.

On the first day of the outbreak, two sailors and several passengers were stricken. By the second day, the victims were delirious and totally prostrated. But Captain Cagigal was unconcerned with anything that didn't directly affect the ship's operation or its schedule, and he refused to fumigate (with woodsmoke) to try to stop the epidemic. The ship's unnamed surgeon, whom Humboldt found an ignorant, unpleasant man, ascribed the disease to a "corruption of the blood" brought on by the tropical heat; he bled the patients, to no avail. Crowded together with no means of escape and without our modern understanding of disease, the other passengers and crew were naturally terrified. Humboldt had particular reason for concern, since his own health had never been robust. He tried to persuade himself that the fever wasn't highly contagious, but he regretted that, in all their boxes of equipment and supplies, he and Bonpland had neglected to pack any quinine bark, on the assumption that the ship's surgeon was certain to have adequate provisions.

By July 8, one of the sailors had become so ill that death seemed imminent. He was taken from his hammock and brought on deck, where he was placed on a small piece of sailcloth in an airy spot near a hatchway, so that the ship's company could gather round while he received last rites. The sacrament was administered, but, away from the heat and stagnant air of the middle deck, the man began to show steady progress and eventually recovered.

A young Spaniard from Asturias contracted a particularly virulent case of the disease as well. Though Humboldt doesn't mention any of the other passengers up to this point, he had clearly gotten to know the young man over the course of the voyage, because he describes him as intelligent and mild mannered. Nineteen and the only son of a poor widow, the boy hadn't wanted to make this journey, but his mother had prevailed upon him to travel to Cuba to work for a wealthy relative, accompanied by a young friend who was also hoping to find a job there. For three days, the Asturian lay in a lethargic, delirious state, his friend always at his side. Then he succumbed, leaving the other not only grief stricken but also despairing of his own future, since his introduction to his prospective employer had expired along with his friend. "It was desperate," Humboldt wrote, "to see this young man abandon himself to deep grief and curse the advice of those who had sent him to a distant land, alone and without support."

That evening found Humboldt in a melancholy frame of mind, watching the hilly, desolate islands slip past. The moon drifted in and out of the clouds, and low waves lapped at the hull as the ship cut through the phosphorescence. The sea was utterly quiet, except for the cries of a few seabirds flying toward the coast. At eight o'clock the dead man's knell was tolled on the ship's bell, and the crew stopped work and knelt to pray. The body was brought on deck in the night, and the next morning the priest recited some prayers and the corpse was buried at sea.

After the young Asturian's death, the master of the *Pizarro* was finally stirred to action. The ship's first stop was to have been Havana, the largest port in the Western Hemisphere and one of the jewels of Spain's New World empire, where Humboldt and Bonpland had intended to disembark. But with the fever still raging, the captain was eager to complete the voyage as soon as possible. He decided to forego their landing in Cuba and to proceed directly to their final destination, Cumaná, in present-day Venezuela. So all the passengers had no choice but to continue across the Caribbean to the coast of South America.

Approaching Cumaná on July 15, the *Pizarro* came upon a low island covered with shimmering sand. The depth of the water continued to drop, and when the soundings reached only three fathoms (eighteen feet), the captain was forced to lower the anchor. Spying two canoes in the distance, the Spaniards fired a cannon as a signal for them to approach, but instead the Indians paddled away to the west. The crew hoisted the Spanish flag and fired the cannon again, and this time the boats cautiously neared. As they came closer, Humboldt had his first glimpse of the Native Americans: "In each canoe there were eighteen Guayqueria Indians, naked to the waist, and of very tall stature. They had the appearance of great muscular strength, and the color of their skin was something between brown and copper. Seen at a distance, standing motionless, and projected on the horizon, they might have been taken for statues of bronze." The captain hailed them in Spanish, and the Indians came on board.

Members of the Guayaquí tribe, they had, according to an apocryphal story, gotten their name from Christopher Columbus. Supposedly, when the first Spanish explorers encountered a group spearfishing and asked them to identify themselves, the Indians, thinking the Europeans were pointing to their tackle, answered, *"Guaike,"* or "spear." In any event, the tribe had long been loyal to Spain. This band was on their way from Cumaná to the cedar forests beyond Cape San José to gather wood.

The sandy, uninhabited island in front of the ship was called Coche, the Indians said, and lay south of the island of Margaretta. British ships frequented the southern channel, and the *Pizarro* wouldn't run aground as long as it steered in that direction. The canoes were loaded with plantains, armadillo, calabash (a gourd used for making drinking vessels), and other products of the land, and the Indians made the captain a gift of some coconuts and fish. It was agreed that the master of the dugouts, a man named Carlos del Pino, would stay aboard the *Pizarro* as pilot.

Toward evening, the *Pizarro* lifted anchor and steered west, following del Pino's directions toward the channel. The ship passed the small desert island of Cubagua, where, shortly after Columbus's landing, Spain had founded a town called New Cádiz. Luxurious homes had been built for the planters, and the island had become famous for its pearl fisheries. But it was now deserted and forlorn, with no trace of its previous prosperity.

Before long, Humboldt spied in the distance the mountains of Cape Macanao, on the western side of Margaretta. With the wind light, the captain decided to wait till morning to approach. On deck, Humboldt spent part of his last night on the *Pizarro* quizzing del Pino in Spanish about the fantastic plants and animals of his country, including crocodiles, boas, electric eels, and jaguars. "Nothing is so exciting to a naturalist," he wrote, "than to hear the wonders of a country he is about to explore." Humboldt was impressed with del Pino's sagacity and disposition, and that night they forged a friendship that would serve the expedition well. In fact, del Pino would later guide Humboldt and Bonpland through the Orinoco Basin, some of the most remote and dangerous jungle in the world. And thus, Humboldt recorded, "By a fortunate chance, the first Indian we met on arrival was the man whose acquaintance became the most useful to us in the course of our researches."

At dawn on July 16, Humboldt and Bonpland had their first glimpse of the verdant mainland, bristling with palms, cacao trees, and all manner of exotic flora. The *Pizarro* anchored in port at nine A.M., forty-one days out of La Coruña.

THREE

Cumaná

THE SKY WAS AZURE, THE SUN DAZZLING. BEYOND THE harbor, the coastline lay green and lush, and in the distance, half veiled in mist, soared the mountains of New Andalusia. From the rail of the *Pizarro*, Humboldt struggled to take it all in—the chalky hills, the picturesque castle, the towering palms, the cacti, mimosas, tamarinds, the pelicans, egrets, flamingos. It was a landscape worthy of a long ocean voyage. "The splendor of the day," Humboldt wrote, "the vivid coloring of the vegetable world, the forms of the plants, the varied plumage of the birds, everything was stamped with the grand character of nature in the equinoctial regions." After all the years of dreaming and preparing, he must have felt a rush of incredulity that this day had finally arrived.

A mile inland, shrouded in tropical foliage, lay Cumaná. Set on a fertile delta between the Manzanares and Santa Catalina rivers, Cumaná was capital of the autonomous *capitanía general* of the same name, which, under the byzantine Spanish system, was included in the province of New Andalusia, which in turn was part of the Viceroyalty of New Granada. Nominally ruled from Bogotá, New Granada took in the huge swath of territory including present-day Venezuela, Colombia, Panama, and Ecuador.

Cumaná had been founded in 1523, making it the oldest continuous European settlement in South America. Columbus, on his third voyage in 1498, was the first European to spy these green shores, and was sufficiently enchanted to christen the nearby Gulf of Paria, Tierra de Gracia, or Land of Grace. The following year, the Spaniard Alonso de Ojeda (perhaps mischievously)

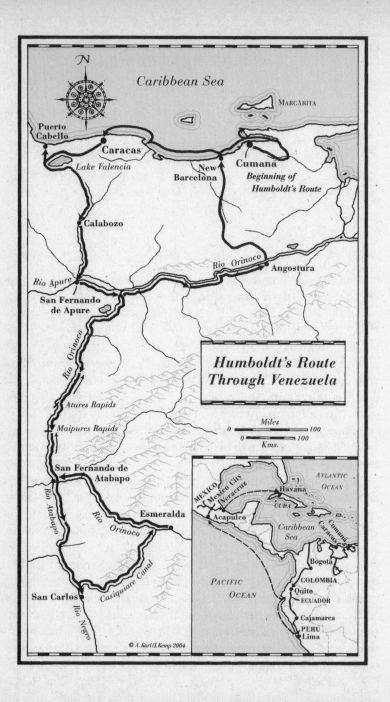

N

Caribbean Sea

MARGARITA

Puerto
Cabello

Caracas

Lake Valencia

New
Barcelona

Cumaná
*Beginning of
Humboldt's Route*

Calabozo

Rio Orinoco

Angostura

Rio Apure

**San Fernando
de Apure**

Rio Orinoco

Atures Rapids

Maipures Rapids

**Humboldt's Route
Through Venezuela**

Miles
0   100
0   100
Kms.

**San Fernando de
Atabapo**

Rio Atabapo

Rio Orinoco

Esmeralda

San Carlos

Casiquiare Canal

Rio Negro

© A. Karl/J. Kemp 2004

*ATLANTIC
OCEAN*

MEXICO Mexico City
Veracruz
Havana

CUBA

Acapulco

*Caribbean
Sea*

Cumaná
Caracas

Bogotá

COLOMBIA

*PACIFIC
OCEAN*

Quito
ECUADOR

Cajamarca
PERÚ
Lima

dubbed the country Venezuela, or Little Venice, because the Indians' houses around Lake Maracaibo, built on stilts, supposedly called to mind that grand city. Originally calling their capital Nueva Toledo, the Spanish later reverted to the Indian term for the place, Cumaná. But the name proved easier to appropriate than the land. The native Caribs (from whose name derives the word *cannibal*, as well as *Caribbean*) resisted fiercely, and English pirates, including Francis Drake, Henry Morgan, and William Hawkins, prowled the coast for the better part of the next two centuries. At the time of Humboldt's arrival, Cumaná was still half entombed in debris from a catastrophic earthquake two years before.

As soon as the *Pizarro* dropped anchor, the passengers fled the infected ship. The ill among them began their long convalescence, but even on land, the fever would claim a second victim—an eighteen-year-old black servant from Guinea, who fell into dementia and succumbed despite the scrupulous attentions of his elderly master. Accompanied by Carlos del Pino, Humboldt and Bonpland set out for town, crossing the stifling plain known as El Salado (The Salty) that stretched between the coast and the capital. On the way—before they had even unpacked their bags—they collected their first botanical specimen from the New World, a species of mangrove (*Avicennia tomentosa*) common along the South American shoreline.

On the outskirts of Cumaná the trio came to the Indian settlement, whose small houses, freshly rebuilt after the earthquake, were laid out in a neat grid. In town, the travelers presented their passports to the governor, Vicente Emparán, who gave them a cordial welcome. As it happened, Don Vicente was interested in scientific matters himself, descending from a seafaring family and having been a naval captain before assuming his current post. In fact, two of his brothers had been involved in a bizarre and tragic incident in the war with Britain. Commanding different Spanish warships, the brothers had come upon each other one night outside the port of Cádiz, near Gibraltar. Each mistaking the other for a British vessel, the two ships had battled all night, till both were eventually sunk. The two brothers, so the story went, had finally recognized each other just moments before expiring.

In Cumaná, Humboldt rented a big, airy house well situated

for astronomical and meteorological observations. For the next few weeks, he and Bonpland busied themselves testing their instruments and inspecting the damage from the earthquake. Although they tried to work methodically, they were overwhelmed by the incredible richness of life in the tropics. "We are here in a divine country," Humboldt wrote his brother. "Wonderful plants; electric eels, jaguars, armadillos, monkeys, parrots; and many, many, real, half-savage Indians, a handsome and interesting race. . . . What trees! . . . And what colors in birds, fish, even crayfish (sky-blue and yellow!). We rush around like the demented; in the first three days we were quite unable to classify anything; we pick up one object to throw it away for the next. Bonpland keeps telling me that he will go mad if the wonders do not cease soon."

Darwin, landing in Brazil in 1832, had a similar reaction, colored by his reading of his predecessor. "Humboldt's glorious descriptions are and will for ever be unparalleled; but even he with his dark blue skies and the rare union of poetry with science which he so strongly displays when writing on tropical scenery, with all this falls far short of the truth," he averred. "The delight one experiences in such times bewilders the mind; if the eye attempts to follow the flights of a gaudy butterfly, it is arrested by some strange tree of fruit; if watching an insect one forgets it in the stranger flower it is crawling over; if turning to admire the splendor of the scenery, the individual character of the foreground fixes the attention. The mind is a chaos of delight, out of which a world of future and more quiet pleasure will arise. I am at present fit only to read Humboldt; he like another sun illuminates everything I behold."

Yielding to this botanical seduction, Humboldt and Bonpland plunged into the nearby plains to gather the first of what would be tens of thousands of plant specimens shipped back to Europe. Some of the most challenging collecting took place in the nearly impenetrable thickets of cacti that the locals called *tunales;* more than once the pair were caught unawares by the brief tropical twilight and had to rush to extricate themselves before the rattlers and coral snakes emerged from their dens for their evening's hunting.

As the visitors flitted from telescope to microscope and back again, their intense activity attracted the curiosity of the residents, who started dropping in to have a look through the instruments themselves. With these human interruptions now piled on top of the myriad natural distractions—plus the rainy winter weather—it was nearly three weeks before Humboldt could begin a regular schedule of astronomical observations. But when the clouds finally cleared, what a profound beauty they revealed. "How can I describe to you the beauty of the sky . . . ?" he wrote. "Venus plays here the role of the moon. It shows big, luminous haloes, two degrees in diameter, of the most beautiful rainbow colors even when the air is completely pure and sky quite blue. I believe that here the stars offer the most magnificent spectacle." Eventually, he was able to place the town at 10 degrees, 27 minutes, and 52 seconds of latitude and 66 degrees, 30 minutes, and 2 seconds of longitude—half a degree, or nearly thirty-five miles, farther north than the previous best measurements had located it. In fact, for three centuries, Humboldt discovered, the entire continent of South America had been positioned too far south on the world's maps, as dead-reckoning mariners, misled by strong currents out of the Caribbean, had overestimated the distance to terra firma.

Humboldt and Bonpland's quarters afforded an unwelcome view of another spectacle, less natural and less innocent than astronomical and meteorological phenomena: The house was located on the same covered square as Cumaná's bustling slave market. Spain had been importing African slaves to their New World colonies since the beginning of the sixteenth century, after it had become apparent that the native population could never provide enough labor for the Europeans' ever-expanding mines and plantations. Though the slave trade had peaked in the eighteenth century, it wouldn't cease for another seventy years, by which time more than eight million Africans would have been transported against their will to South America and the Caribbean (the great majority to non-Spanish colonies), constituting more than eighty percent of all slaves introduced to the New World.

Every morning at the market in Cumaná, young African men aged fifteen to twenty would be forced to rub their bodies with coconut oil to make their black skin gleam. Then prospective buyers would shoulder their way through the captives, examining their teeth and sometimes searing their purchases with a branding iron. The market was Humboldt's first direct experience of slavery, and it incited him to fury. "This," he seethed, "is the treatment bestowed on those 'who [in the words of French moralist Jean de La Bruyère] save other men the labor of sowing, tilling, and reaping.' " Of the rapacious, supposedly Christian Europeans, he demanded rhetorically, "What are the duties of humanity, national honor, or the laws of their country, to men stimulated by the speculations of sordid interest?"

TILL THEIR FORCED LANDING at Cumaná, Humboldt had no plan to explore South America. But now that fate had placed him on the verge of one of the globe's most magnificent—and scientifically uninvestigated—regions, he characteristically embraced the new circumstances. Beyond the coast, beyond the great savannahs to the south, lay the world's premier river system and rain forest. And though a few Europeans had ventured into the Amazon, none had ever lingered to study it in depth. There was a wealth of botanical specimens begging to be discovered, and exotic creatures waiting to be captured. Rocks and minerals to collect and the rivers to be traced on the world's maps. A vast, seething web of life waiting to reveal its secrets.

And not only was the Amazon a sublime temptation for the naturalist, it was the subject of one of the great geographic controversies of the day. On his journey downriver in 1743, La Condamine had heard stories of a natural canal, the Casiquiare River, that supposedly connected the Río Negro to the Amazon via the Orinoco. Such a natural junction between two great river systems exists nowhere else on earth, and when La Condamine reported what he had heard, European scientists rejected the idea as either absurd rumor or Spanish propaganda. "This is not the first time that what is positive fact has been thought fabulous, that the spirit of criticism has been pushed too far, and that this communication has been treated as chimerical by those who ought to have been better informed," La Condamine fumed. As late as 1798, the French geographer Philippe Buache refused to include the river on his new map and even added a dismissive note: "The long supposed communication between the Orinoco and the Amazon is a monstrous error in geography. . . ."

The controversy was not a strictly scientific one. There were also important economic and strategic issues at stake, since such a communication would greatly facilitate settlement, trade, and political and military control over a huge territory. Humboldt decided that, in addition to all the other scientific investigations to be made, the resolution of this pressing geographic question would be the focus of his expedition through the Amazon. His pulse must have quickened as he realized the unique opportunity that had been handed him—thanks to the bit of infected louse feces that had provoked the typhus epidemic on board the *Pizarro.*

There was just one impediment to this scheme. The rainy season had begun, making travel to the interior impossible, and it would be months before the weather would clear. Not wanting to be holed up in Cumaná all that time, the travelers made two excursions along the coast, which in themselves yielded a wealth of scientific information and insight.

The first, and shorter, of these excursions began on August 19, to the rugged peninsula of Araya, which, jutting north of Cumaná, was celebrated for its ruined castle, saltworks, and pearl fisheries. At two o'clock in the morning, Humboldt and Bonpland embarked with their guide in a large canoe on the Río Manzanares, south of the city, near the Indian suburb. Though the plains along the coast were arid and dusty, this region, then as now, was home to dairy farms and plantations. The riverbanks were pleasantly shaded with mimosas, ceibas, and other huge trees, and in the daytime the local children played in the cool, clear water. In the evening, their parents would join them, setting up chairs in the river and sitting for hours in their light clothing, smoking cigars, chatting about the weather, and gossiping about the supposed extravagances of the ladies of Caracas and Havana. Ignoring the small, apparently harmless local crocodiles, the bathers would occasionally be startled by a mischievous dolphin spouting water.

This night, fireflies swarmed in the trees lining the river. Passing plantations, the travelers saw bonfires kindled by the African workers; it was early Monday morning, and the slaves, determined to drain every vestige of enjoyment from their one day of leisure, were dancing to the bittersweet music of a guitar. Jaguar skins were spread in the bottom of the canoe, but the night was unseasonably cool, and the Europeans, already accustomed to the tropical climate, found it too cold to sleep.

At eight o'clock Tuesday morning, the canoe landed at the tip of the peninsula. The largest known salt deposits in the world had been discovered here by the Spanish in 1499, and at a time when the mineral was still the principal method of food preservation, the works had once excited the jealousy of the English, Dutch, and other colonial powers. But now the salt pits were guarded only by a forlorn battery of three guns, plus the house of the inspector and the huts of a few Indian fishermen. In fact, there was a sense of faded glory about the entire peninsula, and the inhabitants spoke wistfully about the splendors of days gone by. In 1726, the same rare hurricane that had destroyed the Spanish fort had transformed the salt lake into an arm of the sea, making it useless for salt production. Afterward, new pits had been dug in the nearby hills. Eight inches deep and up to nearly an acre in size, the pits were filled with a mix of rain and seawater, which left behind the salt deposits as it evaporated.

The party set out for the ruined Spanish fort, Castillo de Santiago, several miles away. The going was slow, with Humboldt stopping often to make detailed geological observations. As night overtook them, the party was navigating a slender path hemmed in by the sea on one side and a vertical rock wall on the other. Worse yet, the incoming tide was narrowing the trail at every step. Moving as quickly as the half-light would allow, the travelers finally gained the demolished castle, built of coral and once boasting sixteen guns. Rising on a barren mountain, overgrown with agave, cactus, and mimosas, the fort was "lugubrious and romantic" in the twilight, with a primordial character. To Humboldt, it bore "less resemblance to the works of man, than to those masses of rock which were ruptured at the early revolutions of the globe." The ruins still stand today.

The naturalists longed to watch Venus set over the decrepit castle, but their guide, parched with thirst and eager for his night's rest, tried to hurry them along with tales of jaguars and rattlesnakes. Relenting, the Europeans allowed themselves to be led through a thicket of prickly pears to a nearby hut, where an Indian family received them graciously. After a supper of fish, plantains, and, most appreciated of all, excellent water, they slung their hammocks for the night.

At dawn, Humboldt saw that they were in a cluster of tiny houses, the remains of a village that had once occupied the site.

There had been a church, whose carcass was half buried in sand and brushwood. When the castle had been dismantled, nearly forty years before, most of the villagers had moved away. But a few families had stayed, and they now made their living catching fish, which they took to Cumaná to trade for plantains, coconuts, and cassava.

In the village, they had been told, was a white shoemaker of Spanish descent. How could a cobbler support himself in a country where everyone went barefoot? Humboldt wondered. And indeed, when they came upon the man he was stringing his bow and sharpening his arrows to go hunting for birds. Still, the shoemaker received them with the self-conscious gravity of someone blessed with a rare talent. He was something of a know-it-all, and Humboldt wryly dubbed him "the sage of the plain." After haranguing them on the manufacture of salt, the warning signs of earthquakes, the techniques of prospecting for gold and silver, the uses of various medicinal plants, the Bible, and the illusion of human greatness, the cobbler drew from a leather pouch a few very small opaque pearls of poor quality, which he insisted that the visitors accept. Refusing all payment, he instructed them simply to relate how a poor shoemaker of Araya, a white man of noble Castilian race, had given them something that in their own country was considered very precious. "I here acquit myself of the promise I made to this worthy man," Humboldt duly concludes in the *Personal Narrative*.

In the sixteenth century, Araya and the surrounding islands had been renowned for their pearls. The Spanish had found the native inhabitants bedecked with necklaces and bracelets made from them, and, perhaps not coincidentally, the peninsula had been one of the first New World territories to be colonized, with Indian and African slaves put to work diving to satisfy the European luxury market. But by the end of the sixteenth century, the pearl industry had declined, and by the 1630s it was defunct, a victim of overfishing, the Venetians' introduction of fine artificial pearls, and the growing fashion for cut diamonds. No wonder the current inhabitants felt their best days were behind them.

Humboldt, Bonpland, and their guide left the village the next morning and, crossing the arid hills nearby, detected the acrid odor of petroleum, carried on the wind from the coast. Gaining the source, they waded out eighty feet from shore and, standing on the sandy bottom in waist-deep water, located a circle about three feet in diameter, from which a transparent yellow fluid bubbled up and slicked the surface of the sea for a thousand feet. It would be another fifty-five years until the invention of the kerosene lamp gave the world its first practical use for petroleum and sixty years before the first oil well was drilled (in Pennsylvania). At the dawn of the nineteenth century, the odiferous liquid was still a curiosity, and not even the prescient Humboldt could have guessed that he was standing over one of the largest oil reserves on earth—which some distant day, after the invention of the internal combustion engine, would become more precious than all the pearls, gold, and silver the Spanish ever took out of the New World.

That night, the travelers set out for Cumaná in a native fishing boat. The vessel was the best they could find, but as they left shore, Humboldt was disturbed to see that the little craft remained afloat only through the constant bailing of the fisherman's son. It was a nerve-wracking nighttime sail, but eventually the boat pulled into port—which Humboldt took as testimony of the extreme serenity of the seas in this region.

ON THE FOURTH OF SEPTEMBER, the travelers began a second, longer excursion from Cumaná, this time south and east to the highlands inhabited by the Chayma Indians. To a naturalist, it was an intensely promising area, with a cool, delicious climate, mountains, caverns, and majestic forests of palms and tree ferns. Humboldt was also eager to get his first prolonged look at the native people, "lately nomadic, and still nearly in a state of nature, wild without being barbarous."

The only route available called for two demanding crossings of the coastal mountains, one to reach the interior missions and another to regain the gulf. Owing to the difficulty of the roads, the travelers pared their baggage down to the bare minimum—two mules' worth, including food, a sextant, a dipping needle, a magnetometer, a few thermometers, a hygrometer, and some paper for drying plant specimens. Traveling southeast, they snaked along a narrow path on the bank of the Manzanares. At the Capuchin hospital of Divina Pastora, they turned northeast and entered a treeless plain where only cacti, spurge, portulaca, and other scrubby plants grew. After two hours more, they came to the foot of the mountains and abruptly found themselves in a magnificent forest, where enormous, vine-strewn trees towered above a floor of lush tropical plants bearing leaves several feet long.

On this excursion through the highlands—and throughout his five-year journey—Humboldt was determined to document as many plants as possible, especially those new to science. No doubt this was partly a reaction to the intoxicating abundance of exotic species. But this impulse to collect and classify lay squarely in the mainstream of botany at the time, when the great voyages of discovery had introduced thousands of previously unknown plants to Europe. However, as with so much of Western science, this enterprise had actually begun in ancient Greece, with a student of Aristotle named Theophrastus. Considered the father of botany, Theophrastus was the author of *Historia plantarum*, which described thousands of plant species and divided them by size into trees, shrubs, undershrubs, and herbs. He had also been the first to distinguish between monocots (plants with one seed leaf) and dicots (plants with two seed leaves), which even today remain the two principal divisions of flowering plants. By the sixteenth century, though, European naturalists were having trouble identifying new plants according to the Greek system. As a result, botanists began to compile books called herbals, which named and categorized species based on a fresh look at the plants themselves. These efforts represented a step forward, though they were still suffused with superstition—such as the "doctrine of signatures," which held that God had placed a mark on each plant to indicate its usefulness (heart-shaped leaves hinting that a plant was beneficial to that organ, for instance).

The following century saw rudimentary advances in plant anatomy (such as the discovery of cells by Robert Hooke) and physiology (such as discoveries in plant respiration and reproduction). But the emphasis remained on identifying and classifying plants, as all those exotic species collected by explorers continued to land on naturalists' worktables. In 1753, the great Swedish botanist Carl Linnaeus published his *Species plantarum*, which described and classified most plants known at the time, and introduced the binomial (genus and species) system of classification still used today. Linnaeus's method provided desperately needed standards, but its criteria for classification were somewhat arbitrary and not necessarily the most botanically significant. Linnaeus also considered species immutable and believed that he had discovered nothing less than God's unchanging plan of creation. Ironically, then, though the Swede's system is a cornerstone of modern botany, it can also be seen as the last hurrah of a rigid, outdated order that would soon give way to the more empirical spirit of nineteenth-century science.

In 1789, just ten years before Humboldt sailed, Antoine-Laurent de Jussieu offered a refinement of Linnaeus's system. The descendant of distinguished French botanists (his uncle Joseph, traveling with La Condamine in South America, had reportedly gone mad after the inadvertent destruction of his specimen collection), de Jussieu organized genera into broader categories called families. Not only did this render classification infinitely easier and less arbitrary (most of de Jussieu's families are still in use today), later, as the theory of evolution gained acceptance, similarities underlying genera and families would come to be seen as evidence of those plants' descent from a mutual forbear.

But *On the Origin of Species* was sixty years in the future. In 1799, botany was still immersed in the hunt for new species, and Humboldt was eager to add what he could to that trove of knowledge. And he had grander ambitions for his journey as well. Not content to deliver even thousands of new plants to Europe, he was intent on discovering "how the forces of nature interact upon one another and how the geographic environment influences plant and animal life." He realized, though, that that underlying unity would reveal itself only through painstaking observation and measurement, collection, and classification. Therefore, he would first focus on the specific, which, besides its intrinsic value, would also allow him to tease out more general laws. Throughout the journey, de Jussieu's system of families would be Humboldt's constant guide as he sought to fit each new genus or species into the emerging pattern of nature.

As the party climbed away from the coast, the flora began to change, and even assume an eerie familiarity. In fact, it improbably began to remind Humboldt of the marsh plants of northern Europe, thousands of miles away in a radically different climate. In the tropics, he realized, plants growing at that altitude resemble vegetation found in temperate regions at lower elevations. Or, as he would later phrase it in *Aspects of Nature*, "The great elevation attained in several tropical countries . . . enables the inhabitants of the torrid zone—surrounded by palms, bananas, and the other beautiful forms proper to those latitudes—to behold also those vegetable forms which, demanding a cooler temperature, would seem to belong to other zones. . . . Thus it is given to man in those regions to behold without quitting his native land all the forms of vegetation dispersed over the globe, and all the shining worlds which stud the heavenly vault from pole to pole."

Humboldt also noted how the geology of the area influenced plant life and even human culture. Following a trail along the stream, he saw that the vegetation was more lush wherever the underlying limestone was topped with a layer of sandstone, which served to trap more water near the surface. In these areas they would also find Indian huts, surrounded by gardens of plantains, papaws, sugarcane, and maize. At first, Humboldt doubted that such small plots could feed a whole family, till he recalled that plantains produce twenty times as much food per acre as grain. But the tropical climate that generated this fecundity also limited human progress, he believed, by the very isolation and ease of living it afforded: The necessity of food, which in a harsher climate would have excited man to labor, was very simply fulfilled in this fertile soil and mild weather. "We may easily conceive why, in the midst of abundance, beneath the shade of the plantain and breadfruit tree, the intellectual faculties unfold themselves less rapidly than under a rigorous sky, in the region of wheat, where our race is engaged in a perpetual struggle with the elements . . . ," he suggested. "Without neighbors, almost unconnected with the rest of mankind, each family of settlers forms a separate tribe. This insulated state arrests or retards the progress of civilization, which advances only in proportion as society becomes numerous, and its connections more intimate and multiplied. But, on the other hand, it is solitude that develops and strengthens in man the sentiment of liberty and independence; and gives birth to that noble pride of character which has at all times distinguished the Castilian race." Thus does the physical environment shape not only the vegetation of a region, but also the character of the human settlers.

As the party continued south, the soil became dry and sandy, and they ventured into a range of forbidding mountains isolating the coast from the savannahs and the rain forest beyond. True to its name, this so-called Imposible was nearly impenetrable. Still, the only road connecting the area to Cumaná twisted through the steep, barren terrain, and the inhabitants of the plains below were forced to send their maize, leather, cattle, and other produce over this rugged country to reach the markets in the capital. At the crest of the Imposible was a spectacular view of the entire peninsula, taking in pastures, the rocky coast, and the natural basin forming the port of Cumaná. In fact, the view was so striking, so unlikely, that it reminded Humboldt of the fanciful landscape that Leonardo da Vinci had created as the background for his *Mona Lisa*.

The travelers slept at a military outpost atop the mountain, then, leaving before sunrise, began the harrowing descent down the other side, tracing a path fifteen inches wide and bordered by steep precipices. At the foot of the mountains, they reentered a dense wood webbed with rivers, where orange and papaw trees gave evidence of *conucos*, or Indian plantations, that had been abandoned. For hours, they walked in half-light under the vaulting forest canopy. The tree trunks were everywhere encrusted by a carpet of vines, orchids, figs, and other parasitic plants so intertwined that the excited botanists could scarcely distinguish the individual species. "It might be said," Humboldt wrote, "that the earth, overloaded with plants, does not allow them space enough to unfold themselves. . . . By this singular assemblage, the forests, as well as the flanks of the rocks and mountains, enlarge the domains of organic nature." From a great tree fern hung the incredible, baglike nests of the orioles, whose cries mingled overhead with the calls of parrots and macaws.

Humboldt was overwhelmed by this first encounter with the rain forest, this realization of his boyhood dream to "travel in those distant lands which have been but rarely visited by Europeans." Awed as he was by the profligacy of life there, the forest's most startling feature was its staggering immensity. "When a traveler newly arrived from Europe penetrates for the first time into the forests of South America, he beholds nature under an unexpected aspect. He feels at every step that he is . . . on a vast continent where everything is gigantic—mountains, rivers, and the mass of vegetation," he found. "If he feels strongly the beauty

of picturesque scenery he can scarcely define the various emotions which crowd upon his mind; he can scarcely distinguish
what most excites his admiration, the deep silence of those solitudes, the individual beauty and contrast of forms, or that vigor
and freshness of vegetable life which characterize the climate of
the tropics."

At the village of San Fernando, near the junction of the Manzanares and Lucasperez rivers, Humboldt visited his first mission in the Americas. Constructed of clay reinforced with vines,
the neat huts of the hundred Chayma families were arranged on
wide, straight streets. The church was located on the village's
main square, along with the priest's house and a humble building used for lodging travelers known, as in all the Spanish
colonies, by the grand name of "the King's House." The missionary was an aged Capuchin friar from Aragon, obese but hale.
Sole representative of Church and Crown, he didn't let his re-
sponsibilities weigh too heavily on him.

"His extreme corpulency, his hilarity, the interest he took in
battles and sieges, ill accorded with the ideas we form in northern countries of the melancholy reveries and contemplative life
of the missionaries," Humboldt noted. "Though extremely busy
about a cow which was to be killed the next day, the old monk received us with kindness and allowed us to hang up our hammocks in a gallery of his house. Seated, without doing anything,
the greater part of the day, in an armchair of red wood, he bitterly complained of what he called the indolence and ignorance
of his countrymen." The sight of the scientists' instruments, books,
and specimens drew a sarcastic smile, and their rotund host
averred that of all the pleasures of life, none was comparable to a
good piece of beef. "Thus does sensuality obtain ascendancy,
where there is no occupation for the mind," Humboldt cautioned.

Three centuries earlier, Columbus had been the first to describe the New World's native peoples, after encountering the
Tainos on Hispañola. The Indians, he wrote his patron, Queen
Isabella, were "very well built, with very handsome bodies and
very good faces" and "very lovely" eyes. "They [had] the sweetest
voices in the world, and . . . [were] always smiling." They were
"simple," "gentle," "artless and generous." "They love[d] their neighbors as themselves," and showed "as much lovingness as though
they would give their hearts." They were a people ready to "be

delivered and to be converted to our holy faith" but "by love rather than force." More to the point, they were pliable—"so cowardly that a thousand of them would not face three [Spaniards]," making them "fitted to be ruled and set to work, to cultivate the land and to do all else that may be necessary."

Humboldt, being of a more scientific and humanitarian turn of mind, and having no need to convince his patrons of the economic or religious potential of the Indians, focused his first detailed portrait of the native peoples on their physical attributes rather than their perceived moral qualities. The description is lengthy and rather clinical, though not unflattering. Through the details, one can discern the obsessive observer, intent on capturing every data point for future study and comparison.

The Chaymas, Humboldt wrote, were short, thickset, and muscular, with tawny skin, broad shoulders, a flat chest, plump belly, and well-rounded, fleshy limbs ending in large feet and delicate hands. Their foreheads were small, and it was said of a beautiful woman, "She is fat and has a narrow forehead." The Chaymas had black, deep-set, almond-shaped eyes, which they had the habit of casting downward. Their cheekbones were high and their nose long and prominent, with broad nostrils. The jaw was wide, ending in a short, rounded chin. Their hair was dark and straight, and the men had very little beard. Humboldt didn't find the women attractive, but the young girls had a soft, melancholy look that was appealing. The women carried their hair in two long braids and wore necklaces and bracelets made from shells, bird bones, and seeds. Overall, Humboldt noted, the Chaymas resembled the Mongol race, suggesting a possible genetic link between Asians and Native Americans.

Showing the limits of his open-mindedness, Humboldt found that all the Chaymas tended to look alike: There were no distinguishing signs of age in individuals from twenty to fifty years old—no gray hair, no facial wrinkles—with the result that he often couldn't tell father from adult son or mother from grown daughter. In addition, all the Chaymas had "a sort of family look," which Humboldt ascribed to inbreeding as well as the vapidity of mission life. "Both in men and animals the emotions of the soul are reflected in the features; and the countenance acquires the habit of mobility, in proportion as the emotions of the mind are frequent, varied and durable. But the Indian of the Missions, being remote from all cultivation . . . drags on a dull, monotonous life . . . [and] this uniformity, this sameness of situation, is pictured" on the face.

In his letter to the queen, Columbus had suggested that it would be well "to teach [the Indians] to go clothed." But, three centuries later, the Chaymas still had not taken to the habit, despite the urging of the missionaries. In fact, the Indians expressed a sense of shame at being covered, and men and women alike remained naked in their houses, where they wore only a *guayuco*, a strip of cloth two or three inches wide and tied to the waist with a string. On trips to the village, they put on a short cotton tunic for the benefit of the padre, but out of the boundaries of the mission, they would carry the tunics rolled up under their arm, especially during the rainy season. The tunic was the extent of their wardrobe, for, as Humboldt reports, "Among the Chaymas, as well as in all the Spanish Missions and the Indian villages, a pair of drawers, a pair of shoes, or a hat, are objects of luxury unknown to the natives."

Leaving the mission, the party pressed farther south, where the road ran through a humid valley filled with small plantations. In the village of Arenas, the chief attraction was a man of Spanish descent named Francisco Lozano, who had reportedly suckled his child with his own milk. The child's mother had fallen ill shortly after the birth, and when the thirty-two-year-old father had held the boy to his breast to comfort him, he'd had been amazed to see that the irritation of the sucking had caused thick, very sweet milk to flow. His breasts had grown, and he had breastfed his son two or three times a day for the next five months, during which the child had had no other nourishment.

Lozano wasn't in the village when the visitors passed through, but later the man journeyed to Cumaná to give the Europeans a look at him, along with his son, now thirteen or fourteen years old. Bonpland, who had trained as a physician, examined the man and found that his breasts were indeed enlarged and wrinkled, like those of an old woman who had suckled a child. The left breast was especially prominent, and Lozano explained that that one had been the more productive.

Though Lozano's case was rare, it wasn't unique. Aristotle

and the Talmud had both reported instances of men suckling children, and Humboldt had heard similar stories from Russia, Syria, and Ireland. He also recalled seeing a he-goat in Germany that had produced more milk than any female. Other human cases were reported throughout the nineteenth and twentieth centuries, leading medical science to verify that under certain circumstances, stimulation of a man's nipples can in fact cause milk to flow, just as Humboldt reported in the strange case of Francisco Lozano.

At the town of Cumanacoa, Humboldt heard of a great crevice in the mountains from which flames were said to sometimes shoot several hundred feet in the air. During the earthquake that had destroyed Cumaná two years before, a dull subterranean rumble had reportedly issued from the same location. This promise of geologic pyrotechnics was too much to be resisted, and he and Bonpland set out to have a look at the site, accompanied by Indian guides eager to show the visiting expert what they hoped were gold deposits nearby. "In these colonies," Humboldt lamented, "every Frenchman is supposed to be a physician, and every German a miner." He neglected to mention that he and Bonpland, by virtue of their respective training, did nothing to belie those stereotypes.

The Indians hacked a path through the woods, and on September 10, Humboldt, Bonpland, and their guides set out for the crevice. Entering the gorge carved by the Río Juagua, they came across a porcupine that had recently been disemboweled by a jaguar, and the Indians ran back to the farm to fetch some small dogs—they wouldn't fend off the jaguars, they explained, but in the event of an attack, the cats would be apt to spring on the animals rather than on the more imposing men.

The party inched along a precipice two or three hundred feet above the river. When the path became too narrow, they picked their way down to the river, forded it, and clambered up the opposite bank. The farther they advanced, the thicker the vegetation grew, till the way became overhung with vines. But the terrain didn't prevent Humboldt and Bonpland from collecting botanical specimens, which they carried with some difficulty. The Europeans particularly admired the abundant mosses and lichens; the ten-foot-tall amomums (aromatic shrubs resembling ginger); the tree ferns; a twenty-foot-tall plant with brilliant purple flowers (*Brownia racemosa*); a previously unknown type of dragon's blood (a croton, or type of spurge); and the striking red and gold tropical woods, which Humboldt foresaw would one day be sought out by European cabinetmakers.

The travelers followed the Río Juagua till they came to the reputed gold mine, which had been dug into the walls of blackish limestone common in the area. As expected, Humboldt saw that the "gold" was worthless pyrite. But the Indians weren't convinced, and they bent over and scooped the glittering stones out of the water when they thought no one was watching.

Tired and soaked from crossing and recrossing the river, the party finally arrived at the cavern of Cuchivano, whence the subterranean flames were said to emerge. But Humboldt was crestfallen to see that the cave was totally inaccessible, set in the middle of an eight-hundred-foot-high, sheer rock wall. Resting at the foot of the cliff, he mused over the source of the pyrotechnics. Could they be caused by hydrogen gas issuing from the cavern? Probably not, he decided, since sulfur was more likely to issue from volcanoes than hydrogen was. Ultimately he concluded that the accompanying earthquakes were probably the result of events taking place not near the surface at all but at enormous distances belowground. It was not a viewpoint that would have been endorsed by Abraham Gottlob Werner, Humboldt's neptunist mentor back at Freiberg, and it was leading toward a radical new approach to geology.

On September 12, the travelers set out for the monastery of Caripe, the principal mission among the Chayma. After an arduous climb back up the coastal mountains, they stopped for three days to rest at the modest farm of a transplanted Spaniard, where Humboldt felt a deep pang of homesickness: "Nothing can be compared to the majestic tranquility which the aspect of the firmament presents in this solitary region," he lamented. "When tracing with the eye, at nightfall, the meadows which bounded the horizon,—the plain covered with verdure and gently undulated, we thought we beheld from afar, as in the deserts of the Orinoco, the surface of the ocean supporting the starry vault of Heaven. The tree under which we were seated, the luminous insects flying in the air, the constellations which shone in the south; every object seemed to tell us how far we were from our native land," Humboldt found. "If amidst this exotic nature we heard from the depth of the valley the tinkling of a bell, or the lowing of herds, the remembrance of our country was awakened suddenly. The sounds were like distant voices resounding from beyond the sea, and with magical power transporting us from one hemisphere to the other. Strange mobility of the imagination of man, eternal source of our enjoyments and our pains!"

To reach the monastery, the party needed to cross a ridge called the Cuchilla de Guanaguana, whose name (*cuchilla* is Spanish for "kitchen knife") hints at the ruggedness of the terrain. The path was only fourteen inches wide in places, and was bordered on either side by a steep, grassy slope, so that if a traveler slipped, he would career straight down for seven or eight hundred feet. Not yet experienced climbers, Humboldt and Bonpland found the crossing difficult, but the mules were surefooted and inspired confidence. When the animals felt themselves in danger, they would stop and look to the right and the left, during which they were oblivious to any attempt to hurry them, before choosing their path. "In proportion as a country is wild," Humboldt noted, "the instinct of domestic animals improves in address and sagacity."

At the foot of the ridge, the party plunged into a dense forest. Owing to its coolness, Caripe was the only one of the high valleys of New Andalusia that was much inhabited. The monastery was magnificently situated, in front of an enormous wall of white rock covered with thick vegetation and studded with numerous springs. The travelers were received with great hospitality by the monks, and as at every other mission they visited throughout their journey, Humboldt detected no hint of intolerance on account of his Protestantism: "No mark of distrust, no indiscreet question, no attempt at controversy, ever diminished the value of the hospitality they exercised with so much liberality and frankness." The superior gave up his own room to Humboldt, who was impressed by the extent of the library he found there. The patio of the convent proved ideal for setting up their instruments.

The mission had a large *conuco*, with five thousand coffee trees plus maize, sugarcane, and other crops, which the Indians were required to work from six to ten every morning, under the watchful gaze of the alcaldes, the native foremen appointed by the superior. The alcaldes, Humboldt notes, "are looked upon as great state functionaries, and they alone have the right of carrying a cane. . . . The alcaldes came daily to the convent, less to treat with the monks on the affairs of the Mission, than under the pretense of inquiring after the health of the newly arrived travelers. As we gave them brandy, their visits became more frequent than the monks desired."

Other than its cool climate, the foremost natural feature of the Valley of Caripe was the Cave of the Guácharo, now known to be the largest cave in Venezuela, with more than five miles of underground passageways. Meaning "one who cries or laments" in Spanish, *guácharo* was the name given to a species of nocturnal bird residing in the huge cavern. About the size of a chicken, with a wingspan of three and a half feet, it had dark, bluish-gray plumage streaked by black, with white, heart-shaped spots on the head, wings, and tail. At nightfall, the birds would leave the cavern to feed on fruits. The creature was previously unknown to naturalists, and Humboldt classified the *guácharo* creature (*Steatornis caripensis*) in a genus of its own.

On September 18, Humboldt and Bonpland, accompanied by the alcaldes and most of the monks, set out for the cavern, located nine miles from the monastery in the base of a high mountain. The entrance to the cave was immense, eighty feet wide and seventy-two feet high, surmounted by spectacular foliage and with a river flowing from the mouth. So huge was the opening that the party penetrated more than four hundred feet before they had to light the resin torches they carried. Entering the cave, they could make out the hoarse, far-off cries of the birds. As the men advanced, the shrieks rose to a piercing crescendo, as the birds were startled by the light from the torches. Fixing a torch to the end of a long pole, the guides illuminated the funnel-shaped nests in the ceiling, fifty or sixty feet above.

Once every summer, the local Indians would come to the cavern armed with such poles. As the *guácharos* hovered overhead, defending their young and making the terrible cries for which their kind was named, the nestlings would be knocked to the ground and gutted on the spot. The area from the abdomen to the anus was loaded with fat, which would be cut out and carried to huts built from palm fronds at the entrance to the cave, where it was rendered in clay pots over fires of brushwood. The half-liquid, clear, odorless fat, called *manteca* ("lard") or *aceite* ("oil") *del guácharo*, was used for cooking and illumination, and was so pure that it could be kept for more than a year without becoming rancid. The crops and gizzards of the birds were full of hard, dry fruits called *semilla del guácharo* ("guacharo seed"), which were carefully collected and circulated across the valley as a treatment for fever. Though several thousand young were killed each year in this way, the species was saved from extinction because many more birds roosted deep in the cavern, where the Indians would not venture.

Eager to explore the cave, Humboldt persuaded the guides to penetrate farther than their annual hunting expeditions had taken them. But even the monks couldn't cajole them to pass beyond a sharp incline with a waterfall. The Indians believed that the souls of their ancestors inhabited the cavern, and in fact, "to go and join the *guácharos*" was a local euphemism for death. Though disappointed, Humboldt also confessed to being "glad to be beyond the hoarse cries of the birds, and to leave a place where darkness does not offer even the charm of silence and tranquility." Two hundred years later, the *guácharos*, now legally protected, still roost in the cave that bears their name.

Despite Humboldt's misgivings about the Church's treatment of the Indians, the following days at the monastery passed in a pleasant routine. The daylight hours the travelers would spend out of doors in the mountains and forests, collecting plants and returning to the monastery only when the bell tolled for meals. The evenings were devoted to making notes and sketches and drying their specimens. The only disappointment in the Valley of Caripe was that the winter weather prevented any astronomical observations.

On rainy days, Humboldt and Bonpland would visit the huts of the Indians or the community garden, or drop in on the alcaldes' evening councils. Sometimes they would attend the early-morning religious lessons, where the Indians' tenuous grasp of Catholic doctrine was evident, along with their imperfect knowledge of Spanish. Because the monks didn't speak Chayma, for instance, it was surpassingly difficult to convey the idea that *infierno*, "hell," and *invierno*, "winter," were as different as hot and cold: Since the Chaymas knew no winter but the rainy season, they imagined the padres' hell as a place where the wicked were exposed to frequent showers.

Though he was surrounded by tropical plants, strange animals, and foreign peoples, Humboldt's predominant impression was not how alien the monastery seemed but rather how much the place reminded him of Europe. "The aspect of this spot presents a character at once wild and tranquil, gloomy and attractive," he wrote. "In the solitude of these mountains we are perhaps less struck by the new impressions we receive at every step, than with the marks of resemblance we trace in climates the most remote from each other. The hills by which the convent is backed, are crowned with palm-trees and arborescent ferns. In the evenings, when the sky denotes rain, the air resounds with the monotonous howling of the alouate apes [howler monkeys], which resembles the distant sound of wind when it shakes the forest," he found. "Yet amid these strange sounds, these wild forms of plants, and these prodigies of a new world, nature everywhere speaks to man in a voice familiar to him. The turf that overspreads the soil; the old moss and fern that cover the roots of the trees; the torrents that gush down the sloping banks of the calcareous rocks; in fine, the harmonious accordance of tints reflected by the waters, the verdure, and the sky; everything recalls to the traveler, sensations which he has already felt."

So taken were the visitors with the charms of the valley that they unintentionally overstayed their welcome. Belatedly, they realized that two extra places at the table were straining the meager resources of the monks, who were too kind to object. The brothers had already eliminated their own shares of wine and bread, which were considered luxuries, and Humboldt and Bonpland's portion of bread had been cut to a quarter of its original level—yet heavy winter rains prevented the visitors from leaving for another two days. "How long did this delay appear!" Humboldt lamented. "It made us dread the sound of the bell that summoned us to the refectory." At last, the rain stopped on September 22, and the travelers were able to say their abashed farewells, accompanied by their Indian guides and their mule train, now grown to four animals laden with instruments and plants.

Continuing their descent toward the coast, the party crossed an uninhabited plateau, then encountered a daunting slope the missionaries called Baxada del Purgatorio, or Descent of Purgatory. Consisting of loose sandstone and clay, the Baxada was so steep and so treacherous that the mules were forced to draw up their hind legs and slide down in stages, while the riders dropped the reins and allowed the animals' instinct to do its work.

At the foot of the Baxada, they entered the dense forest called Montaña de Santa María, where the going was even rougher. "It is difficult to conceive of a more tremendous descent; it is absolutely a road of steps, a kind of ravine, in which, during the rainy season, impetuous torrents dash from rock to rock. The steps are from two to three feet high, and the beasts of burden, after measuring with their eyes the space necessary to let their load pass between the trunks of the trees, leap from one rock to another. Afraid of missing their mark, we saw them stop a few minutes to scan the ground, and bring together their four feet like wild goats." The local Creoles had enough confidence in the animals to stay in the saddle during this process, but "fearing fatigue less than they did," Humboldt and Bonpland dismounted.

The forest was dominated by trees of stupendous height and breadth, draped with orchids and succulents, and the warm, wet air was permeated with the aromas of flowers, fruits, and exotic woods. Though the day was cloudy, the heat and humidity were oppressive. Thunder rolled at a distance, and the howling of the monkeys warned of an advancing storm. Stopping to observe a troop of the animals, Humboldt's party encountered a group of Indians, traveling single file on the road. Naked, the Indians moved silently through the forest, their eyes on the ground. The women carried heavy loads, while the men, and even the youngest boys, bristled with bows and arrows.

Exhausted and overcome with thirst, the Europeans were eager to know how much farther it was to the mission of Santa Cruz, their destination for the night. The Indians smiled and gestured helpfully, but to every question answered only, "Sí, Padre," or "No, Padre" (since to them every white man was a missionary). Unable to get a meaningful answer, Humboldt and the others continued on, with the forest seeming to grow denser and more forbidding by the moment. The mules were stumbling at nearly every step, and as the muleteers struggled with the animals, Humboldt and Bonpland went on ahead. After a torturous descent of seven hours, the forest suddenly gave way to a great green savannah, stretching beyond the horizon. Crossing several terracelike plateaus, the travelers finally arrived at the mission of Santa Cruz, in the center of the plain. They settled into the "King's House" there for only one night before pressing on through the dense forest.

Humboldt had planned to trek eastward along the Paria Peninsula, but at the mission of Catuaro he learned that the rains had washed out the roads there. Not wanting to risk losing the plant specimens they had already gathered, the travelers decided to take the forest trail to Cariaco, on the coast, where they could find a boat back to Cumaná. The corregidor, or governor, of the region dispatched three Indians to clear the path of vines and branches. And, to Humboldt's annoyance, the missionary insisted on accompanying them, as he was on his way to Cariaco on the coast to give last rites to a black man who had been condemned to death for his part in an unsuccessful slave rebellion the year before. En route, the rest of the party was subjected to the missionary's tirades on the innate wickedness of the blacks, the necessity of the slave trade, and the benefits that the slaves derived from their bondage among the Christians. Humboldt, of course, demurred. Though Spanish laws were "less harsh than some other nations'," he wrote, "such is the state of the negroes, that justice, far from efficaciously protecting them during their lives, cannot even punish acts of barbarity which cause their death."

The descent through the forest of Catuaro was no better than the Baxada del Purgatorio, especially over the slippery, wet clay of the Saca Manteca (Lard Sack). Finally, they reached Cariaco, set on the gulf in a broad plain and surrounded by plantations. Notorious for its fevers, Cariaco had nevertheless grown rapidly in recent years, to over six thousand people, as a result of the colony's liberalized trade policy. Cotton was the primary crop, having supplanted coconut, which was still grown for its oil (Humboldt called coconut the "olive" of the tropics). Early the next morning, the travelers boarded a narrow canoe the Spanish called a *lancha* to cross the Gulf of Cariaco to Cumaná. They followed the River Carenicuar, which cut a line straight as a canal through the surrounding plantations. On the banks of the river, Indian women were washing their laundry with soapberry, whose rubbery fruit produces a thick lather; since the coming of the missions, the small, round fruits were also used for rosary beads.

In the gulf, the waves were high and the winds contrary. Rain began to fall in torrents, and thunder rolled very near. Flocks of flamingos, egrets, and cormorants sought shelter, while the pelicans continued to fish, oblivious to the storm. Though Cumaná lay only thirty-six nautical miles across the gulf, the lancha was forced to pull into a small farm on the lightly settled south shore. After dark the rain abated, and sailing all night, the travelers finally arrived in Cumaná the following morning.

BESIDES GIVING HUMBOLDT his first glimpse of the South American rain forest, this excursion to the country of the Chayma offered him his first extensive contact with the New World's native peoples. Numbering about seven million by 1800, the Indians of South America and the Caribbean had fared miserably at the hands of the supposedly civilized Europeans, and Humboldt was deeply moved by their fate. It was in this region that Columbus had first spied terra firma, and by the beginning of the sixteenth century the coastal Indians had already fallen victim to slaving expeditions. Devastated by violence, cultural dislocation, and European diseases, the native peoples had suffered horribly, their civilization had collapsed, and their population had dwindled. With the arrival of the missionaries, the Church had pushed through laws that, though hardly enlightened by our standards, did protect the Indians from some of the worst depredations. Perhaps motivated by a desire to please their new masters and a belief that their own gods had failed to protect them, the Indians had adopted the white man's religion, at least superficially. And thus the Chaymas, like native peoples throughout the Americas, had been absorbed, village by village, into Spain's vast network of missions, stretching thousands of miles from California to the tip of South America.

It's difficult to exaggerate the crucial role played by the missions in the New World—or, for that matter, the influence exerted by the Catholic Church in both Spain and its overseas empire. In the fifteenth century, as King Ferdinand and Queen Isabella had consolidated power in the Spanish monarchy, they had incorporated the Church into the apparatus of state. As an intimate partner of the Crown, the Church—though not without its own internal rivalries and conflicts—controlled tremendous wealth at home and abroad and, along with other institutions such as the university and the crafts guilds, even convened its own courts, which operated outside secular law. In fact, the Spanish monarchs (among whose ceremonial titles was "Most Catholic King") derived their very authority to colonize the New Continent from Pope Alexander VI, whose 1493 Bull of Demarcation, amended by the Treaty of Tordesillas the following year, had divided the Americas between Spain and Portugal (first along a north-south line drawn 100 leagues west of the Azores and Cape Verdes, then along a line another 260 leagues farther west, or 46 degrees west longitude, slicing through the eastern hump of South America). In many areas of the New World, the clerics were better organized and more plentiful than the secular officials, and by the time of Humboldt's journey, the Church's property in the Americas was worth more than that of the Crown itself. As the foundation stones of this vast enterprise, the missionaries were charged not only with propagating the faith and introducing rudimentary education but also with exercising political control over the native peoples. In their far-flung districts, they were confessor, teacher, governor, judge; their power was absolute, divinely granted and royally sanctioned. In fact, it was the missionaries, rather than the bureaucrats or the soldiers, who from the beginning had been entrusted with disseminating Spanish culture throughout the New World.

Though the missions had imposed a semblance of peace, by gathering the Indians into artificial settlements they had also sapped the native people of their independence. "By subjecting to invariable rules even the slightest actions of their domestic life . . . ," Humboldt wrote, "[the Indians] have been rendered stupid by the effort to render them obedient. Their subsistence is in general more certain, and their habits more pacific, but subject to the constraint and the dull monotony of the government of the Missions, they show by their gloomy and reserved looks that they have not sacrificed their liberty to their repose without regret." Conquered by the sword, the Indians had been subdued by the Cross.

The Spanish believed that they were carrying civilization to the native peoples, but Humboldt didn't see the results in such simple terms. All Indians not subjected to European rule were commonly supposed to be nomadic hunters, but in fact agriculture had been practiced in the Americas long before the arrival of the Europeans. The Indians of the forest, settled in villages, ruled by chiefs and raising plantains, cassava, and other crops,

were, Humboldt argued, "scarcely more barbarous than the naked
Indians of the Missions, who have been taught to make the sign
of the cross." Moreover, as Humboldt witnessed during the cate-
chism lesson at Caripe, the padres tended to overestimate the effect of their teachings. "The reduced [missionized] Indian is often
as little of a Christian as the independent Indian is of an idol-
ater," he notes. "Both, alike occupied by the wants of the moment,
betray a marked indifference for religious sentiments, and a secret tendency to the worship of nature and her powers."

Though the missionaries had debased the native culture, they
had not succeeded in obliterating it. The Indians maintained
their own language, and they showed an independent spirit and
a reverence for tradition that the Spanish had not been able to
expunge. "The missionaries may have prohibited the Indians
from following certain practices and observing certain ceremonies," Humboldt noted; "they may have prevented them from painting their skin, from making incisions on their chins, noses, and
cheeks; they may have destroyed among the great mass of the
people superstitious ideas, mysteriously transmitted from father
to son in certain families; but it has been easier for them to pro-
scribe customs and efface remembrances, than to substitute new
ideas in the place of the old ones." By failing to educate the Indians to the secular aspects of European civilization, the Spanish
had not given the native peoples the tools to raise themselves up.

Indeed, the Spanish had stripped native culture of far more
than they had replaced. The monotonous way of life imposed on
the mission Indians wasn't likely to inspire the industry and ex-
ertion seen in independent Indians, yet the resulting inactivity
was taken for idleness and lack of intelligence. Thus the Europeans had created a cultural vacuum—which they then cited as
evidence of their own supposed superiority. On the contrary,
Humboldt suggested that the Indians that the Spanish considered "savages" were in fact the descendants of a highly advanced
civilization. If Europe had surpassed the New Continent, he argued, it wasn't due to any barbarity or inherent inferiority on the
part of the native cultures. Rather, their current condition was
the tragic result of degradation from a state of high civilization—
a debasement the Spanish had done little to ameliorate.

This was a novel, even outlandish, notion at the time, but

Humboldt's later exposure to the Inca, Aztec, and other ancient civilizations would only confirm his view. Though Creoles such as Francisco Clavijero had studied pre-Conquest indigenous peoples, Humboldt was the first European scholar of his stature to investigate these cultures—their history, architecture, language, religion—to recognize in them "those family features by which the ancient unity of our species is manifested."

FOUR

Caracas

IN CUMANÁ, HUMBOLDT BEGAN CHOOSING INSTRUMENTS, engaging guides, and otherwise preparing for the expedition into the Orinoco and Amazon basins. He intended to leave at the end of October, after an eclipse of the sun that he was keen to observe from the coast, where viewing conditions were apt to be better.

The evening of October 27, the day before the eclipse, was overcast. The nightly breeze never arose, and the humidity was stifling. For the past two weeks, the weather had been unseasonably hot, and a strange reddish-brown haze had spread over the sky, sometimes waning to reveal the craters of the moon and other times waxing to conceal all but the brightest stars. The phenomenon alarmed the locals, who considered the haze and stillness an infallible omen of disaster.

At about eight o'clock on the evening of the 27th, Humboldt and Bonpland went, as usual, down to the gulf to take the air and to mark the time of high tide. As they crossed the beach between the Indian suburb and the embarcadero, Humboldt suddenly became aware of footsteps in the sand behind them. Turning, he was startled to see a tall, half-naked Indian brandishing a macana, a bulbous club made of palm wood. Dodging to the left, Humboldt managed to avoid the blow. Bonpland, taken unawares, wasn't so fortunate. The club caught him above the right temple, and he collapsed on the ground. The clout had knocked off the Frenchman's hat, and instead of pressing the attack, the assailant inexplicably went after it. Humboldt rushed to his friend's side and helped him to his feet. Having lost the element of surprise, the Indian ran toward a thicket of cacti and mangrove. But he slipped in the loose sand, and Bonpland, gal-vanized by pain and anger, was the first to reach him. As he threw his arms around the Indian, the man drew a long knife. Alone, un-armed, a mile and a half from the nearest house, the two Euro-peans were seriously outmatched. But just then a group of Spanish merchants, who also happened to be strolling the beach, came to their aid. Outnumbered, the assailant fled through the cacti with the white men in pursuit. After a long chase, the Indian tired and took shelter in a cowshed, where he ultimately surrendered.

Taken into custody, the attacker told the authorities that he was from the area around Lake Maracaibo, in northwestern Venezuela, and had served on a privateer out of Santo Domingo. In fact, he was not a full Indian but what was called a zambo, with a mixture of Indian and black blood. As historian William Lytle Schurz points out, such mixed-bloods were "considered to be the one insoluble—and completely undesirable—ingredient in the racial melting pot of the colonies. Bitter and truculent, and spurned by the two peoples responsible for his hybrid soul, as well as by the Spaniard, he was a hopeless pariah and a poten-tial enemy of the society that would have none of him." Having quarreled with his captain, the man had been put ashore at Cumaná when the ship left port. But what was the motive for the attack? Why, after knocking Bonpland down, was he content with stealing only a hat? The authorities questioned him in his prison cell, but the man's answers were confused. Sometimes he seemed to imply that he had meant to rob them; other times he claimed he'd flown into a rage on hearing them speak French, the lan-guage of his erstwhile captain.

Humboldt helped Bonpland back to their house. During the night, he contracted a fever, and Humboldt sat up with him. Even so, the Prussian was out on the patio at five o'clock the next morning, preparing for the eclipse. The day was clear, and he had an unobstructed view. "Being endowed with great energy and for-titude, and possessing that cheerful disposition which is one of the most precious gifts of nature," Bonpland also insisted on work-ing as usual that day. Still, he became lightheaded whenever he bent over, and it was obvious that he was in no condition for an arduous journey through the rain forest. The travelers had no choice but to postpone their expedition into the Amazon and to wait in Cumaná in the hope that his symptoms would clear.

A few days later, on the night of November 3–4, the strange red mist that had been hanging over the city grew so dense that the moon disappeared save for a smudge in the sky. The next day, about two in the afternoon, heavy black clouds enveloped the inland mountains, and at four o'clock, a hollow, sporadic thunder was heard, seemingly coming from a great height. A few minutes later there was a violent gust of wind, followed by great drops of rain. At 4:12, the precise moment of the loudest peals of thunder, Cumaná was struck by an earthquake. Bonpland, hunched over a table of plant specimens, was nearly thrown to the floor. Humboldt, lying in a hammock, began to sway from the strong shock. Still, he didn't lose his scientific detachment. The quake ran north to south, he noted, and didn't seem to be a rolling motion, but a powerful up-and-down jolt. Another shock followed fifteen seconds later. People ran into the street, screaming.

For the rest of the afternoon, the sky remained cloudy and deadly calm. The sunset that evening was spectacular, with golden clouds parting near the horizon to reveal an indigo sky and a huge, distorted sun throwing off rainbow-hued rays. At nine o'clock, there was a smaller aftershock, accompanied by more subterranean rumblings.

The townspeople attributed the earthquake, the underground thunder, and the odd red mist all to the eclipse the week before. The tremor hadn't done great damage, but the people were still traumatized by the earthquake that had leveled the city just twenty-two months before. That evening they collected in the zócalo, afraid to return to their homes, and Humboldt and Bonpland were frequently interrupted by people wanting to know whether their instruments predicted another quake. There was great alarm the following day, when again at two in the afternoon, there was a violent gust of wind, followed by thunder and a few drops of rain. The phenomenon repeated itself at the same hour for the next week, to the continuing distress of the population.

A few years later, an unidentified Chilean tried to describe to British naval captain Basil Hall the psychological power of the temblores. "These earthquakes are very awful. . . . Before we hear the sound, or, at least, are fully conscious of hearing it, we are made sensible, I do not well know how, that something uncommon is going to happen: everything seems to change colour; our thoughts are chained irrevocably down; the whole world appears to be in disorder; all nature looks different from what it was wont to do; we feel quite subdued and overwhelmed by some invisible power, beyond human control or comprehension," he explained. "Then comes the horrible sound, distinctly heard; and, immediately, the solid earth is all in motion, waving to and fro, like the surface of the sea. Depend upon it, Sir, a severe earthquake is enough to shake the firmest mind."

The temblor at Cumaná was the first Humboldt experienced, and though relatively minor, it left a profound impression. Many years later, he wrote in *Cosmos*, "We are accustomed from early childhood to draw a contrast between the mobility of water and the immobility of the soil on which we tread. . . . When, therefore, we suddenly feel the ground move beneath us, a mysterious and natural force, with which we are previously unacquainted, is revealed. . . . A moment destroys the illusion of a whole life; our deceptive faith in the repose of nature vanishes, and we feel transported, as it were, into a realm of unknown destructive forces. . . . We no longer trust the ground on which we stand."

The aftershocks gradually subsided. On November 7, the strange red mist vanished, and the sky took on the brilliant blue often seen after a violent storm. That night, Humboldt set his telescope on one of the satellites of Jupiter, and the planet's atmospheric belts appeared more distinct than he had ever seen them before. Viewing conditions remained superb for the next several days, and he took advantage of the extraordinary clarity to measure the comparative brightness of various stars.

In the early hours of November 12, Bonpland happened to get up for a breath of air, when he noticed a meteor shower in the eastern sky. He awakened Humboldt, and the two went out onto the patio to watch. There were no clouds, and for the next four hours thousands of white streaks traced in a north-to-south pattern. The extraordinary storm gradually diminished after four o'clock, but was still visible even a quarter hour after sunrise. The meteors didn't go unnoticed by the townspeople, who customarily arose at four to prepare for morning mass. Coming on the heels of the temblor, the meteor storm wasn't a matter of indifference to the locals, who recalled that the great earthquake of 1766 had been preceded by a similar shower.

For the rest of their journey, Humboldt would inquire about

the storm and discover that it had been visible throughout Latin America. On his return to Europe, he would be astounded to learn that the massive meteor shower of 1799, now famous in the annals of astronomy, had been observed over nearly a million square miles, as far away as Greenland and his native Germany. These November meteor storms are now known as the Leonids, because they seem to originate in the constellation Leo. In fact, Humboldt's detailed observation of the storm of 1799 helped to demonstrate the regularity of meteor showers, and he was the first to suggest that they are caused by the earth passing through the orbiting debris fields of ancient comets—which we now know to be the case.

By the middle of November, Bonpland's symptoms had improved enough that he and Humboldt thought it safe to leave Cumaná for Caracas, the capital. After the end of the rainy season, they planned to strike out from there, across the vast plains known as the Llanos, toward the Amazon. On the sixteenth, at eight in the evening, they sailed for La Guaira, 180 miles to the west, which then, as now, was the seaport for Caracas and the principal port of Venezuela. Their vessel was a thirty-foot trader, low to the water, with an enormous triangular sail. Though these seas could be rough, the Indian pilots were skilled, and they navigated, like their forebears, without benefit of chart or compass. In any event, the alternative to sailing would have been a difficult, four-day journey over an arduous road through fever-infested territory.

Humboldt was leaving with some regret. "We quitted the shore of Cumaná as if it had long been our home," he wrote. "This was the first land we had trodden in a zone, towards which my thoughts had been directed from earliest youth. There is a powerful charm in the impression produced by the scenery and climate of these regions; and after an abode of a few months we seemed to have lived there during a long succession of years . . . ," he found. "In the lower regions of both Indies, everything in nature appears new and marvelous. In the open plains and amid the gloom of forests, almost all the remembrances of Europe are effaced. . . . Cumaná and its dusty soil are still more frequently present to my imagination, than all the wonders of the Cordilleras [Andes]. Beneath the bright sky of the south, the light, and the magic of the aerial hues, embellish a land almost destitute of vegetation." Moreover, the "sun does not merely enlighten, it colors the objects, and wraps them in a thin vapor, which, without changing the transparency of the air, renders its tints more harmonious, softens the effects of the light, and diffuses over nature a placid calm, which is reflected in our souls."

A few days after their departure, Humboldt later learned, the man who had attacked them managed to escape from jail. Far from being outraged at this supposed miscarriage of justice, Humboldt was relieved. Neither he nor Bonpland seems to have harbored any ill will toward their assailant, who clearly was suffering from some type of derangement. In fact, they worried for the man, since at that time and place, an accused could be held for as long as eight years before even facing trial. Still, one wonders whether the "escape" was engineered by the local authorities, who, after the departure of their distinguished European guests, were just as happy to forget the whole incident.

As their boat passed the mouth of the Río Manzanares outside Cumaná, the evening breeze raised gentle waves on the Gulf of Cariaco. The moon hadn't risen yet, and the Milky Way was reflected in the placid sea. Castle San Antonio peeked out from behind the coconut palms, and the torches of the native fishermen twinkled onshore. As the trader approached Cape Arenas, a pod of dolphins frolicked alongside, leaving a swirl of phosphorescence in their wake. From midnight to dawn, the boat drifted amid the barren, geologically tortured Caracas Islands. In the morning, it anchored in the crocodile-infested harbor of New Barcelona, a busy port of trade with Cuba, especially for beef. Humboldt took the opportunity to fix the town's latitude and longitude and to scramble up to the nearby fort to take in the scenery and do some geologizing. The boat hoisted sail again at noon and by late in the day had reached the Piritu Islands, which rose only eight or nine inches above sea level. The islands were covered with grass and wildflowers, and except for the huge red sun setting behind them, Humboldt would have sworn he was gazing out on a European meadow.

By dawn they had made such good progress that it seemed they'd be in La Guaira by dusk. But, fearing privateers working the area, the Indian pilot dropped anchor in the little harbor of Higuerote to await the shelter of night. This part of the coast had

a reputation as being unhealthy, and judging from the condition
of the inhabitants, Humboldt was inclined to agree. Since popular opinion blamed the insalubrity on the mangrove thickets
guarding the shore, Humboldt collected some of the wood for
later analysis. In experiments at Caracas, he'd discover that the
yellowish-brown tint to the water around mangroves was due to
the leaching of tannin and that moistened mangrove bark absorbed oxygen from the air.

The road from Higuerote to Caracas was wild and prone to
flooding. But though Bonpland still suffered sporadic dizziness
from his injury, he was determined to disembark and prospect for
new plant specimens. Humboldt must have been concerned when
his friend insisted on taking the overland route, but he himself
had no choice but to stay on board and safeguard the instruments. At dusk the trader set sail into choppy seas and contrary
winds, and by sunrise on November 21 they had rounded Cape
Codera. The coast here was wild and picturesque, dominated by
perpendicular mountains three or four thousand feet high. But
the land was also green and fresh, and the valleys were cultivated with sugarcane, corn, and coconut palms. The tallest of the
mountains were Niguatar and the double-domed Silla ("Saddle")
de Caracas, jutting into the sky like snowless Alps.

With its back to the steep rock wall, the town of La Guaira reminded Humboldt of Santa Cruz on Tenerife—stifling, solitary,
dreary. In fact, analyzing the detailed readings taken by a Spanish physician, Humboldt pronounced the port one of the hottest
places on earth. With about seven thousand residents, the town
consisted of just two streets running parallel to shore. Though
the waters were rough, the anchorage bad, and the tides inconvenient, La Guaira had grown into an important hub for transporting coconuts to market. With the waves precluding the use of
mules, free blacks would wade into the shark-infested waters up
to their waists, lugging their heavy loads. No one ever seemed to
be attacked, a fact they attributed to the bishop having given his
benediction to the fish.

For the past two years, La Guaira had been suffering from an
epidemic of yellow fever (called *vómito negro* in Spanish, after
the dark blood clots spit up in severe cases), characterized by a
high temperature, jaundice, and vomiting. Humboldt puzzled over
the possible causes. The epidemic seemed to have begun with the

arrival of more ships from northern Europe; did the vessels bring the disease, or were the arriving Europeans just more susceptible? It also seemed to have coincided with flooding of the nearby Río de la Guaira; had the resulting pools of stagnant water released some pestilential miasma? The disease scarcely ever penetrated into the cooler mountains, and it was widely considered to be, as Humboldt said, "developed by a corrupted air, destroyed by cold, conveyed from place to place in garments, and attached to the walls of houses." The fever was generally thought to be contagious, yet outbreaks didn't spread from quarter to quarter of the city, as one would expect in that case. And its progress didn't seem to increase with contact or decrease with quarantine. The disease was confined mainly to shore, where epidemics seemed to break out during the hot season, then die down during cooler weather. Despite all this data, Humboldt was stymied. "The more I reflect on this subject," he admitted, "the more mysterious it appears to me. . . ."

But the remarkable thing here isn't Humboldt's failure to solve the riddle; it's the accuracy of his description. All the clues are present—the stagnant water, the outbreak in warm weather and improvement with cooler temperatures, the lack of infection at sea and in the mountains—but Humboldt was lacking the vital context that would tie them all together. To do this he would have to have made his journey sixty years later—after Louis Pasteur had proven the preposterous idea that human health could be affected by creatures too small to be seen by the naked eye—to have known that diseases such as yellow fever are not caused by unhealthy air but by tiny parasitic organisms called germs. And it wouldn't be until 1881 that Cuban physician Carlos Juan Finlay would suggest that yellow fever is caused by a virus transmitted by a mosquito, which, yes, is "conveyed from place to place in garments, and attached to the walls of houses."

Warned not to spend the night at La Guaira because of the epidemic, Humboldt continued on to nearby Maiquetía. The road to Caracas was good, and the next day a five-hour uphill walk through an area renowned in Europe and North America for its spectacular vistas brought Humboldt to the capital city. Bonpland didn't fare so well, arriving four days later, after slogging along the flooded road from Higuerote. But his health had held up, and his specimen cases were packed, including several plants previously unknown to science, such as *Brownea racemosa* (a scarlet flower resembling a rose) and *Bauhinia ferruginea* (a small tree with large pale flowers).

Whereas La Guaira suffered in constant summer, Caracas enjoyed perpetual spring, owing to its elevation of about three thousand feet. Named after the Caracas Indians, the city had been founded in 1567, only to be sacked by the English buccaneer Francis Drake in 1595. Built on a precipitous slope along a narrow valley, the city had about forty thousand residents, eight churches, five monasteries, and a fine stone bridge. The streets were wide and straight and set out in a grid, like most of Spain's New World cities. The houses were spacious, though Humboldt feared they were built a bit high for this earthquake-prone country. He and Bonpland rented a large, pleasant house with a view of the surrounding landscape to wait out the rains. It would be another four months before the dry season would begin.

With the mountains pressing in on it, Caracas was a gloomy place, particularly in these cool, rainy months. At this time of year, the mornings would be clear, but toward evening wisps of vapor would form over the peaks, then gather into big, fleecy clouds. Again a pang of homesickness crept into Humboldt's journal. "Beneath this misty sky, I could scarcely imagine myself to be in . . . the torrid zone," he wrote, "but rather in the north of Germany, among the pines and larches that cover the mountains of the Harz."

The elevation also lent the climate a variability unusual in the tropics, and the city's inhabitants would complain of experiencing several seasons in a single day. The fickle weather was blamed for a surfeit of catarrh, an inflammation of the nose and throat, but the residents especially dreaded those occasions when the wind would rebound off the mountains and, filled with moisture, swing around from the west. Known as the *catia*, after the place where it seemed to originate, this wind was said to cause headaches among the susceptible, and people sometimes shut themselves up in their houses while it was blowing. The "wind of Petare," on the other hand, hailing from the east and southeast, was dry, pleasant, and invigorating—except during the rainy season, when breezes from this quarter brought frequent storms.

Varied though it was, the climate around Caracas was well suited to sugarcane, coffee, bananas, pineapples, and even strawberries, grapes, and fruit trees such as apples and quince. Looking back on all the capitals of Spanish America that he had visited, Humboldt would later reflect that, whereas he discovered the greatest passion for science in Mexico City and Bogotá and the most pronounced interest in literature in Quito and Lima, he found the strongest European flavor in Cuba and Venezuela, owing to their extensive commerce with the Old World and their ready access to the Caribbean—made all the more piquant by the addition of Native American and African influences.

The city also boasted a theater of more than fifteen hundred seats. By all accounts, Humboldt had no ear for music. But the theater was the hub of the city's social life, and he would often attend the concerts there. On those occasions, he appreciated the fact that the orchestra seats were open to the air, which allowed him to glance up during the performance and gauge the conditions for astronomical viewing later that night.

Humboldt found in Caracas a few families well versed in literature, music, and mathematics, but he was disappointed to see that natural history and the visual arts had been neglected. There was no newspaper in the city, and curiosity about the natural world was so tepid that he couldn't find a single person who had ever ventured to the top of the neighboring Silla de Caracas, the saddle-shaped mountain towering over the valley. Concerned only with their crops, the weather, and other minutiae of daily existence, the people seemed "to live not to enjoy life, but only to prolong it." The truth is that by the time of Humboldt's visit, Caracas was a deeply divided city—racially, socially, and politically. In another few years, it would become the locus of the South American independence movement.

The instigators of independence were not the most oppressed segments of the population, the Native Americans or the African slaves, but rather the whites known as Creoles. Elsewhere this term denotes a variety of racial and ethnic mixes, but in the Spanish colonies it referred to supposedly purebred Spaniards who had been born in the New World. Constituting ninety-five percent of the white population, the Creoles were themselves a diverse group, including descendants of the conquistadors, who had grown rich in service to the Crown over a period of centuries, as well as families who had arrived only recently from Spain, where they had enjoyed a level of prominence and wealth that they were eager to continue in the New World. Though there were bitter rivalries among the Creoles, they were nearly unanimous in their hatred of the Gachupines, the five percent of the white population that had been born in Spain. Also known as *peninsulares*, the Gachupines were given precedence in everything, while the Creoles were shut out of the best administrative and military positions after the Bourbon Reforms of the late eighteenth century. Well aware of their own superiority, the Gachupines adopted a gallingly supercilious manner. "The most miserable, uneducated, uncultured European," Humboldt pointed out, "thinks himself superior to the whites born in the New World. . . ."

Increasingly restive under Spanish rule, some Creoles had begun to take pride in the land of their birth and to see themselves as a people apart. "One often hears people saying proudly, 'I am not Spanish, I am American,'" Humboldt reports. In fact, the Creoles' resentment was fanned by the presence of the Indians and Africans. For, however low he happened to be on the social scale, every white could comfort himself that he was at least superior to every Indian and every black, as reflected in the oft-heard expression, *Todo blanco es caballero*, "Every white is a gentleman." With their common skin color blurring the less obvious distinctions among Creole and Gachupine, the former found it easier to challenge the reputed superiority of their so-called betters. Humboldt would often hear a barefoot white ask in indignation, "Does that rich man think himself whiter than I am?" And thus, the spirit of independence was fostered in part by the subjugation of the nonwhite races.

These simmering resentments, exacerbated by an economic downturn, had bubbled over into an abortive Creole revolt in Venezuela in 1797. That same year, Great Britain seized the nearby island of Trinidad, which it started using as a base for illicit trade with the Spanish colony. This was not the last of the meddling by Spain's longtime rival, who would later play a key role in Venezuela's struggle for independence. Several years after

Humboldt and Bonpland's departure, Caracas would take the lead in the revolt against Spain.

THE TERRACE of Humboldt and Bonpland's house in Caracas had a tantalizing view of the Silla, and it wasn't long before the naturalists resolved to scale the peak, rainy season or no. The night of January 2, 1800, they spent at a coffee plantation outside town, so as to get an early start the following morning. But the weather was unexpectedly clear, and instead of resting up for the climb, the travelers stayed up all night watching for three predicted eclipses of the moons of Jupiter, since comparing the exact times of the eclipses to published tables would allow them to determine their longitude. However, owing to an error in the charts, the pair ended up missing all three occlusions, much to Humboldt's disgust.

The party of eighteen climbers set out before dawn, including local dignitaries as well as slaves to bear the instruments. The path was known to be frequented by smugglers, and though their black guides were somewhat familiar with the route, no one in the party had ever climbed as far as the eastern, or higher, spur of the Silla that was their objective. For the next two hours they hiked through the clear, cool morning, one behind the other on the narrow trail. On reaching the Puerta de la Silla, where the grade increased to thirty degrees, they felt the lack of crampons and iron-tipped sticks, since the grass was too short to provide a handhold and the underlying rock was too hard to be dug out into steps. One of their companions, a young Capuchin monk and professor of mathematics, had been boasting of his superior strength and courage. But now he and the other townfolk found the going particularly hard. Frequently, the fitter naturalists were forced to stop and wait for the rest of the party—till, peering down the mountain, they watched their erstwhile companions, including the young Capuchin, give up and begin their descent without them. The desertion was all the more unfortunate since the Capuchin had been entrusted with the entire supply of food and water. After that, Humboldt and Bonpland positioned the slaves bearing the most important instruments in front of them on the trail, to prevent the porters from abandoning them as well.

They had been advised not to make their climb after a day of fine weather, since at this time of year it was unlikely to find two clear days in a row. Now the wisdom of that advice was revealed. The weather was deteriorating, and slender streaks of vapor rose up the mountain in the morning breeze. Still ascending, the climbers became enveloped in mist so thick that they could barely find their way, even on their hands and knees. After four more hours of hard going, the grade moderated at a little wood called the Pejual, and, despite the bad weather and difficult terrain, Humboldt "felt an indescribable pleasure in examining the plants of this region. Nowhere, perhaps, can be found collected together, in so small a space, productions so beautiful, and so remarkable in regard to the geography of plants," he wrote.

In fact, on the Silla de Caracas, the very first of many mountains he would climb in the New World, Humboldt confirmed some of his ideas about plant distribution and made several observations that Charles Darwin would later marshal as evidence for the theory of natural selection. Here on the Silla, just as in the hills outside Cumaná, Humboldt was struck by the similarity between the local flora and plants he knew from Europe. The rhododendrons bore a striking resemblance to those he had seen in the Alps. In the valley below was a willow (later named *Salix humboldtiana*) comparable to ones growing at home. Other travelers had made similar findings: Grasses like those in Switzerland had been found in the Straits of Magellan; more than forty flowering plants known in Europe had also been discovered in Australia; a violet once thought unique to Tenerife had been seen in the Pyrenees. And so on.

What could account for such similar plants growing half a world apart? Though soil and atmospheric pressure no doubt played a role, the key factor was temperature, Humboldt realized. Similar plants tended to grow where the average temperatures were comparable, which sometimes meant at corresponding latitudes—in Pennsylvania and central France, for instance. But average temperature wasn't determined only by distance from the equator; height above sea level must also be factored in. How to keep track of all these variables of latitude and altitude and to organize the data in a way that would let the underlying patterns emerge? His solution was as elegant as it was functional—lines could be drawn on a map connecting those places with the same mean temperature. It would be along such lines that one would expect to find similar species of plants. Invented as a tool for discovering "the unity of nature," Humboldt's isotherms proved a cornerstone of climatology, allowing an immediate visual comparison of climate from region to region. The technique, still universal in meteorology, is also familiar to anyone who has ever watched the Weather Channel or checked the weather map in the daily newspaper.

But climate only created the conditions for plants to grow. How did similar species actually arrive at these far-flung locales? How could the plants have traveled across the ocean? Humboldt, like Darwin after him, concluded that simple migration couldn't account for the spread of species around the globe. Or even across a single continent. In South America, for instance, the Andes provided an impassable barrier, yet the same plants were found on either side. How could they have crossed some of the highest mountains in the world? (Today, we know that the geographical distribution of plants is determined by a complex interplay of geologic and climatic factors—the shifting of tectonic plates, the rise of mountains, the advance and retreat of glaciers—but at the beginning of the nineteenth century such concepts were still well in the future.)

True, Humboldt allowed, all these widely distributed plants might not be exactly the same, but they were definitely related: "When nature does not present the same species, she loves to repeat the same genera." Thus Humboldt, like Darwin, also noted differences in geographically isolated species. But whereas both Humboldt and Darwin asked why the plants were so similar, only Darwin went on to ask why they weren't *exactly* the same. That is, how had the various species come to diverge? It was by framing the question the other way around that the Briton would remake the science of biology and revolutionize our understanding of man's place in the natural world. Still, Humboldt's obsessive cataloging would provide Darwin with important evidence for his theory. *On the Origin of Species* cites the variety of plants growing on the Silla de Caracas, as detailed in Humboldt's *Personal Narrative*, as exactly the type of biological diversity that the theory of natural selection sought to explain.

Engrossed in their botanizing, Humboldt and Bonpland lingered on the Silla, though the sky was growing blacker and the temperature had dropped to barely fifty degrees. Eventually they crossed the western, lower hump of the saddle and entered the hollow between the two peaks, where they hacked a path through a dense forest. The fog was now so thick that they could guide their way only by compass—so that at every step they were in danger of plummeting over the sheer, six-thousand-foot ledge into the sea below. Finally they were forced to stop and, while waiting for the weather to clear, were rejoined by some slaves sent up with provisions. Whether the Capuchin had miscalculated or whether the slaves had helped themselves along the way, all that arrived were some olives and a little bread. "Horace, in his retreat at Tibur, never boasted of a repast more light and frugal," Humboldt commented dryly, "but olives, which might have afforded a satisfactory meal to a poet, devoted to study, and leading a sedentary life, appeared an aliment by no means sufficiently substantial for travelers climbing mountains." And it had been nine hours since they'd come across a spring.

Their duty done, the slaves were intent on leaving again, and Humboldt and Bonpland had a hard time stemming a mass desertion. Eventually they sent down half the servants, with orders to rejoin them in the morning with some salt beef. No sooner had they made these arrangements than a strong, warm wind began to blow from the east and, within two minutes, dissolved the mist and revealed the Silla's peaks again. The party resumed the climb, leaving the forest and entering an area of grass and low shrubs leading to the summit. Their route led very close to the precipice, but with typical understatement, Humboldt reassured those who would seek to follow him: "This part of the way is not dangerous, provided the traveler carefully examines the stability of each rock fragment on which he places his foot."

After another forty-five minutes, the party reached the summit, about eight thousand feet above sea level. The view was spectacular, encompassing the ocean, the Valley of Caracas, even the Ocumare Mountains, beyond which were hidden the Orinoco and the mighty Amazon. But the climbers didn't have long to enjoy the vista before a milky-white fog rolled up from the valley, partially obscuring the sun. Humboldt rushed to make as many measurements as he could—with sextant, dip needle, barometer, thermometer, cyanometer—but was harassed by small, hairy bees clinging to his hands and face. The guides assured him that the bees wouldn't sting unless grasped by their legs, but, for once, Humboldt wasn't tempted to try the experiment.

With the fog thickening, the party inched down the mountain, gathering a previously unknown genus of grass along the way. Back in the hollow between the peaks, Humboldt was surprised to discover quartz pebbles that had been rounded by running water. Clearly the Silla had once been submerged. His mentor, Werner, would have explained that the primordial seas had reached this high before subsiding, but Humboldt found it hard to believe that the earth had ever been covered in water so deep. More likely, he thought, the peaks of the Silla and all the mountains along the coast had been heaved up by volcanic forces. The more he saw of South American geology, the further Humboldt seemed to be moving toward the vulcanist camp.

By now it was four-thirty in the afternoon. The party retraced their steps to the enchanting Pejual, where they collected yet more specimens. "When, in these climates," Humboldt confesses, "a botanist gathers plants to form his herbal, he becomes difficult in his choice in proportion to the luxuriance of the vegetation. He casts away those which have been first cut, because they appear less beautiful than those which were out of reach. Though loaded with plants before quitting the Pejual, we still regretted not having made a more ample harvest." In fact, Humboldt and Bonpland were so bewitched by the botanical abundance that they lost track of the hours. By the time they reached the savannah below, the fleeting tropical twilight had passed, leaving them in absolute darkness still at an altitude of five thousand feet and buffeted by a rough, cold wind. The moon drifted in and out of clouds, and the resulting play of light and shadow created the illusion of precipices in the grass. Pulling off their boots for better traction, the climbers had to make their way on all fours to keep from slipping in the slick vegetation. One by one the guides deserted them to sleep on the mountain, but among those who remained was a slave from the Congo, who effortlessly balanced the large dip needle on his head, seemingly oblivious to the difficult terrain.

As they descended, the mist gradually dissipated, and lights from the valley came into view. Voices and even guitar music wafted up to them. Yet the longer they walked, the farther the valley seemed to recede. Toward the end, the remaining guides lost their way trying to find a shortcut to the farm where the party was to spend the night, and the resulting detour presented the steepest descent of all. It was ten o'clock when they finally reached the floor of the valley, after an arduous six-hour climb down. The Europeans' feet were bloody from the rocks and dry grass stalks, and everyone in the party was consumed by hunger and thirst, having hiked for fifteen hours with scant food or water. One can imagine how deeply Humboldt and Bonpland must have slept that night, not just from exhaustion but also from satisfaction at all those botanical specimens in their collecting cases, waiting to be dried, classified, and, someday, exhibited to their colleagues in science.

THE WEATHER HAD CHANGED. It was time to push into the Amazon. On the evening of February 7, Humboldt and Bonpland left Caracas for the Llanos, the great savannahs extending between the coast and the rain forest. A dozen years later, the city that Humboldt knew ceased to exist. Caracas had suffered several previous earthquakes, including one just two years after Humboldt's visit, but none was as violent as the one of March 26, 1812, during Venezuela's war of independence. Beginning the year before, the area around Caracas experienced unusually violent tremors. In fact—counter to Werner's theory that local causes, such as burning subterranean coal, were responsible for volcanic activity—the years 1811 to 1813 were particularly active, geologically speaking, throughout a wide region of the Western Hemisphere, with new islands created in the West Indies and earthquakes as far away as Tennessee and Kentucky.

March 26, 1812, the day of the Caracas quake, was clear and unusually warm. As it was Ascension Thursday on the Christian calendar, most of the residents were in church when the first shock struck, at seven minutes after four in the afternoon. Though the tremor lasted just five or six seconds, it was powerful enough to ring church bells throughout the city. The townspeople thought the danger had passed, when a tremendous subterranean roar was heard, resembling a peal of thunder but longer and louder. The second shock, measured between fifty and seventy-two seconds in duration, leveled the city and buried nine to ten thousand people in the rubble. The churches of La Trinidad and Alta Gracia, a hundred fifty feet tall and supported by pillars twelve to fifteen feet in diameter, were reduced to a heap of rubble scarcely six feet high. The military barracks were swallowed by the earth, taking their occupants with them. Ninety percent of the city was demolished, and much of what remained was uninhabitable.

"The night of the Festival of the Ascension witnessed an awful scene of desolation and distress," Humboldt would later write, drawing on contemporary accounts.

The thick cloud of dust which, rising above the ruins, darkened the sky like a fog, had settled on the ground. No commotion was felt, and never was a night more calm or more serene. The moon, then nearly at the full, illumined the rounded domes of the Silla, and the aspect of the sky formed a perfect contrast to that of the earth, which was covered with the bodies of the dead, and heaped with ruins. Mothers were seen bearing in their arms their children, whom they hoped to recall to life. Desolate families were wandering through the city, seeking a brother, a husband, or a friend, of whose fate they were ignorant, and whom they believed to be lost in the crowd. The people pressed along the streets, which could be traced only by the long line of ruins. . . .

Wounded persons, buried beneath the ruins, were heard imploring by their cries the help of the passers-by, and nearly two thousand were dug out. Never was pity more tenderly evinced; nor was it more ingeniously active than in the efforts employed to save the miserable victims whose groans reached the ear. Implements for digging and clearing away the ruins were entirely wanting; and the people were obliged to use their bare hands, to disinter the living. . . . Beds, linens to dress the wounds, instruments of surgery, medicines, every object of the most urgent necessity, was buried in the ruins. Everything, even food, was wanting; and for the space of several days water became scarce in the interior of the city. . . .

There was a duty to be fulfilled to the dead, enjoined at once by piety and the dread of infection. It being impossible to inter so many thousand bodies, half-buried under the ruins, commissioners were appointed to burn them: and for this purpose funeral pyres were erected between the heaps of ruins. . . . Amidst so many public calamities, the people devoted themselves to those religious duties which they though best fitted to appease the wrath of heaven. Some, assembling in processions, sang funeral hymns; others, in a state of distraction, made their confessions aloud in the streets. . . . Children found parents, by whom they had never till then been acknowledged; restitutions were promised by persons who had never been accused of fraud; and families who had long been at enmity were drawn together by the tie of common calamity.

FIVE

The Llanos

LLANOS TRANSLATES FROM THE SPANISH AS "PLAINS," but the word fails to capture the rugged essence of the land. Formed over millions of years by the erosion of the Andes, South America's steppes are forbidding, unsuited to raising crops, sparsely inhabited. Yet these bleak grasslands extend southward from Venezuela's coastal mountains, and to journey from Caracas to the great rain forest beyond, Humboldt had no alternative but to trek across this rugged territory. In fact, he found the stark landscape, so different from the lush coast, one of the truly awe-inspiring experiences of his five-year odyssey (which took in some of the planet's geographical superlatives): "I know not," he wrote, "whether the first aspect of the Llanos excite[s] less astonishment than that of the chain of the Andes."

The Llanos' size alone is daunting. Sprawling from the mountains of Caracas to the forests of Guiana to the delta of the Orinoco, the savannahs occupy some 225,000 square miles—an area larger than France. But to Humboldt, raised in mountainous, populous Germany, the desolation and absolute flatness were even more unsettling. "There is something awful, as well as sad and gloomy, in the uniform aspect of these steppes," he found. "Neither hill nor cliff rises, like an island in the ocean, to break the uniformity of the boundless plain. . . . Like the ocean, the Steppe fills the mind with the feeling of infinity. . . . Yet the aspect of the clear, transparent mirror of the ocean, with its light, curling, gently foaming, sportive waves, cheers the heart like that of a friend; but the Steppe lies stretched before us dead and rigid, like the stony crust of a desolated planet. . . ." Moreover, here "no Oasis recalls the memory of earlier inhabitants; no carved stone, no ruined building, no fruit tree . . . speaks of the art of industry of former generations. As if estranged from the destinies of mankind and riveting attention solely to the present moment, this corner of the earth appears as wild theater for the free development of animal and vegetable life. . . ."

In May, as the Llanos' rainy season begins, the grasses put out tender green shoots, mimosas and aquatic plants blossom, and boas and crocodiles emerge from the mud where they've been estivating. Then, as the downpours continue, water collects in muddy, malarial sloughs, and the myriad rivers, all tributaries of the Orinoco, begin to swell. Huge tracts of the Llanos flood, forming an inland sea navigable by oceangoing ships. The fauna—native species such as jaguars, agoutis (nocturnal rodents about the size of a rabbit), deer, antelope, armadillos, hares, capybaras (the largest of the rodents, growing up to four feet long), as well as domesticated animals such as horses, cattle, oxen, and mules—are forced to swim from island to island in search of pasturage or prey, struggling to avoid the crocodiles and electric eels en route.

But Humboldt crossed the Llanos in March, toward the end of the dry season, when the plains exude a menacing, unearthly sterility. "Everything seems motionless," he wrote; "scarcely does a small cloud, passing across the zenith . . . cast its shadow on the earth. Under the vertical rays of the never-clouded sun, the carbonized turfy covering falls into dust, the indurated soil cracks asunder as if from the shock of an earthquake. . . . Everywhere the death-threatening drought prevails, and yet, by the play of the refracted rays of light producing the phenomenon of the mirage, the thirsty traveler is everywhere pursued by the illusive image of a cool, rippling, watery mirror."

Dead palm trees, shorn of their branches, jut from the sand like the masts of sunken ships, and the traveler is tormented by dust storms: "Like conical shaped clouds, the points of which descend to the earth, the sand rises through the rarified air in the electrically charged center of the whirling current; resembling the loud waterspout dreaded by the experienced mariner. The lowering sky sheds a dim, almost straw-colored light on the desolate plain. The horizon draws suddenly nearer; the Steppe seems to contract, and with it the heart of the wanderer . . . ," Humboldt admitted. "Half concealed by the dark clouds of dust, restless with the pain of thirst and hunger, the horses and cattle roam around, the cattle lowing dismally, and the horses stretching out their long necks and snuffing the wind, if happily a moister current may betray the neighborhood of a not wholly dried up pool. More sagacious and cunning, the mule seeks a different mode of alleviating his thirst," by carefully scraping the thorns off the melon cactus with its hooves to get at the cool juice inside. Though temperatures abated when the sun went down, at night the hapless animals were preyed upon by enormous vampire bats, who opened wounds in the sleeping creatures' backs. The bats also hovered over the hammocks, threatening at any moment to affix themselves to the faces of the human travelers.

For their first two nights in this godforsaken territory, Humboldt's party rode only after dark, to avoid the blistering daytime sun. Late on the third afternoon, they arrived at a little farm called Hato del Caimán (Alligator Ranch), a house surrounded by a few rude huts constructed from reeds and animal skins. There were no corrals at Alligator Ranch, and the cattle, oxen, horses, and mules wandered at will, herded when necessary by half-naked men on horseback. Some of these *peones llaneros* were slaves and others were free, but all had been hardened by the unforgiving land. Living in the saddle, they subsisted on dried, salted meat, which they sometimes even fed to their horses.

Stepping wearily from their mounts, Humboldt and Bonpland let the mules lead them to a muddy yellow pool, the sole source of water, which an old slave suggested they drink through a handkerchief to minimize the grit and odor. Having slaked their thirst as best they could, the travelers, hot, dusty, and wind burned, stripped naked and dived into the fetid pool. But they had barely begun to savor their bath when they heard a splash on the opposite bank—an alligator slithering into the murky water. The men made a precipitous retreat, grabbing their clothes as they ran.

Dusk was approaching, and the pair headed back toward the farmhouse where they were staying. But after walking for more than an hour, they were forced to admit that they had lost their way. By now the sun had set, and in the dark they were unable even to retrace their steps to the water hole. We may find it ironic that the greatest scientific explorers in South American history could lose their way less than a mile from camp, but in deadly fact, the Llanos were no place to go wandering, whether by day or night. The two meandered across the savannah for a long time before finally taking shelter under a palm, where they kept watch for snakes, jaguars—and bandits, who reportedly made a practice of stripping unlucky travelers and tying them to the trunks of trees.

Eventually, hoofbeats sounded across the steppes, and with some trepidation, the two strained through the dark to see who was approaching. A rider appeared, lance in hand—a *llanero*, making his nightly rounds among the cattle. Relieved, Humboldt and Bonpland explained their predicament. But the sight of two strange white men claiming to be lost only heightened the horseman's suspicions, and it was some time before he deigned to let them chase his trotting horse back to the farmhouse.

The party set off again at two o'clock that morning, hoping to reach the small trading town of Calabozo before the worst of the afternoon heat. Incongruous though it seemed, the dark Llanos reminded Humboldt of his days back on the *Pizarro*. "The solemn spectacle of the starry vault, seen in its immense expanse;—the cool breeze which blows over the plain during the night;—the waving motion of the grass, wherever it has attained any height; everything recalled . . . the surface of the ocean." (Indeed, an oft-heard observation was *Los Llanos son como un mar de yerbas*, "The Llanos are like a sea of grass.") As the sun rose, so did the eerie mirages, making a herd of oxen appear to be floating above the earth, or causing phantom towers and burial mounds to shimmer in the distance, then abruptly vanish. A hot, dust-laden wind kicked up, and as they continued on, the men filled their hats with leaves as insulation against the vicious sun.

Calabozo, situated between the Guárico and the Uritucu rivers in the center of the Llanos, was a prosperous town of five thousand people and twenty-five thousand cattle. The muddy hollows in the surrounding countryside were infested with *tembladores*, or electric eels (actually not eels at all but a type of fish, *Electrophorus electricus,* related to carp). The animals had been described by other Europeans but never scientifically examined, and Humboldt, long fascinated by Luigi Galvani's concept of "animal electricity," was impatient to find some of the creatures.

Around 1750, Benjamin Franklin had proved, via his famous (and perilous) experiment with the kite, that lightning was a form of electrical discharge. But by the late eighteenth century, the predominant researchers on the topic were two Italians, Luigi Galvani and Alessandro Volta. Galvani, who had trained as a physician, discovered that a frog's muscles would contract when connected in a simple circuit with two different metals, such as brass and steel or copper and zinc. Publishing his results in 1791, he suggested that this "animal electricity" was produced by the frog's nerves and transmitted through the metal to the muscle.

Volta, however, had trained as a physicist and saw the phenomenon in starkly different terms. The nerves didn't produce the electricity, he argued; it was an interaction between the two metals that created the current, with the animal's tissues simply acting as a conductor. To prove his point, Volta substituted a chemical solution for the frog and was able to produce an electric current without the use of any animal tissue at all.

Not to be outdone, Galvani repeated his experiments without the metal electrodes and managed to create a contraction in the frog's muscle just by contact with the nerve. What to make of these seemingly contradictory results? It was a classic case of two branches of science being blinded by their own prejudices. For Galvani and Volta were actually studying two different aspects of the same phenomenon: The former, a physician, had stumbled upon electrical conduction through the nerves, while the latter, a physicist, had discovered a more general type of electrical phenomenon independent of living cells. But at the time, neither researcher nor any of their contemporaries could see the distinction.

Back in Germany, Humboldt had refined Galvani's laboratory techniques and performed thousands of experiments with animal electricity. In the end, he concluded that the nerves produced a substance that caused the muscles to contract, and that, although the metal electrodes increased this effect, they clearly didn't produce it. But by focusing his efforts on Galvani's side of the problem and never really turning his attention to Volta's more general phenomenon, Humboldt, too, never adequately distinguished between the physiological and purely physical manifestations of electricity. Although Humboldt was tantalizingly close to the solution, it was Volta who ultimately discovered the electrical storage device that would be named the voltaic pile in his honor—the electricity-generating configuration of fluids and metal plates that was the forerunner of today's ubiquitous batteries.

Humboldt had made some original discoveries concerning the interplay of nerves and muscles, but after Volta's breakthrough, both his and Galvani's work was brushed aside. Humboldt was mortified to have been on the verge of the same discovery but never to have made the critical leap. As he despaired to a friend, "I have observed a great many things, as I am not without ability, but I have achieved nothing." Had he thought to substitute nonorganic material for organic in some of his thousands of experiments, today we might be measuring electric potential not in units called volts but rather in humboldts.

Humboldt was eager to continue his research on animal electricity in a natural environment, and immediately after his arrival in Calabozo he set about trying to procure some electric eels, without success. He inquired again, promises were made, but still no eels materialized. He offered cash, but even that did no good. As Humboldt discovered, "Money loses its value as you withdraw from the coast; and how is the imperturbable apathy of the ignorant people to be vanquished, when they are not excited by the desire of gain?"

Meanwhile, he was astonished to meet in Calabozo a self-taught electrical experimenter named Carlos del Pozo. Del Pozo had never even heard the names Galvani and Volta. But he had pored over books on the subject, including Benjamin Franklin's memoirs, and, with that slight introduction, had managed to assemble two capacitors (devices for storing electricity)—a Leyden jar and a more sophisticated apparatus fashioned from two large metal plates he'd had shipped from Philadelphia. He'd also constructed electrophori for charging the capacitors and electrometers for measuring their potential, with the result that his one-man laboratory, hidden in these remote plains of South America, was virtually as complete as anything to be found in the great European centers of learning.

The Leyden jar, developed about fifty years before by Pieter van Musschenbroek at the University of Leyden (or Leiden) in the Netherlands, consisted of a glass bottle partly covered on both its outer and inner surfaces with metal foil; a conducting rod passed through an insulating material in the neck of the jar

and contacted the foil on the inside. With the outside grounded, a
charge was given to the inside surface (as by rubbing a glass rod
over a piece of fur and touching it to the conductor in the jar's
neck), which gave the outside surface an equal but opposite
charge. When the inside and outside surfaces were joined by a
conductor, the stored electricity would be released and a spark
would be produced. Del Pozo's more sophisticated capacitor used
the large metal plates, separated by a nonconducting material
such as glass or wax, to produce an even more impressive (and
potentially hazardous) jolt of electricity.

Until the unexpected appearance of these European visitors,
del Pozo had had no audience for his demonstrations except his
neighbors, who, though impressed, didn't have the scientific training to fully appreciate his accomplishment. Accordingly, he was
delighted to display his apparatus to the distinguished naturalists and in turn to examine the Leyden jar and the electrometers
they carried. Humboldt was pleased to encounter a kindred
spirit in this out-of-the-way place. And for his part, "Señor del
Pozo could not contain his joy on seeing for the first time instruments which he had not made, yet which appeared to be copied
from his own."

Humboldt hadn't given up his quest for the electric eels. But
he'd concluded that if he were to procure some of the *tembladores*, he would have to find them himself. Early on the morning of
March 19, he and Bonpland hiked to the village of Rastro, where a
group of Indians led them to a muddy basin in which eels were
known to congregate. Because the creatures buried themselves in
the mud, nets were useless for their capture. The roots of certain
trees were sometimes used to poison them, but this method perma-
nently enfeebled the fish, so that their powers couldn't be demonstrated adequately. The only other alternative was the method
called *embarbascar con caballos*, "to excite [the eels] with horses."

Rounding up about thirty wild horses from the savannah, the
Indians drove the animals into the pond. Their hooves aroused
the *tembladores*, who wriggled to the surface and initiated a
cruel and extraordinary contest. Shocked repeatedly around their
bellies, the terrified horses strove to escape, but were blocked by
Indians standing on the banks, crying and brandishing reeds.
Stunned into unconsciousness, two horses disappeared beneath

the surface, and at first it appeared that the others would share
their fate. But the ferocity of the eels' shocks diminished in time,
and the surviving horses were finally allowed to stumble out of
the water and throw themselves, exhausted, in the sand. Meanwhile, the eels, having discharged their stored electricity, were
easily taken by the Indians, using small poles fitted with long
cords.

The *tembladores* were about five feet long, with olive-green,
scaleless skin, yellowish-red heads, and two rows of yellow spots
along the back. Even in an enfeebled state, their two- or three-
second charge produced a painful numbness, similar to the shocks
that Humboldt had administered to himself during his physiology experiments. After four hours of handling the enfeebled animals, the naturalists experienced muscle weakness, joint pain,
and a general lethargy that lingered till the next day. And once
the eels had a chance to recharge themselves, they delivered a
severe shock—up to six hundred volts—whose "pain and numbness," Humboldt discovered when he imprudently stepped on
one, "are so violent that [they are] impossible to describe. . . ."

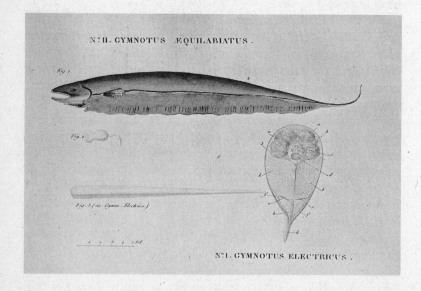

Electric eels.

From *Receuil d'observations de zoologie* by Humboldt and Bonpland.

No wonder the locals had been so reluctant to search out the creatures.

Over the course of a typically exhaustive investigation, Humboldt and Bonpland determined that the eels were able to discharge their shock at will. The creatures could also electrocute their prey without contact, and could even aim the shock. Cutting an eel in two, Humboldt discovered that only the front half was capable of producing a discharge. Also, although the eels' skin was covered with mucus, an electrical conductor, the animals were immune from shocks both from other eels and from themselves. Indeed, the only way to observe the shocks was by the obvious effect on their victim: No sparks were produced during the discharges, even in the dark; the bursts didn't register on the electrometer; and they produced no magnetic reading. Over the next several days, Humboldt and Bonpland subjected themselves to every imaginable abuse in the name of science. Among their discoveries: By gripping the eel with both hands or by holding the animal in one hand and a piece of metal in the other, they could magnify the charge; also, when two people held hands and one touched the eel, both felt the jolt simultaneously. Conversely, glass, wax, horn, dry wood, and bone served as effective insulators.

The first to scientifically study electric eels, Humboldt greatly increased our understanding of the animal and its peculiar abilities. However, he realized that a great deal more work remained to be done. Accordingly, he called on other scientists to continue the investigations, which he believed could even elucidate the mechanism by which all animals move. "It will perhaps be found," he presciently suggested, "that, in most animals, every contraction of the muscular fiber is preceded by a discharge from the nerve into the muscle; and that the mere simple contact of heterogeneous substances is a source of movement and of life in all organized beings." Two centuries later, scientists have built on these observations to develop a thorough knowledge of nerve-muscle interactions.

Their researches concluded, the scientists left Calabozo on the twenty-fourth of March. As they trekked south through the Llanos in the coming days, the ground grew even drier and more lifeless. The palm trees petered out, and the temperature hovered at a sun-scorched ninety-five degrees from eleven in the morning until sunset. If losing their way at the water hole had left any doubt concerning the Llanos' hazards, the point was driven home to them on their first day out of Calabozo. Around four o'clock in the afternoon, the travelers came upon an Indian girl sprawled in the brittle grass. About thirteen or fourteen years old, the girl was naked, exhausted, disoriented, and dangerously dehydrated. Her eyes, nose, and mouth were caked with earth, her breathing was reduced to a rattle in her throat, and she was unable to speak. A clay pitcher, half filled with sand, lay at her side.

The men managed to rouse her by washing her face and forcing a few drops of wine down her throat. Overcoming her initial alarm, she eventually told them her story. She had been working at a nearby farm, she said, till an illness had reduced her capacity for work. Losing her position, she'd set out for a nearby mission but had run out of water and collapsed in the unrelenting heat. From the position of the sun, she estimated she'd been unconscious for several hours. The travelers tried to convince her to mount one of the mules and accompany them as far as the neighboring town of Uritucu, but she could not be persuaded. All they could do was empty the sand from her pitcher, fill it with water, and watch her straggle off in a plume of dust.

Three scorching but uneventful days later, the party reached the Capuchin mission of San Fernando on the Río Apure, the principal tributary of the Orinoco. Stopping in San Fernando for three days, just long enough to hire a canoe and crew and to take on provisions, Humboldt prepared, at last, to venture into the greatest forest on the planet.

IT HAS BEEN SAID that a journey through the South American rain forest is a journey through time as well as space. Over the past hundred million years, while continents have drifted and mountains have risen, the forest has remained essentially unchanged. (The great forests of Europe and North America, by comparison, are just eleven thousand years old, having grown up only after the last ice age.) The most abundant, diverse ecosystem on earth, the Amazon constitutes only two percent of the planet's landmass but is home to fifteen percent of the world's plant mass, including an estimated five million botanical and animal species (most still unidentified by science today). Paradoxically, despite this biological profusion, the rain forest is not particularly fertile, since the nutrients have already been absorbed by the abundant plant life. With up to 150 inches of rain per year and relative humidity at a near-constant eighty percent, the forest is one of the wettest places in the world, and the unremitting dampness, combined with the high temperature, has long since leached the nutrients from the soil. Only the tiny fraction of forest regularly flooded by the silt-bearing rivers has soil that could be considered fertile.

Consequently, competition for nutrients in the rain forest is fierce. Since only ten percent of the sun's energy penetrates to the ground, due to the density of the foliage, most botanical life is found in the canopy, where plants battle for precious sunlight. Throughout the forest, no scrap of food goes wasted, and each species of plant and animal must carve out a highly specialized niche in order to endure. This produces the second paradox of the rain forest: Although there is unparalleled diversity of species, there is extremely low density of any particular species. Because there simply isn't enough food to support a large number of identical individuals in a given locale, members of each species must fan out through the forest to secure the resources they need. As a result, there are no groves of like trees in the rain forest, as there are in temperate forests, only isolated individuals. Animal and human populations are similarly limited. Though there are more than 1,500 species of fish, for example, compared with about 150 species in all of Europe, they are abundant only during the wet season. Similarly, food-bearing trees such as the Brazil nut are widely scattered, and having discovered one, a band of foragers could travel for days without sighting another. The only human food that could be called relatively abundant in the rain forest is the cassava, cultivated for its roots, which are ground into a flour that is the traditional staple of the region.

For someone with Humboldt's scientific training and relentless curiosity, the rain forest must have held an irresistible attraction. At last he would see for himself the exotic plants and animals that he had dreamed about since boyhood. Surely amid this living profusion he and Bonpland would discover many species never before recorded by naturalists—an incalculable scientific booty just waiting to be harvested. And where better but in the most prolific environment on earth to explore the myriad interconnected strands forming the great web of life itself? Where better to study the effects of soil and precipitation on plants, the interdependence of the plant and animal kingdoms, the myriad ways in which the physical environment shapes human culture? If the rain forest had nothing to teach about the unity of nature, then he would have to admit that the concept was pure fantasy.

In addition to his biological researches, Humboldt was determined to test La Condamine's hearsay report that the Orinoco connects with the Amazon through the Casiquiare River and the Río Negro, forming an extensive natural highway into the interior of the continent. Toward this end, though still a neophyte in the tropics, he mapped out an arduous itinerary through some extraordinarily demanding terrain. His party would sail down the Río Apure to its mouth in the Orinoco, then upstream on that great river, past the notorious cataracts at Ature and Maipures, and up its tributaries the Atabapo, the Temi, and the Tuamini in turn. From there they would make the five-day portage to the Caño Pimichin, which emptied into the Río Negro, part of the vast watershed drained by the Amazon. They wouldn't follow the Negro as far as the Amazon itself, but only to its junction with the Casiquiare, which (assuming La Condamine were correct) they would trace back to the Orinoco. Then they would descend the Orinoco to Angostura (present-day Ciudad de Bolívar), the capital of Spanish Guiana, located in the river's massive delta. Thus Humboldt hoped to prove once and for all the existence of the so-called Casiquiare Canal, the only natural connection on the planet between two great river systems.

On March 30, at four P.M., the explorers pushed their boat into the Apure and let the current carry it downstream. The craft was one of the long canoes known as lanchas. A cabin thatched with palm fronds had been constructed in the stern, fitted with a table and chairs made of stretched oxhides, and the canoe had been stocked with provisions for a month, including eggs, plantains, cassava, cacao, sherry, oranges, and tamarinds (the pulp of which was mixed with water and sugar to make a refreshing drink). Preferring to put their trust in the bounty of the river, the Indian guides had also brought fishhooks and nets. Some guns for hunting small game were shipped as well, along with extra fishing tackle, firearms, and casks of brandy for trading en route. In addition to their Indian pilot and crew of four, Humboldt and Bonpland were joined by a large stray mastiff they had adopted.

Completing the party was a Spaniard named Nicolás Soto, brother-in-law of the provincial governor, who had recently arrived from Cádiz and was eager to see the interior of the country.

Soto was notable as the first in a series of young European men to attach themselves to the party for weeks or months at a time. Though Humboldt was circumspect about these companions, from his few comments in the *Personal Narrative* we can infer that he enjoyed their company. One, in fact, would remain his fast companion for years to come, long after his return to Europe. It seems likely that Humboldt—loquacious, sociable, and the financial sponsor of the expedition—would have been the one to extend these invitations. Did Bonpland, the junior partner, find these supernumeraries a welcome diversion from Humboldt's exclusive company? (For that matter, can any two men, however compatible, be content to spend virtually every day and night together for five long years?) Or did Bonpland see them as interlopers who fulfilled no purpose except to occupy precious space and to consume limited provisions? Humboldt said nothing to suggest the latter, but this is not the type of personal information he would have divulged in any case. And since Bonpland didn't write his own record of the journey, we will never know his point of view.

Their first night on the river, the travelers slept at a sugar plantation called Diamante (Diamond). The next morning a contrary wind kept them ashore till noon, but when they pushed off, they quickly passed into a wild, exhilarating territory. In some

Humboldt and Bonpland's lancha.
From *Vom Amazonas und Madeira* by Franz Keller-Leuzinger, 1874.

places, the forest came right down to the water's edge, while in others the river's periodic flooding had created a sandy beach. Low bushes called *sausos* formed a natural hedge along the bank, broken only by the paths made by forest animals coming down to the water to drink.

Humboldt's sense of wonder and excitement are palpable. "You find yourself in a new world," he wrote, "in the midst of untamed and savage nature. Now the jaguar . . . appears upon the shore; and now the *hocco* [peacock pheasant], with its black plumage and tufted head, moves slowly along the *sausos*. . . . We saw flocks of birds, crowded so closely together as to appear against the sky like a dark cloud which every instant changes its form." Not that Humboldt imagined a real-life Peaceable Kingdom in the passing tableau. While the pilot, an old Indian of the missions, commented that the river was "just as it was in Paradise," the naturalist showed that from the outset he was aware of the dangers to be found in the forest. "In carefully observing the manners of animals among themselves, we see that they mutually avoid and fear each other," Humboldt responded. "The golden age has ceased; and in this Paradise of the American forests, as well as everywhere else, sad and long experience has taught all beings that benignity is seldom found in alliance with strength." Nature is "red in tooth and claw," Alfred, Lord Tennyson, would later write. And later still Charles Darwin would posit such competition as an integral component of "natural selection," the amoral, unforgiving means by which species improve and perpetuate themselves.

Everywhere, it seemed, crocodiles basked onshore, sometimes in groups of eight or ten, and so numerous that half a dozen individuals were nearly always in view. When excited by prey onshore, the huge reptiles—sometimes over twenty feet long—would raise themselves up on all fours and dart after it with astonishing swiftness. Though their diet generally consisted of *chiguires* (the local name for capybaras, the world's largest rodents), they would attack other animals when given the opportunity. Once Humboldt's mastiff went for a swim, and as the men watched helplessly from shore, a crocodile slithered straight at it. But just as the creature came within striking distance, the canny dog made a tight turn and began to paddle upriver. Though a much faster swimmer, the crocodile was far less maneuverable, and while it was momentarily swept downstream by the current, the mastiff was able to gain dry land.

On occasion, the crocodiles also went after larger prey, each year taking two or three people living along the river. But the inhabitants, Humboldt says, had "marked the manners of the crocodile, as the torero has studied the manners of the bull. When they are assailed, they put in practice . . . the counsels they have heard from their infancy." The Indians told the story of a young girl who successfully fought off one of the creatures and, despite having lost the lower part of her left arm and a great deal of blood, managed to swim to shore. "I knew," she explained coolly, "that the cayman lets go his hold, if you push your fingers into his eyes."

Jaguars were also very common along the riverbank, where they came to drink and to prey on other creatures, including *chiguires*, birds, snakes, deer, fish, turtles, and even crocodiles. Taking its common name from the Indian word for it, *yaguar*, magnificent *Felis onca* is the largest cat in the New World and the third largest on earth (after the lion and the tiger), reaching a length of up to six feet and a weight of up to four hundred pounds. Near a bend in the Río Apure, Humboldt spied one of the big cats, lying in the shade of a huge zamang tree (a type of mimosa), its paw resting on a fresh-killed *chiguire* to guard it from querulous vultures gathering nearby. Humboldt and Bonpland climbed into the lancha's small skiff to get their first close look at the magnificent predator, on the assurance of the crew that jaguars rarely attack boats—at least not unless they're exceedingly hungry. As the humans approached, the jaguar rose reluctantly and crouched behind the *sauso* bushes. But when the vultures ventured too close to the *chiguire*, the great cat "leaped into the midst of them, and in a fit of rage . . . carried off his prey to the forest." At night the explorers had to be constantly on guard against jaguars, lighting fires on the perimeter of camp to keep them at bay. Despite these precautions, they would often hear the creatures circling in the darkness.

In the middle of the day, as the temperature rose, the rain forest would assume an unnatural quiet. As famed British naturalist Henry Walter Bates wrote, "When the paddlers rested for a time, the stillness and gloom of the place became almost painful; our voices waked dull echoes as we conversed, and the noise made by fishes occasionally whipping the surface of the water was quite startling. . . . The few sounds of the birds are of the pensive or mysterious character which intensifies the feeling of solitude rather than imparts a sense of life and cheerfulness."

Humboldt noted this paradoxical quiet too. But always sensitive to the entire scene before him, he also detected the nearly subliminal buzz of tropical life: "How vivid is the impression produced by the calm of nature, at noon, in these burning climates! The beasts of the forests retire to the thickets; the birds hide themselves beneath the foliage of the trees, or in the crevices of the rocks. Yet, amidst this apparent silence, when we lend an attentive ear to the most feeble sounds transmitted through the air, we hear a dull vibration, a continual murmur, a hum of insects, filling . . . all the lower strata of the air. Nothing is better fitted to make man feel the extent and power of organic life," he found. "Myriad insects creep upon the soil, and flutter round the plants parched by the heat of the sun. A confused noise issues from every bush, from the decayed trunks of trees, from the clefts of the rocks, and from the ground undermined by lizards, millipedes, and *cecilias* [wormlike amphibians]. There are so many voices proclaiming to us that all nature breathes; and that, under a thousand different forms, life is diffused throughout the cracked and dusty soil, as well as in the bosom of the waters, and in the air that circulates around us."

A little downriver from the jaguar guarding its prey, the lancha passed a herd of *chiguires*. The animals watched unconcernedly as the canoe landed, their upper lips quivering like rabbits', until the mastiff finally scattered them. Though Humboldt found the creatures' smell disagreeable, the Indians made hams from them, which the missionaries ate during Lent, conveniently classifying the *chiguire* as an amphibian and not proper meat at all. Humboldt never acquired a taste for the animal, finding that the odor justified the traditional name for it, "water hog." Still, food was scarce on the river, and lacking anything more appetizing, that evening the travelers planned to roast a *chiguire* they had killed.

However, when they beached the lancha at a ramshackle riverside farm for the night, the proprietor had another idea. "Don" Ignacio, as he styled himself, using the traditional title of respect, claimed to be of pure Spanish blood despite his dark complexion. Insisting that *chiguires* were not fit food for *"nosotros caballeros blancos"* ("white gentlemen like ourselves"), he feasted the travelers on venison that he had killed with his bow and arrow the day before. Despite his airs, the man and his wife, "Doña" Isabella and his daughter, "Doña" Manuela, lived virtually naked and out of doors, with no roof whatsoever. Slinging their hammocks under the trees, Humboldt and the other travelers "soon had reason to complain of a system of philosophy which is indulgent to indolence, and renders a man indifferent to the conveniences of life."

After midnight, a furious thunderstorm broke, soaking them all without regard to skin color. The fierce wind carried Doña Isabella's cat into the hammock of one of the Indians, who awoke screaming that he was being attacked by a jungle animal—much to the others' amusement. Their host, meanwhile, congratulated the travelers on their luck in being able to weather the storm "in his domain among whites and persons of respectability." But the irony was not lost on Humboldt: "Wet as we were, we could not easily persuade ourselves of the advantages of our situation. . . . We were struck with the singularity of finding in that vast solitude a man believing himself to be of European race and knowing no other shelter than the shade of a tree, and yet having all the vain pretensions, hereditary prejudices, and error of longstanding civilization!"

By sunrise the next day, April 1, the rain was gone. The explorers bid farewell to Don Ignacio and Doña Isabella and steered the lancha through the perilous maze of floating trees uprooted in the storm. Later, after passing a low island called Isla de Aves (Bird Island), so densely packed with flamingos, rose-colored spoonbills, herons, and moorhens that the birds seemed unable to move, the canoe pulled into a wide beach for the evening. "The night was calm and serene," Humboldt wrote, "and there was a beautiful moonlight." The tracks of three jaguars were visible in the sand, and the crocodiles positioned themselves along the shore so as to be able to see the campfire. With no trees nearby, the men stuck their oars in the beach and tied hammocks to them. "Everything passed tranquilly till eleven at night," Humboldt related, "and then a noise so terrific arose in the neighboring forest, that it was almost impossible to close our eyes. Amid the cries of so many wild beasts howling at once, the Indians discriminated such only as were at intervals heard separately. These were the little soft cries of the sapajous [spider monkeys], the moans of the alouate apes [howler monkeys], the howlings of the jaguar and cougar, the peccary, and the sloth, and the cries of the curassow, the parraka, and other . . . birds." When a jaguar approached the camp, the mastiff, "which till then had never ceased barking, began to howl and seek for shelter beneath our hammocks. . . . We heard the same noises repeated, during the course of whole months, whenever the forest approached the bed of the river." When the Indians were quizzed as to the cause of the tremendous racket issuing from the forest, they explained that the animals were "keeping the feast of the full moon."

The next day, the party was on the river before dawn. The air was cool, and freshwater dolphins frolicked alongside the lancha. Water birds perched on shore and on the snags, watching for un-suspecting fish. The canoe hung up several times on submerged trees, any of which had the potential to stave in the craft's deli-cate hull; but each time the Indians managed to extricate the boat without apparent damage. That night, after a supper of roasted iguana (which Humboldt found, along with armadillo, one of the more palatable foods of the region) the men slept, as usual, in the open air. There were some Indian huts nearby, and the guides assured the Europeans that there would be no trouble with jaguars that night, since the cats avoided areas inhabited by humans—or as the Indians put it, "Men always put the jaguar out of humor."

Since their departure from San Fernando, the travelers hadn't encountered a single other boat on the river. In fact, human pres-ence seemed beside the point in these vast solitudes. "In that interior part of the New Continent one may almost accustom oneself to regard men as not being essential to the order of na-ture," Humboldt considered. These desolate surroundings couldn't help but accentuate the travelers' sense of isolation from the wider world. Recognizing the uncertainty of his correspondence reaching Europe, Humboldt made a practice of repeating the same information in several different letters, in the hope that at least one would survive.

In fact, by the summer of 1801, Wilhelm, then in Paris, would still have received no word of his brother. Over the first three years of his journey, Alexander would receive only six letters from his family. He must have yearned to know how they were faring, and how Europe was surviving the years of almost continual warfare. Yet the people the explorers encountered along the river depended on the travelers for information of the outside world; they had no news to share. And even in the cities, in this pretelegraphic era, the newspaper reports were likely to be months out of date. No wonder Humboldt was so sensitive to the forlorn aspects of the passing landscape. "The earth is loaded with plants," he wrote, "and nothing impedes their free development. . . . Crocodiles and boas are masters of the river; the jaguar, the peccary, the *dante* [tapir], and the monkeys traverse the forest without fear and without danger; where they swell as in an ancient inheritance. This aspect of animated nature, in which man is nothing, has something in it strange and sad. . . . Here, in a fertile country, adorned with eternal verdure, we seek in vain the traces of the power of man; we seem to be transported into a world different from that which gave us birth."

The next morning the Indians dropped their hooks in the water and caught some tasty little fish the Spanish called caribes, or "cannibals"—piranhas. Related to the carp and the catfish, the piranha was dreaded by the Indians for the vicious damage it could inflict with its razorlike teeth. Sometimes only a few inches long, the fish would congregate unseen at the bottom of streams, swarming at the barest trace of blood and stripping a carcass in minutes. Several of the Indian guides bore deep scars on their legs left by encounters with the voracious fish.

Humboldt was the first naturalist to describe the caribe, concluding that "no other fish has such a thirst for blood." A century later, when Theodore Roosevelt ventured into the Amazon, *piranha* was—and still is—synonymous with "rapacious." Roosevelt considered them "the most ferocious fish in the world. Even the most formidable fish, the sharks or the barracudas, usually attack things smaller than themselves. But the piranhas habitually attack things much larger than themselves. They will snap a finger off a hand incautiously trailed in the water; they mutilate swimmers—in every river town in Paraguay there are men who have been thus mutilated; they will rend and devour alive any wounded man or beast; for blood in the water excites them to madness," he wrote. "They will tear wounded wild fowl to pieces; and bite off the tails of big fish as they grow exhausted when fighting after being hooked. . . . The rabid, furious snaps drive the teeth through flesh and bone. The head with its short muzzle, staring, malignant eyes, and gaping, cruelly armed jaws, is the embodiment of evil ferocity; and the actions of the fish exactly match its looks. . . . They are the pests of the waters. . . . The only redeeming feature about them is that they are themselves fairly good to eat, although with too many bones."

At noon, the crew put the boat ashore, and while the Indians were preparing the midday meal, Humboldt went off on his own to observe some crocodiles that were dozing in the sun, their plated tails curled one atop the other. The reptiles' leathery skin, greenish-gray and caked with dried mud, lent the animals the appearance of bronze; in fact, some small white herons were striding along the creatures' backs, and even on their heads, as if stepping over so many statues.

As Humboldt strolled along the shore, his eyes were trained on the riverbank. But just now his attention was caught by some bits of the mineral mica gleaming in the sand at his feet. As he bent down to examine them, he noticed some fresh tracks leading into the forest. Turning in that direction, he was startled to see a huge jaguar lying under a ceiba tree, a scant eighty paces away. From the safety of the lancha, Humboldt had seen many jaguars prowling the riverbank, but none had ever seemed so huge as this specimen. The Indians had given instructions on how to respond in such an event, and Humboldt, his heart pounding, now did his best to follow their advice. Keeping his arms at his side, he made a wide turn toward the water and away from the jaguar. Then he began to walk ever so slowly away. The Indians had warned not to look backward on any account, lest eye contact make the cat aggressive. So Humboldt inched down the riverbank, his back to the animal, not knowing whether he was being stalked or not. When he finally permitted himself a furtive glance, he saw the jaguar still lounging under the ceiba, its attention fixed on a herd of capybara fording the river. Redoubling his pace, Humboldt arrived, out of breath, at the canoe. He told the others what had happened, but by the time the guides loaded their rifles and returned with him, the jaguar had vanished.

That evening, the lancha passed the mouth of the Caño del

Manatí (Manatee Creek), named for the many sea cows that congregated there. Growing ten to twelve feet long and weighing up to eight hundred pounds, the manatees were so huge that, after harpooning one of the creatures, the only way the Indians could load it into a canoe was to flood the boat, position it under the animal, then refloat the vessel by bailing it out with calabashes. Given the opportunity to examine the lungs of one of the manatees, Humboldt was astonished to see that the organs were three feet long, with a volume of more than a thousand cubic inches. The meat, he found, was "savory" and resembled pork; when salted and dried, it would keep for up to a year. Though the manatee is a mammal, the missionaries considered it a fish, and its meat was particularly prized during Lent. Besides the meat, the Indians extracted from each carcass nearly forty pounds of fat, used in lamps and for cooking, and stripped the inch-and-a-half-thick hide, which was cut into bands and fashioned into cords and whips to be used on recalcitrant slaves and impious Indians.

That night the guides lit campfires as usual, and Humboldt again noticed that, though the fires appeared to keep the jaguars at a distance, they seemed to attract the crocodiles. Porpoises, also drawn to the light, made playful noises in the river, leaving the travelers sleepless till the fires finally died out. But even then it didn't prove a restful night. First a female jaguar and cub approached close to camp to drink from the river. The Indians succeeded in chasing the cats away, but mother and young were separated in the process, and the cub's plaintive mews filled the camp for a long time, till the two were finally reunited. Soon after, Humboldt's dog let out a yelp when it was bitten on the tip of the nose by one of the enormous vampire bats that hovered over camp each evening.

But even more troublesome than the vampires were the voracious insects that appeared every night after sundown and, able to pierce through clothing and even hammocks, covered the explorers with painful bites. Every visitor to the rain forest—not to mention the Indians and missionaries who made it their home— cursed the mosquitoes, gnats, flies, ticks, fleas, ants, and myriad other insects, and Humboldt's experience would be no different. Biting, chewing, stinging, burrowing, preying on their fellow creatures, the most numerous class of animal made life hell for every other species that came into unfortunate contact with it. And the insects would only get worse as the party penetrated deeper into the forest, ever closer to the putative Casiquiare Canal.

The next evening, as the travelers were slinging their hammocks at their chosen campsite, they discovered two jaguars lurking behind a locust tree. Prudently returning to the canoe, they paddled upstream to an island near the junction of the Apure and the Orinoco. Since there were no trees here to sling hammocks from, the men passed an uncomfortable night on oxhides spread on the ground. In the rainy season, the mouth of the Apure would have spread across the savannahs as far as the eye could see, but at this time of year the river was reduced to a width of only four hundred to five hundred feet and a depth of just twenty to twenty-five feet. In fact, the water was so low that the next morning even the shallow-draft canoe had to be towed along the bank in several places. It was April 5. The journey down the Apure, some two hundred miles, had taken six days. More than a thousand miles of river lay before them, but they had reached the first milestone on their expedition into the rain forest. "It was not without emotion," Humboldt wrote, "that we beheld for the first time, after long expectation, the waters of the Orinoco. . . ."

SIX

The Orinoco

ONE OF THE GREAT WATERWAYS OF THE NEW WORLD, the Orinoco carves a thirteen-hundred-mile arc through the heart of Venezuela, from its source in the Guiana Highlands, in the southeastern tip of the country, to its vast delta on the Atlantic Ocean, just below the island of Trinidad. En route, the river and its four hundred-plus tributaries drain some 360,000 square miles—more than half the size of Europe. It's thought that Columbus was the first European to spy the Orinoco's mouth (on his third voyage, in 1498) and that the river's massive, silty discharge convinced him that he had finally found tierra firma. Indeed, so fragrant was the evening air and so clear the starry sky that the great navigator, thinking himself in Asia, fancied he had found one of the four biblical rivers descending from Eden.

Thirty-three years later, Diego de Ordáz entered the labyrinthine delta and christened the river after its Indian name, Orinucu. In the sixteenth and early seventeenth centuries, several explorers ventured upstream, but, bent on conquest or pillage, all met with disappointment or disaster. In 1553, Alfonso de Herrera led an ill-fated expedition far up the Orinoco and into its tributary, the Meta, before being killed by an Indian arrow dipped in curare, the black, resinous nerve poison concocted from certain jungle plants. Seven years later, the Spanish adventurer Lope de Aguirre is believed to have traveled down much of the river during his quixotic, traitorous campaign to proclaim himself emperor of Peru.

In 1595, Sir Walter Raleigh sailed three hundred miles up the Orinoco in search of El Dorado, the fabled country of gold.

Returning with a few samples of the precious ore, he assured Queen Elizabeth I "that from Inglaterrra those Ingas should be againe in time to come restored and delivered from the servitude of the said conquerors. I am resolved that if there were but a small army afoote in Guiana marching towards Manoa, the chiefe citie of Inga, he would yield Her Majestie by composition so many hundred thousand pounds yearely as should both defend all enemies abroad and defray all expences at home, that he woulde besides pay a garrison of 3000 or 4000 soldiers very royally to defend him against other nations. The Inga will be brought to tribute with great gladness." But despite a second expedition in 1616, Raleigh never discovered the riches he believed were waiting there.

The first European explorers found significant villages along the river, where the indigenous peoples supported themselves by fishing, hunting, and raising cassava, and over the next two centuries, the Lower Orinoco was thinly settled by missionaries and ranchers. But at the time of Humboldt's journey, the so-called Upper Course of the river, beyond the great rapids at Ature and Maipures, was still a wild, unknown country visited by only a handful of white men—a trackless forest rumored to be filled with savage animals, cannibals, and races with a single eye in the center of their forehead, or a dog's head, or a mouth below their stomach. "He who goes to the Orinoco," an old Spanish saying warned, "either dies or comes back mad."

Humboldt had come to the Orinoco because it was the shortest route to the Río Negro. The Negro was known to intersect with the Casiquiare. And, if La Condamine could be trusted, the Casiquiare rejoined the Orinoco to form the natural canal whose existence was roundly denied by European geographers. Whatever the outcome, Humboldt was determined to settle one of the great geographic controversies of the age. He would need every shred of that resolve for the journey that lay ahead.

Humboldt saw that the country changed dramatically at the junction of the great river. During the rainy season, the Orinoco would have been nearly six miles wide here, but even in April it stretched before the lancha like a huge lake, well over two miles across and whipped with whitecaps several feet high. The flamingos, herons, and other water birds that had been the explorers' constant companions on the Apure vanished, and even the ubiquitous crocodiles were reduced to a few individuals slicing obliquely through the waves. Forest stretched to the horizon, but along the riverbanks was only a vast, empty beach. Even the distant hills appeared forlorn. "In these scattered features of the landscape, in this character of solitude and of greatness," Humboldt wrote, "we recognize the course of the Orinoco, one of the most majestic rivers of the New World."

Despite the high waves, the north-northeast wind was ideal for sailing upstream. The canoe made good progress, arriving later that day at the mission of Encaramada, which was spectacularly situated before a chain of weathered granite mountains that resembled ancient ruins swallowed by forest. At the adjacent port, Humboldt had his first, fascinated glimpse of the Caribs, the people once renowned for their ferocity (and cannibalism—the tribe's original name, Calibi, had been corrupted by the Spanish into Canibal). In the century before Columbus, the Caribs had come to dominate the Caribbean Basin and, utilizing their prodigious sailing skills, had spread as far as the South American mainland. Though the Caribs had put up a fierce resistance against the Spanish, they had proved no match for European technology, and by the time of Humboldt's arrival the tribe existed only in isolated, missionized pockets.

"These Caribs are men of an almost athletic stature," Humboldt noted; "they appeared to us much taller than any Indians we had hitherto seen. Their smooth and thick hair, cut short on the forehead like that of choristers, their eyebrows painted black, their look at once gloomy and animated, gave a singular expression to their countenances." The Caribs' features were more regular than those of the other native peoples Humboldt had seen, their noses smaller and less flattened, and the cheekbones not so high. "Their eyes, which are darker than those of the other hordes of Guiana, denote intelligence, and it may even be said, the habit of reflection. The Caribs have a gravity of manner, and a certain look of sadness which is observable in most of the primitive inhabitants of the New World." The women, "less robust and good-looking than the men," were also "very tall, and disgusting from their want of cleanliness." Infants' legs were wrapped at prescribed locations by tight strips of cotton cloth, with the flesh bulging out between the ligatures. "It is generally to be observed," Humboldt explained, "that the Caribs are as attentive to their exterior and their ornaments as it is possible for men to be, who are naked and painted red. They attach great importance to certain configurations of the body; and a mother would be accused of culpable indifference toward her children, if she did not employ artificial means to shape the calf of the leg after the fashion of the country."

Despite the Caribs' loss of their empire to the white men, "the remembrance of their ancient greatness, has inspired them with a sentiment of dignity and national superiority, which is manifest in their manners and their discourse. 'We alone are a nation,' say they proverbially; 'the rest of mankind are made to serve us.' " So great was their contempt of other peoples that Humboldt once saw a ten-year-old boy fly into a rage at mistakenly being called a Cabre—another local people—though he had never even seen a member of that tribe.

Traveling upriver for the annual turtle egg harvest, the Carib cacique, or chief, sat in his canoe, attended by a smaller boat. Seated beneath a canopy of palm leaves, his "cold and silent gravity, the respect with which he was treated by his attendants, everything denoted him to be a person of importance. He was equipped, however, in the same manner as his Indians. They were all equally naked, armed with bows and arrows, and painted with *onoto* [a vegetable pigment]. . . . The chief, the domestics, the furniture, the boat, and the sail, were all painted red."

Among the Caribs and certain other tribes, body paint served as the principal "clothing." Generally, in fact, it was considered less immodest to present oneself without a *guayuco*, the strip of cloth covering the genitals, than without paint. "Thus," Humboldt reports, "as we say, in temperate climates, of a poor man, '[H]e has not enough to clothe himself,' you hear the Indians of the Orinoco say, '[T]hat man is so poor, that he has not enough to paint half his body.' " The missionaries permitted the practice, and some made a handsome profit retailing the pigments to their charges.

The type of body paint used even indicated the wearer's relative affluence. To make *onoto*, the more common variety, Indian women would mix with water the seeds of a plant called *achote*, beat the mixture for an hour, and allow a sediment to form. The residue would be collected, mixed with oil from turtle eggs, then shaped into little round cakes for later use. A more costly type of paint was made from a species of Bignonia, a family of woody vines, which Bonpland named *Bignonia chica*. The vine's leaves became red when dried, and when soaked in water they produced a powder that was collected and patted (without added oil) into small cakes. So highly valued was the *chica* that a man would need to work for two weeks just to earn enough money to paint his body once.

Ever since the practice had been first reported, there had been speculation in Europe that the body paint helped to keep voracious insects at bay. But Humboldt proved that the intent was purely decorative and that painted skin was just as subject to bites and stings as unpainted. Not always satisfied to spread the pigment uniformly over their bodies, the Indians sometimes whimsically imitated European clothing. "We saw some [Indians] . . . ," Humboldt wrote, "who were painted with blue jackets and black buttons. . . . [Others] are accustomed to stain themselves red with *onoto*, and to make broad transverse stripes on the body, on which they stick spangles of silvery mica. Seen at a distance, these naked men appear to be dressed in lace clothes. If painted nations had been examined with the same attention as those who are clothed, it would have been perceived that the most fertile imagination, and the most mutable caprice, have created the fashions of painting, as well as those of garments."

After four centuries of oppression and dislocation at the hands of the Spanish—enduring forced labor, banishment from traditional homelands, the imposition of an uncompromising new religion, and even unheard-of diseases—the indigenous cultures of South America existed only as faded reflections of their former selves. Yet Humboldt seemed to be the only observer to recognize this insidious process of debasement, and now, as he studied the Indians of the Orinoco, he was struck again by how little they resembled Rousseau's "noble savages." "How difficult to recognize in this infancy of society, in this assemblage of dull, silent, inanimate Indians, the primitive character of our species!" he concluded. "Human nature does not here manifest those features of artless simplicity, of which poets in every language have drawn such enchanting pictures. . . . We are eager to persuade ourselves that these natives, crouching before the fire, or seated on large turtle-shells, their bodies covered with earth and grease, their eyes stupidly fixed for whole hours on the beverage they are preparing, far from being the primitive type of our species, are a degenerate race, the feeble remains of nations who, after having been long dispersed in the forests, are replunged into barbarism."

In the area around Encaramada were ancient petroglyphs of animals and symbolic figures carved on unscalable rock cliffs, often at great heights. When Humboldt asked the guides how their forebears could have created these images so far above-ground, "they answer[ed] with a smile, as if relating a fact of which only a white man could be ignorant, that 'at the period of the great waters, their fathers went to that height in boats.'" Humboldt would discover that nearly every tribe along the Orinoco had a traditional belief in a devastating flood. According to their guides' version, a man and a woman had saved themselves by landing on a tall mountain and had afterward repopulated earth by casting over their heads the seeds of the Mauritia palm.

Humboldt concluded that these flood myths were not simply local adaptations of the biblical story of Noah and the Ark, because the stories were current even in places that had never seen a missionary. "These ancient traditions of the human race, which we find dispersed over the whole surface of the globe, like the relics of a vast shipwreck, are highly interesting in the philosophical study of our own species," he wrote. "Like certain families of the vegetable kingdom, which, notwithstanding the diversity of climates and the influence of heights, retain the impression of a common type, the traditions of nations respecting the origin of the world, display everywhere the same physiognomy, and preserve features of resemblance that fill us with astonishment. How many different tongues, belonging to branches that appear totally distinct, transmit to us the same facts!"

On the morning of April 6, paddling upstream on the Orinoco, the travelers came upon an encampment of more than three hundred Indians on a sandy island in the middle of the river. It was the *cosecha*, or annual harvest of turtle eggs. Several tribes had converged for the event, each distinguished by the color of their body paint, and had constructed temporary huts of palm fronds. They were all mission Indians, but Humboldt found them "as naked and rude as the 'Indians of the woods'" even though "they go to church at the sound of the bell, and have learned to kneel down during the consecration of the host." Among the Indians were several white traders, or *pulperos*, who had come up-river from Angostura to buy turtle oil. Clear, pale yellow, and odorless, the oil was used in lamps and for cooking (as well as for making *onoto*, the more common form of body paint), and by buying at the source the traders were able to turn a profit of seventy to eighty percent.

A Franciscan priest was on hand to say Mass each morning and to set down the rules of the harvest. The Jesuits had organized the annual turtle hunt before they had been expelled from Spain's New World colonies in 1767 as a result of ecclesiastical infighting and, by making it more efficient and less wasteful, had maximized the yield while ensuring that some nests went undisturbed each year. However, the Jesuits' successors in the missions, the Franciscans, now allowed the entire beach to be dug, with the result that the harvest had become less productive from year to year.

Astounded to see strange white men in the area, the missionary was skeptical that Humboldt and Bonpland had come to verify the existence of the Casiquiare Canal. "How is it possible to believe," he asked, "that you have left your country, to come and be devoured by mosquitoes on this river, and to measure lands that are not your own?" Fortunately, Humboldt had brought a letter of recommendation from the superior of the Franciscan Missions, which served to allay the priest's suspicions. The missionary invited the travelers to share his frugal meal of plantains and fish, then, after admiring their scientific instruments, warned them of the dangerous wilderness they would face in ascending the Orinoco beyond the cataracts at Ature and Maipures.

Undeterred by this friendly advice, the explorers took on some provisions of turtles, dried turtle eggs, fresh meat, rice, and even wheat biscuits, and pushed off at four in the afternoon. The wind had freshened and was now blowing in squalls. Since entering the open waters of the Orinoco, the travelers had discovered that their light canoe did not carry sail very well. However, as they left shore, the Indian pilot, eager to impress the spectators onshore, insisted on heaving as close to the wind as possible so as to reach the middle of the river on a single tack. But even as he was boasting of his prowess, the boat was struck by a violent gust and the leeward gunwale was forced beneath the surface. Before anyone could react, water swept over the table in the stern where Humboldt was writing, and the passengers found books, papers, and plant specimens floating about their knees. As the water continued to fill the heavily laden canoe, it began to sink.

Bonpland had been napping in the floor of the boat but was awakened by the deluge and immediately assumed his characteristic coolness under pressure. Because the canoe from time to time managed to right itself enough to draw the rail out of the water, he concluded that the craft might be saved. If not, he thought it might be possible to swim to shore, as no crocodiles were in immediate sight. Humboldt was a poor swimmer, and Bonpland offered to carry him on his back.

But even if they survived the long swim, what then? How would they replace their instruments, books, and provisions? Without a canoe, how would they continue their journey or even rescue themselves from the wilderness? As they struggled with the unsavory possibilities, the boat was struck by another squall, which snapped the mainsheet, the line regulating the mainsail. Emptied of wind, the sail began to shake, and the little craft popped unexpectedly upright. Thus Humboldt writes, "The same gust of wind, that had thrown us on our beam, served also to right us." Within half an hour, the boat was bailed out and the mainsheet repaired. With the sail set at a more prudent angle, the canoe got under way once more, the only casualty one book lost overboard, Johann Christian Daniel von Schreber's *Genera Plantarum.*

In the *Personal Narrative* Humboldt generally went out of his way to minimize the risks and discomforts of the journey. But realizing how close they had come to catastrophe on the river, he couldn't conceal his deep concern over this incident. "We had escaped as if by a miracle," he felt, though the pilot, whether out of insouciance or bravado, didn't seem to share their dismay. "To the reproaches that were heaped on [him] for having kept too near the wind, he replied with the phlegmatic coolness peculiar to the Indians, observing 'that the whites would find sun enough on those banks to dry their papers.'"

That night, seated on turtle shells on a moonlit beach, Humboldt and Bonpland were still reliving the events of the afternoon. "What satisfaction we felt on finding ourselves thus comfortably landed!" Humboldt wrote. "We figured to ourselves the situation of a man who had been saved alone from shipwreck, wandering on these desert shores, meeting at every step with other rivers which fall into the Orinoco, and which it is dangerous to pass by swimming, on account of the multitude of crocodiles and caribe fishes. We pictured to ourselves such a man, alive to the most tender affections of the soul, ignorant of the fate of his companions, and thinking more of them than of himself. . . . Our minds were full of what we had just witnessed."

The near accident had driven home to Humboldt the dangers of the rain forest in a way that even his encounter with the jaguar had not. Behind the rapturous beauty, the fantastic plants, the marvelous creatures, lay an implicit violence that threatened to engulf the unwary at any moment. Thus, with the Orinoco expedition barely under way, Humboldt uncharacteristically began to question the wisdom of his plan. "There are periods in life when, without being discouraged, the future appears more uncertain," he confides. "It was only three days since we had entered the Orinoco, and there yet remained three months for us to navigate rivers encumbered with rocks, and in boats smaller than that in which we had so nearly perished."

That night was intensely hot, and to add to the explorers' misery, the torments of the mosquitoes had magnified daily. With no trees to hand, the men were forced to sleep on the ground again. And, as if to exacerbate their disquiet, the campfire failed to keep jaguars away from camp. Toward morning, the animals' cries were heard very near. The next day, the canoe passed the mouth of the Río Auraca, renowned for its huge flocks of birds. Though the Orinoco's mouth was some six hundred miles distant, during the rainy season this section of river would have been more than four miles across. As the travelers paddled on, the mountains pressed closer to the eastern bank, constricting the flow and magnifying the force of the current. Progress upstream became proportionately harder, especially since the mountains also blocked the wind—except at narrow passes, where it careened through in dangerous, violent gusts. As the river narrowed, crocodiles became more numerous again.

Early on the morning of April 9, the party beached their canoe at a place called Pararuma, where they found another encampment of Indians collecting turtle eggs. The great rapids at Ature and Maipures were approaching, and, having had no experience

with the notorious cataracts, Humboldt's pilot and crew refused
to venture any farther upriver. Not two weeks into the rain forest,
the explorers were abruptly stranded, without boat or guides.

Their dilemma was partially resolved when one of the missionaries on the beach agreed to sell Humboldt a canoe. Crafted,
like all the Indian boats, from the trunk of a single tree, the fine
lancha had been hollowed out with fire and hatchets. Though
forty feet long, larger than their first canoe, the boat was less
than a yard across, not wide enough for three persons to sit
abreast. Still, the lancha would serve Humboldt's purpose well—
if he could find a crew.

The explorers might have languished at Pararuma for weeks
searching for new guides if another missionary, Bernardo Zea,
hadn't come to their rescue. Having traveled down from his mission at Maipures for the turtle hunt, Father Zea volunteered to
accompany the expedition upriver to the Casiquiare, despite the
intermittent fevers that he'd suffered for years. Thus he became
the second additional member of the party, after Soto. The priest
also volunteered to recruit another crew. But "the missionary
from the cataracts made the preparations for our voyage with
greater energy than we wished," Humboldt found. Finding two
men familiar with the cataracts, Zea ordered them imprisoned in
the *cepo*, a kind of stocks, in case their resolve faded in the night.
And the next morning, Humboldt was awakened by the cries of
another young Indian, named Zerepe, who was mercilessly
beaten with a manatee whip when he declined to join the expedition. When Humboldt protested, the missionary explained,
"Without these acts of severity, you would want for everything. . . .
If left to their own will [the Indians] would all go down the river
to sell their productions, and live in full liberty among the whites.
The Missions would be totally deserted." Zerepe's reluctance may
have stemmed from the fact that he was about to be married.
After his whipping, he joined the party, and his twelve-year-old
fiancée agreed to wait at Maipures for his return. Though Humboldt was appalled by Zea's methods, he was relieved to have
a crew.

Besides the canoe, Humboldt purchased a veritable menagerie
at Pararuma—some two dozen caged birds and monkeys to add
to the books, instruments, provisions, botanical and geological
samples, and myriad other gear heaped onboard. With the excep-

tion of the mastiff, Humboldt didn't consider these creatures
pets, but living natural-history specimens. Privileged to be conducting the first extensive scientific exploration of the New
World, Humboldt felt a deep responsibility not only to observe
and record and interrelate all that he could himself, but also to
make available to his Europe-bound peers—and to posterity—
the physical means of additional study. One never knew whether
an anomalous rock or a previously unknown animal would fill a
gaping taxonomic gap, or overthrow a long-held theory, or lead to
some startlingly original new idea. "I aim at collecting ideas
rather than material objects," Humboldt wrote, but he considered the safe delivery of these many thousands of specimens a
crucial purpose of his journey as well.

However, the ever-growing collection of cases and cages presented a problem in the narrow confines of the lancha. The travelers distributed the cargo as carefully as they could in the
unstable canoe, but even so, if one of the passengers wanted to
move about, he had to first warn the paddlers, seated two by two
in front, so they could lean to the opposite side in compensation.
If Humboldt or Bonpland needed an instrument, the boat had to
be landed while the object was unpacked.

To increase the canoe's capacity, a framework of branches was
constructed over the gunwales at the stern. This was covered
with a thatched roof, or toldo. But to keep down weight and to
present a smaller surface to the wind, the roof was built low and
short. As a result, the Europeans were consigned to spend interminable days stretched out flat on the uncomfortable branches,
their legs exposed, alternately, to burning sun and soaking rains.
To the perennially restless Humboldt, it was a torture. "It is difficult," he comments curtly, "to form an idea of the inconveniences
that are suffered in such wretched vessels."

Cramped though their quarters were, that was nothing compared to the growing misery of the mosquitoes. "We attempted
every instant, but always without success, to amend our situation," Humboldt recounts. "While one of us hid himself under a
sheet to ward off the insects, the other insisted on having green
wood lighted beneath the toldo, in the hope of driving away the
mosquitoes by the smoke. The painful sensations of the eyes, and
the increase of heat, already stifling, rendered both these con-
trivances alike impracticable." As the canoe continued upriver,

Humboldt's characteristic rapture at the mysteries of nature began to flag noticeably, replaced by a grim determination to keep
going. One has to wonder how the four Europeans—Humboldt,
Bonpland, Soto, and Zea—were getting along in such cramped
quarters. Did Soto, newly arrived in South America, regret ever
meeting these naturalists-cum-madmen? Were the mosquitoes
puncturing the travelers' camaraderie as well as their epider-
mis? But Humboldt reported that morale was still good. "With
some gaiety of temper, with feelings of mutual goodwill, and with
a vivid taste for the majestic grandeur of these vast valleys of
rivers," he explains, "travelers easily support evils that become
habitual."

On the tenth of April, at four o'clock in the morning, the lancha left the beach at Pararuma. Near the mouth of the Paruasi
River they passed a ruined Jesuit fortress that once boasted
three gun batteries and a complement of soldiers. The fort had
not been built as a defense against hostile Indians, but as a base
for the *conquista de almas*, or "conquest of souls." The soldiers,
animated by hope of material gain as much as by pangs of religious duty, would use the fortress to launch illegal incursions,
called entradas, into the surrounding territory, during which
they would kill Indians who resisted, burn their huts, destroy
their crops, and carry away their women, children, and old men
to work in faraway missions. Yet such atrocities were reckoned
necessary for the Lord's work. As one missionary explained, "The
voice of the Gospel is heard only where the Indians have heard
also the sound of firearms. By chastising the natives, we facilitate their conversion." After the expulsion of the Jesuits, their
successors in the missions had discontinued the entradas, and
the fort, Humboldt was pleased to note, had been abandoned.

As the party proceeded upstream, the Orinoco continued to
narrow, and the current grew more insistent. Little islands and
shoals began to dot the riverbed, and even small cascades spawned
dangerous eddies. A rapids known as the Raudal de Marimara
stretched across nearly the entire river, passable only via a narrow channel carved through a block of solid granite some eighty
feet high and three hundred feet around. Mile by mile, the river
became more foreboding. None of the Europeans except Father
Zea had ever experienced anything remotely like this terrain.
Though Humboldt had spent a fair amount of time trekking the

mountains of Europe, that was meager preparation for a journey through the South American rain forest. As he watched the landscape literally closing in on them, one wonders whether Humboldt's characteristic optimism wasn't at last deserting him. With so many waterborne miles stretching before them, he must have wondered at times whether the Casiquiare Canal, if it existed, was worth all the misery and danger.

The explorers passed the huge rock known as the Piedra del Tigre (Rock of the Jaguar), where the water was so deep that their line could find no bottom. That afternoon, there was a violent squall, and, hemmed in by rocks, the river began to swell. As the canoe approached the mouth of the Río Meta, one of the principal tributaries of the Orinoco, the vicious current pinned the craft against a rock in the middle of the river. With dusk coming on, the explorers were forced to spend the night on the bare rock, surrounded by raging water. At four o'clock the next morning they pushed off again, struggling for every foot of headway. Wherever the river became too strong to paddle against, the Indians could work the canoe upstream only by leaping into the water, fixing a rope to a rock shelf, and hauling the boat behind them. After twelve hours of brutal labor, without even a break for food, the crew finally maneuvered the craft through a five-foot-wide dam of granite boulders and into the relatively quiet waters beyond.

At nine o'clock the next morning, the canoe passed the mouth of the Meta, on the border of present-day Colombia and Venezuela. The junction was a desolate place. The sandy banks held the forest from the river's edge, and on the eastern shore blocks of granite were stacked atop one another like the ruins of a bygone civilization. A huge boulder in the middle of the river was called la Piedra de la Paciencia, or the Rock of Patience, because canoes were sometimes trapped in its treacherous whirlpool for two days before their crews could extricate themselves. Humboldt's party were lucky that day; it took them only two hours of ferocious paddling to negotiate the eddy.

Above the junction with the Meta, the Orinoco became wider and smoother for a time, and there was no need to tow the canoe. That evening, the explorers stopped just below a cataract known as the Raudal de Tabaje, forced to sleep on a steep rock shelf. Bats swarmed in the crevices of the rock and jaguars prowled very close; sensing the great cats, the mastiff howled for hours. It was a miserable, exceedingly dark night. The rumble of thunder mingled with the roar of the cataract, though it never rained.

Early the next morning the canoe cleared the rapids and pulled into shore to allow Father Zea to say Mass at the mission of San Borja, established just two years before. The Indians here were Guahíbos, a notoriously nomadic, omnivorous people who subsisted on fish, centipedes, worms, and anything else they could forage. (The other Indians had a saying that "a Guahibo eats everything that exists, both on and under the ground.") To Humboldt's eye these Indians' relative independence lent them a more animated manner than he'd seen among the Indians of the longer-established missions. Though the Guahíbos listened dutifully to Father Zea's Mass, they betrayed not the slightest comprehension of the rite.

As the explorers struggled up the Orinoco toward the Casiquiare, the temperatures moderated, but the mosquitoes grew steadily worse. Now at San Borja, the men "could neither speak nor uncover our faces without having [their] mouths and noses filled with insects." Worse yet, the piranha and huge crocodiles, up to twenty-four feet long, prevented them from bathing in the river to gain some relief. After another miserable night, the party broke camp at five o'clock in the morning of April 14, hoping the mosquitoes would be less thick on the river. Just to the southeast lay the infamous cataracts of the Orinoco, the Ature and the Maipures, dividing the so-called Lower Course from the Upper. The travelers were now nearly eight hundred miles from the coast and some five hundred miles from the river's source.

THE SPECTACLE OF THE HUGE RAPIDS, formed by two granite ridges crossing the riverbed at right angles, almost made the arduous journey seem worthwhile. "Nothing can be grander than the aspect of this spot," Humboldt rhapsodized. "Neither the [water]fall of the Tequendama, near Santa Fe de Bogotá, nor the magnificent scene of the Cordilleras [Andes], could weaken the impression produced upon my mind by the first view of the rapids of Ature and of Maipures. When the spectator is so stationed that the eye can at once take in the long succession of cataracts, the immense sheet of foam and vapors illumined by the rays of the setting sun, the whole river seems as if it were suspended over its bed."

The first, northernmost rapid was the Ature, named for the extinct Indian people who had once inhabited its banks. The cataracts, pinched on each side by high mountains, were a five-mile-long jumble of islands, rock shelves, and blocks of granite stretching from one side of the river to the other. "Persons who have dwelt in the Alps, the Pyrenees, or even the Cordilleras, so celebrated for the fractures and the vestiges of destruction which they display at every step, can scarcely picture to themselves, from a mere narration, the state of the bed of the river," Humboldt wrote. "It is traversed . . . by innumerable dikes of rock, forming so many natural dams, so many barriers. . . . The space between the rocky dikes of the Orinoco is filled with islands of different dimensions; some hilly, divided into several peaks, and two or three hundred toises [about thirteen hundred to two thousand feet] in length, others small, low, and like mere shoals. These islands divide the river into a number of torrents, which boil up as they break against the rocks. . . ." In addition, the river "is engulfed in caverns; and in one of these caverns we heard the water roll at once over our heads and beneath our feet. We were struck with the little water to be seen in the bed of the river [owing to all the rocks], the frequency of the subterraneous falls, and the tumult of the waters breaking on the rocks in foam."

Both banks were steep and virtually inaccessible, but the left, generally lower than the right, formed part of a vast plain of meadows, crisscrossed by streams and studded with granite blocks and shelves on which flamingos, herons, and other water birds perched like sentinels. (Their resemblance to an army was so striking that the birds' sudden appearance downriver at the capital of Angostura once panicked the citizens, who thought themselves surrounded by hostile Indians, until the birds suddenly took wing.) So great was the roar of the cataracts that they could be heard more than three miles away. Missionaries reported that Indians living along the river suffered hearing loss from the continuous thunder of the water. In the stillness of the night, the roar seemed even louder, and only served to underscore the remoteness of the place.

The Indians living along the Orinoco had worked out various

methods of passing the rapids. Where the natural dams were only two or three feet high, they could be run downstream in a canoe. To ascend the river in these places, the Indians would swim on ahead of the boat, tie a rope to an outcropping, then pull the craft over the top of the rocks, as Humboldt's crew had done near the junction of the Meta. But this operation was more easily described than accomplished, since the boat would often fill with water or sometimes be dashed against the rocks, in which case the Indians, bruised and bleeding, would have to extricate themselves from the treacherous whirlpools and swim to the nearest island. Where the dams were too high for this technique, the canoes had to be portaged along shore, with tree trunks serving as rollers. At the great falls of Ature, this last method was the only option.

To help with the portage around the rapids, eight Indians were hired from the nearby mission of San Juan Nepomuceno de los Atures. Founded by the Jesuits in 1748, the mission was in a deplorable state by the time of Humboldt's arrival, its population having dwindled from more than three hundred to just forty-seven, as the Indians, tired of mission regulations and fearful of epidemic fevers, had gradually drifted off to their ancestral home in the forest. So few canoes came this way that the missionary kept count: Over the past three years, exactly sixteen boats had passed the rapids—three to fetch the soldiers' annual pay from Angostura, five with Indians bound for the turtle egg harvest, and eight laden with trade goods.

In the two days the Europeans spent at the mission, the insects grew worse than ever, reaching a crescendo of carnivorous misery. "Persons who have not navigated the great rivers of equinoctial America . . . can scarcely conceive how, at every instant, without intermission, you may be tormented by insects flying in the air; and how the multitude of these little animals may render vast regions almost uninhabitable," Humboldt lamented. "Whatever fortitude may be exercised to endure pain without complaint, whatever interest may be felt in the objects of scientific research, it is impossible not to be constantly disturbed by the mosquitoes, *zancudos* [large gnats], *jejenes* [venomous flies], and *tempraneros* [small gnats], that cover the face and hands, pierce the clothes with their long needle-formed suckers, and getting into the mouth and nostrils, occasion coughing and sneezing whenever any attempt is made to speak in the open air. . . ." In fact, "the lower strata of air, from the surface of the ground to the height of fifteen or twenty feet, [is] absolutely filled with venomous insects." And this was all during the dry season. "I doubt whether there is a country on earth were man is exposed to more cruel torments in the rainy season," he ventured.

With customary thoroughness, Humboldt made a study of the six-legged perpetrators, discovering that the infestation followed a daily pattern, an orderly changing of the insect guard, broken only when a downpour would mercifully (but all too briefly) sweep away the pests: From half past six in the morning to five in the afternoon the air was filled with mosquitoes; the *tempraneros* emerged an hour before sunset, disappearing between six and seven; they were followed in turn by the *zancudos*, whose bite produced the worst pain of all, as well as a swelling that lasted for weeks.

In the missions and villages along the river, *la plaga de las moscas* afforded an inexhaustible subject of conversation. When two persons met in the morning, they wouldn't ask, "How did you sleep?" but "How did you find the *zancudos* during the night?" or "How are the mosquitoes today?" "How comfortable must people be in the moon," one Indian exclaimed. "She looks so beautiful and so clear that she must be free from mosquitoes."

Condemned to live among these voracious pests, people along the river devised various means of self-defense—waving objects about the head and hands; erecting mosquito curtains woven from palm fibers; filling the air with the acrid fumes of burning cow dung; sleeping on rocks in the middle of the rapids, or surrounded by cattle (which were thought to attract the pests away from the humans), or buried in sand, with only the head exposed (and that covered with a handkerchief), even retreating to a windowless hut with a smoky fire in the center. (Bonpland found such *hornitos* ["little ovens"] ideal for drying plant specimens.) At his mission at Maipures, Father Zea had built a small, windowless room high in a grove of palm trees, in which Humboldt and Bonpland managed to find some peace from the insects. Still, Humboldt supposed, "it is neither the dangers of navigating in small boats, the savage Indians, nor the serpents, crocodiles, or jaguars, that make Spaniards dread a voyage on the Orinoco; it is . . . 'the perspiration and the flies.' "

The explorers were grateful when the portage was accomplished on April 17 and they were able to leave Ature behind. Owing to its isolation, provisions were hard to come by at the mission, and Humboldt was able to buy only a few bunches of plantains, some cassavas, and several fowl to replenish their stores. Immediately above the rapids of Ature, the river was relatively free of shoals. The canoe easily passed the small Raudal de Garcita, but the swarms of biting insects were so thick that night that they filled the sky and prevented Humboldt from taking an astronomical reading to fix the party's location. At three o'clock the next morning, the travelers got under way, to be sure of reaching the treacherous Raudal de los Guahíbos in daylight.

At five that afternoon, the rapids came into view. Following the prescribed method, one of the Indians swam to the natural dam in the center of the cataract. He fixed a rope to the rock, and the lancha was pulled close enough to allow the Europeans to disembark, along with the animals and all the gear. While the passengers waited for the boat to be hauled up the rapids, Father Zea was stricken by one of his periodic fevers. To offer relief, the others scooped some water from the river and poured it in a large hole in the top of the rock. They added sugar and lime juice, and "in a few minutes had an excellent beverage, which is almost a refinement of luxury, in that wild spot; but our wants rendered us every day more and more ingenious," Humboldt wrote.

After an hour, the canoe was ready to be reloaded. Above the rapids the river was about a mile wide, and still running fast. As the Indians worked against the current, rain began to fall in torrents. Thunder drowned out the roar of the water, and twice lightning struck the water very near the canoe. After twenty strenuous minutes, the paddlers began to lose their battle, and the canoe started slipping back toward the rapids. If the current managed to take the craft over the natural dam, at the least the lancha and its passengers would be dumped into the swirling waters; at the worst the fragile craft would be dashed against the rocks. "These moments of uncertainty appeared to us very long," Humboldt admits; "the Indians spoke only in whispers, as they do always when they think their situation perilous." Redoubling their efforts, the crew finally managed to pull away from the threatening cascade. Two hours later, drenched to the skin, the party arrived safely at the port of Maipures, located just downstream from the second great rapids of the Orinoco.

Onshore, as soon as the rain let up the gnats reemerged with new voracity. At Father Zea's suggestion the party decided to press on to his unfinished house at the nearby mission, in the hope that the insects would be less thick away from the river. Carrying tubes of bark filled with copal resin, which gave off more smoke than light (before they went out completely), the men stumbled over rocks and twice had to cross streams in the dark by balancing on the trunks of fallen trees, as all the while the Indian pilot warned against jaguars and poisonous snakes. At last, the travelers reached the village of San José de Maipures. The inhabitants were asleep, and the only sounds came from a few night birds calling above the distant roar of the rapids. Humboldt must have been glad to leave the river behind, at least temporarily. But even so, "in the calm of the night, amid the deep repose of nature," he found that "the monotonous sound of a fall of water has in it something sad and solemn."

The party stayed for three days at the village while the canoe was portaged around the great *raudal*. Like the rapids at Ature, those at Maipures (named after the native people who still lived beside them) were strewn with small granite islands connected by natural dikes. But Father Zea had chosen his crew well, and the Indian guides could recite the name of every ledge and waterfall. One of the largest, known as el Salto de la Sardina (the Leap of the Sardine), was nine feet high and formed an imposing fall across a wide breadth. But the rapids' danger derived not so much from their height, Humboldt realized, as from the narrow channel and the resulting currents and countercurrents.

During the time the travelers spent at the mission, Humboldt never tired of climbing one of the low mountains and gazing at the spectacle. From the summit, he described, "a wonderful prospect is enjoyed. A foaming surface of four miles in length presents itself at once to the eye: iron-black masses of rock resembling ruins and battlemented towers rise frowning from the waters. Rocks and islands are adorned with the luxuriant vegetation of the tropical forest; a perpetual mist hovers over the waters, and the summits of the lofty palms pierce through the cloud of spray and vapor," he wrote. "When the rays of the glowing evening sun are refracted in these humid exhalations, a magic optical effect begins. Colored [rain]bows shine, vanish, and reappear; and the ethereal image is swayed to and fro by the breath of the sportive breeze." The rocks were covered with clusters of palm trees eighty feet tall. "I do not hesitate to repeat," he averred, "that neither time, nor the view of the Cordilleras, nor any bode in the temperate valleys of Mexico, has effaced from my mind the powerful impression of the aspect of the cataracts. . . . The majestic scenes of nature, like the sublime works of poetry and the arts, leave remembrances that are incessantly awakening, and which, through the whole of life, mingle with all our feelings of what is grand and beautiful."

To the west of the rapids rose a peak known as Keri. The mountain's western face was graced with a prominent white spot, visible from some distance, which to the Indians represented the image of the full moon. Opposite Keri, its twin mountain, Ouivitari, had a similar spot, facing east, which was taken to be the image of the sun. Humboldt regretted that the mountains were too sheer to permit a climb, but he believed the mysterious markings were formed by the conjunction of veins of quartz, often found in granite outcroppings.

In the time of the Jesuits, the mission at Maipures had boasted more than six hundred souls, but, like the mission at Ature, it now numbered fewer than sixty. Humboldt found the remaining Maipures temperate, without the inordinate fondness for alcohol that prevailed among some of the tribes. And their huts were neater than the houses of most of the missionaries he had seen. The Indians cultivated cassava, kept pigs, and made a nourishing beverage from the fleshy fruit of the *seje* tree, which resembled a coconut. They also manufactured pottery, baking the vessels in brushwood fires and painting them with geometric patterns and the figures of crocodiles and monkeys. A number of tribes in the region had made painted pottery for centuries; Francisco de Orellana, the first white man to journey the length of the nearby Amazon (1541–42), had been impressed by the painted pots he'd discovered among the Omagua people there.

On April 21, with the portage complete, the party resumed their journey toward the junction of the Orinoco and the Casiquiare. Though the lancha had been battered by the shoals and portages, it was still intact, and Humboldt hoped it would withstand the insults that still lay in store for it. Leaving the rapids, the party advanced cautiously, for they were entering a wild region that only a few white men had ever seen. "When the traveler has passed the Great Cataracts," Humboldt explained, "he feels as if he were in a new world, and has overstepped the barriers which nature seems to have raised between the civilized countries of the coast and the savage and unknown interior."

Over the next several days, the canoe slipped by the mouths of the rivers Sipapo, Vichada, and Zama, and the peaks of Cerros and Sipapo, which formed an immense rock wall above the surrounding plain. Judging from the rivers' mouths, the Vichada and the Zama appeared to be among the largest tributaries of the Orinoco, but since they had never been explored, one could only guess at their source or length. In truth, the entire region was terra incognita only three miles from the rivers' banks, since no white man had ever ventured farther into the forest. These days the land was believed to be occupied by the fierce Chiricoas people, but they were little seen, since they didn't have the art of boat building. In previous times, when the territory had been inhabited by the Caribs and their enemies the Cabres, no explorers could have safely camped at the junction of these rivers. But since the coming of the whites, these warlike tribes had retreated from the banks of the Orinoco, and now the region was eerily devoid of any human presence whatsoever.

On April 24, a violent storm forced the party to break camp in the middle of the night. By two A.M. the drenched men were back on the river, having abandoned some books they were unable to locate in the darkness. That day the lancha passed the mouths of the Ucata, Arapa, and Caranaveni rivers, and by late afternoon the travelers had reached the outlying Indian plantations attached to the mission of San Fernando de Atabapo. Paddling in the gathering darkness, they left the Orinoco and entered its tributary the Río Atabapo, the first of the three rivers to be ascended before they reached the portage at the Caño (Stream) de Pimichín, where they would cross from the watershed of the Orinoco to that of the Amazon.

The men reached the mission a little after midnight, and the surprised priest received them with hospitality. At daybreak the next morning, the travelers "found themselves as if transported to a new country, on the banks of a river the name of which we had scarcely ever heard pronounced [Atabapo], and which was to conduct us . . . to the Río Negro, on the frontiers of Brazil." In the darkness, they had passed from the white waters to the black.

SOUTH AMERICAN WATERWAYS are divided into two broad classes—the so-called white rivers and the so-called black rivers. The white include the Orinoco, the Casiquiare, and the Amazon itself. But the term is misleading, as these are actually a dirty yellow-green, tinted by the great quantities of silt washing out of the younger, softer Andes to the west, where virtually all the white rivers originate. Nearly all the black rivers, by contrast, flow south out of the Brazilian Highlands. One of the oldest geological formations on the planet, these highlands have long since given up most of their silt, and the basalt that remains is very slow to erode. Whereas the white rivers are cloudy, the black are so clear that even small fish can be seen at a depth of thirty feet or more, despite the dark tint, which is caused by plant matter dissolved over the rivers' long courses. "Their waters," Humboldt found, "seen in a large body, appear brown like coffee, or of a greenish black. These waters, notwithstanding, are most beautiful, clear, and agreeable to the taste. . . . When the least breath of wind agitates the surface of these 'black rivers' they appear of a fine grass-green, like the lakes of Switzerland. . . . These phenomena are so striking, that the Indians everywhere distinguish the waters by the terms black and white." The Indians, in fact, could identify the water of a specific river by taste alone. So striking is the color difference between the two classes of river that at the Río Negro's junction with the Amazon, the white and black waters flow side by side, readily discernible, for some fifty miles before merging.

"Everything changes on entering the Río Atabapo," Humboldt wrote; "the constitution of the atmosphere, the color of the waters, and the form of the trees that cover the shore. You no longer suffer during the day the torment of mosquitoes; and the long-legged gnats (*zancudos*) become rare during the night. . . ." Crocodiles also disappeared, along with manatees, tapirs, and howler monkeys, though the freshwater dolphins were still in evidence. Huge water snakes resembling boas, up to fourteen feet long, swam beside the canoe, making it dangerous for the travelers to bathe. The jaguars on the banks appeared large and well fed, but were reputed to be less aggressive than those on the Orinoco. "Nothing can be compared to the beauty of the banks of the Atabapo," Humboldt found. "Loaded with plants, among which rise the palms with feathery leaves; the banks are reflected in the water, and this reflex verdure seems to have the same vivid hue as that which clothes the real vegetation." So lovely was this new terrain, and so blessedly free of insect pests, that Humboldt "began to regret the Lower Orinoco."

In 1756, Jesuits under Father Solano had founded a mission at the junction of the Orinoco, the Río Negro, and the Atabapo, among the Guaypunave people. The Guaypunaves' chief, Cusero, warned Solano to wait at the cataracts of Maipures for a year before attempting to settle his mission, so that a cassava plantation could be established upstream. But Solano, impatient to begin God's work, refused this advice, with the result that a large proportion of his party, as well as many of the mission Indians, died of starvation in the coming months.

By the time of Humboldt's journey, no one was starving at San Fernando, which had grown into one of the largest settlements along the Upper Orinoco. There were only seven or eight cattle in the entire village, but every Indian family cultivated a grove of coconut palms. Cassava and plantains were also raised, and for several months of the year the Indians supplemented their diet with the fruit of a remarkable palm called the *piritú*. More than sixty feet high, the *piritú* produced a starchy fruit, yellow and slightly sweet. Eaten boiled or roasted like plantains, the fruit of the *piritú* seemed to confirm Linnaeus's belief that the palm had served as the original food of mankind. "Man dwells naturally within the tropics," the great Swedish botanist had written, "and lives on the fruits of the palm tree; he *exists* in other parts of the world [including Linnaeus's native Europe], and there makes shift to feed on corn and flesh."

The missionary at San Fernando lived in a neat house surrounded by gardens. As he conducted his visitors around the mission, he proudly described the depredations still made on the surrounding villages, when, "for the conquest of souls," priests and mission Indians would travel up the Río Guaviare and raid settlements, seizing children above eight or ten years of age and carrying them to San Fernando to be parceled out as slaves among the Indians of the mission. Hearing these stories, Humboldt was outraged by the persistence of a barbaric practice outlawed many years before.

One account in particular excited his fury. Beyond San Fernando, where the Atabapo overspread its banks and merged with the surrounding forest, flat rocks, or *piedras*, protruded above the surface of the water, forming familiar landmarks. One such rock, known as La Madre, or la Piedra de la Guahíba, had been christened after a courageous woman of the Guahíbo tribe. Not three years before Humboldt's arrival, the missionary at San Fernando had led a raiding party up the Guaviare, and among the prisoners taken were a Guahíba and her two small children, surprised at home while her husband and older children were away fishing.

The prisoners were transported to San Fernando but managed to escape repeatedly, only to be recaptured each time. Finally, after ordering the woman mercilessly beaten, the missionary separated mother and children and had the woman transported up the Atabapo to a mission called Javita. En route, she managed to slip her restraints, jump out of the canoe, and climb onto the rock that afterward bore her name. But her captors rowed ashore, tracked her down in the woods, and dragged her back to the rock, where they administered another brutal beating with manatee whips. Bound more securely, the mother was then conveyed to Javita, where she was imprisoned in "the king's house."

But during the night, the Guahíba escaped once again. For days, she trekked through the forest, fording rivers and clawing through vines and underbrush, making her way over territory thought to be impenetrable, all the while living on ants she dug out of the ground. Four days after her escape, she was spied hiding outside the mission at San Fernando, where her young children were still being held. Retaken yet again, the Guahíba was exiled to a faraway mission on the Upper Orinoco, where, her spirit finally broken, she refused all nourishment and soon expired.

Humboldt was deeply moved by this story of courage and resistance in the face of cruelty. "In this relation of my travels I feel no desire to dwell on pictures of individual suffering," he explained, "—evils which are frequent wherever there are masters and slaves, civilized Europeans living with people in a state of barbarism, and priests exercising the plenitude of arbitrary power over men ignorant and without defense. . . . If I have dwelt longer on the Rock of the Guahíba, it was to record an affecting instance of maternal tenderness in a race of people so long calumniated; and because I thought some benefit might accrue from publishing a fact, which I had from the monks of San Francisco, and which proves how much the system of the missions calls for the care of the legislator."

Humboldt's canoe made slow progress over this portion of the river, due in part to the swift current and in part to frequent stops for plant collecting. On the night of April 29, the party put in at the well-tended mission of San Baltasar, where they watched Indians preparing a substance resembling caoutchouc, or India rubber. Produced from the milky sap of several tropical plants, rubber had been reported by Spanish and Portuguese explorers from Columbus onward. The first to study the substance in detail, La Condamine had called it *latex*. (The name *rubber* had been coined in 1770 by the great English scientist Joseph Priestley, who had noticed that it could be used to rub out pencil marks; the "India" in the name was a reference to the East Indies, where it was mistakenly thought to originate.) For the first part of the nineteenth century, the material was used mainly for rubber bands, erasers, and crude waterproof clothing, but in 1839 Charles Goodyear stumbled on vulcanization, the process in which rubber is heated and combined with sulfur, greatly enhancing its strength, elasticity, and resistance to heat and cold.

After that breakthrough, the world rushed to exploit rubber's huge commercial potential, transforming the frontier village of Manaus, Brazil, located where the Río Negro joins the Amazon, into the center of the worldwide rubber market. By the late nineteenth century, Manaus boasted mansions, paved streets, streetcars, an opera house, a racecourse and bullring, twenty-three department stores, seven bookshops, and a cost of living four times higher than that of New York City. The boom, which lasted till the early years of the twentieth century (when Asian rubber and later synthetic rubber came to dominate the world market), would vindicate Humboldt's prophesy that "there [in the Amazon Basin], sooner or later, the civilization of the world will be found."

The material that Humboldt discovered and introduced to Europe under the name *dapicho*, or *zapis*, was similar to but not the same as so-called India rubber. Instead of being collected as a milky liquid, this white, fungous substance was dug from two or three feet deep in the earth, amid the roots of certain trees. The *dapicho* would be placed on a spit and roasted like meat, giving off a resinous odor, till it gradually grew black and elastic. Then it would be formed into balls, to be used in an Indian game resembling tennis, or fashioned into drumsticks for beating hollow trees, or cut into artificial corks. But the genus *Hevea*, whose sap flowed freely from slashes in its trunk, was destined to become the commercial source of the world's natural rubber.

Five miles above the mission of San Baltasar, Humboldt's party left the Atabapo and entered the second of the small rivers that would conduct them toward the Casiquiare—the Río Temi, which, about a mile wide, frequently overspread its banks even during the dry season. Venturing into the flooded forest to avoid the brunt of the current, the pilot attempted to follow narrow, shallow channels between the trees while one of the crew crouched in the bow, chopping at the underbrush with a machete. Finally, at five that evening, the boat regained the main channel. Then the canoe caught between the trunks of two trees, and after the crew finally managed to free it, they immediately reached a confusing intersection of several channels. Choosing one, the pilot led them into forest so thick that navigation was impossible by either sun or stars. It was long after dark when they finally found a place dry enough to camp for the night.

The next morning, May 1, they departed well before sunrise and held to the riverbed till daylight, when they veered once again into the network of channels snaking through the inundated forest. At the junction with the smaller Tuami, the canoe turned up that river, traveling southwest, till it reached the mission of San Antonio de Javita, where the unfortunate Guahíba woman had been transported. Here the explorers planned to hire Indians to help them make the portage to the Caño de Pimichín, which would carry them from the watershed of the Orinoco to that of the Amazon.

It would be a matter of four or five days to maneuver the canoe over the portage. In the meantime, Humboldt, Bonpland, Zea, and Soto stayed at the mission, whose population of about 160 Indians of various tribes were occupied in building dugout canoes. Like all the Spanish missions, San Antonio de Javita combined a saint's name with an Indian name. For the sake of clarity (since there might be several missions named after a particular saint), the missions were generally known by their Indian name, except in cases where that proved too difficult for the Spanish to pronounce. Following this convention, the mission of San Antonio was popularly known as Javita, in honor of a local Indian leader.

Renowned for his courage and energy, Javita had at first been an ally of the Portuguese, who, encroaching into Spanish territory, had authorized him to conduct slaving expeditions in the region under the pretext of making converts. On one such excursion, Javita was captured by the Spanish, who managed to win him over to their camp. With the chief's help, the Portuguese were driven from the area, and the mission of San Antonio was founded. Though the chief was now advanced in age, Humboldt found him "of great vigor of mind and body" and still exerting considerable influence over the neighboring Indian nations.

Javita accompanied Humboldt and Bonpland on their botanizing forays during the time they passed at the mission and regaled them with stories of days gone by, when it had been a common custom to eat enemies captured in battle. Whatever the practice of cannibalism may say about human nature, Humboldt believed that the impenetrable terrain of the rain forest played a crucial role in the gruesome practice, by keeping the various tribes isolated and suspicious. "In Spanish Guiana a mountain, or a forest half a league broad, sometimes separates hordes who could not meet in less than two days by navigating rivers," he wrote. "In open countries, or in a state of advanced civilization, communication by rivers contributes powerfully to generalize languages that appear to us radically distinct and keep up national hatred and mistrust. Men avoid, because they do not understand each other; they mutually hate, because they mutually fear." Was this comment, one wonders, intended as a reflection on the political turmoil in Europe as well as that in South America?

The mission was located in an extremely wet area, and even when it wasn't raining the sky was overcast, making astronomical measurements impossible. In fact, the missionary told them, it was not uncommon to see rain for four or five months without letup. From April 30 to May 4, Humboldt was unable to take a single geodetic reading. During that time, the travelers added to their botanical collection, though the extreme height of the trees— over a hundred feet, with foliage only at the crown—made the work inordinately confusing. Leaves and flowers would fall to the ground, but it was devilishly difficult to determine which materials had dropped from which specimen. "Amid these riches of nature herborizations caused us more chagrin than satisfaction," Humboldt confessed. "What we could gather appeared to us of little interest, compared to what we could not reach."

During their stay at the mission, Humboldt and Bonpland determined to rid themselves of some painful parasites. For the past two days, both men had suffered from an extraordinary irritation on the backs of their hands and in the joints of their fingers, where under a magnifying glass they could make out parallel white furrows beneath the skin. The missionary recognized the culprits as *aradores*, or "ploughers," and he sent for the woman who was the *curandera*, or healer, of the village. Heating the point of a hard bit of wood over a lamp, the *curandera* dug into the furrows and, after a lengthy, excruciating excavation, removed three or four of what appeared to be tiny round sacks. Though painful, the operation did give immediate relief. It had grown late, and since the explorers' hands were covered with the furrows, further treatment was postponed till morning.

However, before the *curandera*'s return the next day, the Europeans met another Indian, who claimed to have a quick and painless cure for their misery. He made a cold infusion of the bark of a shrub called *uzao*, which bore small, glossy leaves. With a bluish color and a taste like licorice, the infusion became frothy when beaten. Humboldt and Bonpland drank it, and found that the irritation from the *aradores* disappeared, without any painful subcutaneous probing. So effective was the cure that the travelers took some of the bark away with them, in case of reinfestation.

Every day Humboldt and Bonpland went to check on the progress of the portage. Altogether, twenty-three Indians, including their own crew, were employed in cutting trees to serve as rollers and in propelling the long, fragile canoe through the forest. It was hazardous as well as exhausting work. On the evening of May 4, one of the Indian workers was bitten by a viper.

Though the man was robust, the bite rendered him comatose, and even after regaining consciousness he continued to suffer such nausea and vertigo that Indians and Europeans alike feared the worst. After he was administered a traditional antidote, he eventually recovered.

The next day the canoe cleared the portage. Reaching the boat at dusk, Humboldt and Bonpland spent the night in a hut recently abandoned by a native family, who had left behind their fishing nets and tackle, pottery, and other domestic accumulations. Before the party settled into the house, the guides killed two vipers that had taken up residence there. With white bellies and brown-and-red-spotted backs, the snakes were beautiful but deadly. Since there was no place to hang their hammocks, the travelers slept on grass spread on the floor, passing a restless night in which every rustle evoked images of unseen snakes. Indeed, in the morning one of the Indians lifted the jaguar skin he'd been sleeping on and discovered that a viper had slithered beneath it in the night.

In the daylight, Humboldt examined the canoe. Though its hull had been worn noticeably thinner during the portage, no cracks were visible. The Caño de Pimichín's banks were low but rocky, and there was only one rapids to cross. The lancha was lowered into the water, and four and a half hours later the travelers entered the Río Negro. After thirty-six days' confinement in a crowded, unstable boat; surviving encounters with crocodiles, jaguars, and poisonous snakes; suffering horribly from hordes of biting, stinging, and burrowing insects; withstanding the discomfort and insalubrity of one of the wettest climates in the world; successfully passing some of the most celebrated rapids on earth; completing the arduous portage between two of the world's great river systems with their fragile canoe intact, the travelers had at last left the Orinoco. "After all we had endured," Humboldt wrote, "it may be conceived that we felt no little satisfaction . . . in having reached the tributary streams of the Amazon."

But there remained another thousand waterborne miles before the party would reach the delta city of Angostura. And Humboldt still hadn't fulfilled "the most important object of our journey, namely, to determine astronomically the course of that arm of the Orinoco which falls into the Río Negro, and of which

the existence has been alternately proved and denied during half
a century" since its putative discovery by Jesuit missionaries.
If all went well, he hoped to be able to report at last whether
the so-called Casiquiare Canal existed, as La Condamine had believed, or whether it was a fiction, as the geographers of Europe
had decided. Either way, the explorers now found themselves in
the basin of the most fantastic river on the planet.

SEVEN

The Amazon

THE AMAZON IS THE QUINTESSENTIAL RIVER OF EARTH. Rising at eighteen thousand feet in the Peruvian Andes, just 120 miles from the Pacific, it first rushes and later meanders across four thousand miles to the Atlantic, on the far side of the continent. It is not the longest river in the world; the Nile nudges it out of that honor by about two hundred miles. However, the Amazon lays claim to every other superlative. It is the deepest river on the planet. And it has by far the greatest volume, disgorging some 3.4 million gallons of water *per minute* into the Atlantic—a staggering one fifth of all the world's fresh water (twelve to fourteen times the amount disgorged by the Mississippi). The Amazon has more than a thousand tributaries—seventeen of which are over a thousand miles long—longer than the Rhine—and six of which are themselves among earth's ten longest rivers. Moreover, because the Amazon flows so slowly in its flat lower course, it is estimated that, at any given time, the river and its nearly fifty thousand miles of tributaries contain a staggering two thirds of the world's river water.

The vast area drained by these waterways encompasses 2.5 million square miles (equal to eighty percent of the continental United States) and takes in two fifths of South America, including half of Brazil and parts of eight other nations. The river is two hundred miles wide at its mouth, where the largest island is the size of Switzerland. In fact, Vicente Yáñez Pinzón, the Spanish captain who first spied the river in 1500, sailed two hundred miles upstream before realizing he'd even left the ocean; he thought he'd discovered some bizarre arm of the Atlantic, which he christened the Freshwater Sea. A thousand miles from the ocean, the river is still seven miles wide. Oceangoing vessels on the Amazon can navigate more than two thousand miles inland.

In 1639, Jesuit missionary and explorer Cristóbal de Acuña, traveling with Portuguese explorer Pedro Teixeira, published the earliest first-person account of the Amazon. "It flows along, meandering in wide reaches," he wrote, seeking to capture the river's majesty, "and, as absolute lord of all the rivers which run into it, sends out its branches, which are like faithful vassals, with whose aid it goes forth, and, receiving from the smaller streams the lawful tribute of their water, they become incorporated into the main channel. . . . In breadth it varies greatly, for in some parts its breadth is a league, at others two, at others three, and at others many more; . . . spread out into eighty-four mouths, it may [be placed] on an equality with the ocean."

The Amazon is known by various names over different segments of its length. Near its source, where it descends an alarming sixteen thousand feet in only six hundred miles, the river is known as the Marañon. From the Brazilian border to the mouth of the Río Negro, it is called the Solimões. It is only in the second half of its length, as it descends a lazy quarter inch per mile, that the river is universally known as the Amazon. So convoluted is its course that there was no definitive map of the river until the latter half of the twentieth century, when it was finally charted from the air. No wonder that for centuries it was called the Amazons, in acknowledgment of its multiple guises.

Hundreds of millions of years ago, when South America nestled into the western bulge of Africa to form (along with Australia, India, and Antarctica) the southern supercontinent of Gondwanaland, the Amazon flowed from east to west. About 150 million years ago, as earth's tectonic plates shifted, Gondwanaland started to splinter and South America began its drift to the west, opening up the South Atlantic Ocean. As the continent drove into the plates beneath the Pacific floor, the Andes were forced upward, eventually damming the Amazon and creating a huge inland lake. Then, perhaps 50 million years ago, the lake breached two of the oldest rock formations on earth, the Guyana Shield to the north and the Brazilian Shield to the south, and the Amazon carved a course to the Atlantic, establishing the west-to-east flow that exists today.

Though this vast river system was still relatively little known at the time of Humboldt's journey, it was actually better explored than the Upper Orinoco. By 1799 five expeditions had sailed the length of the Amazon. The first, led by Francisco de Orellana in 1541–42, was a journey of accident and desperation, not exploration. Orellana had followed the conquistador Gonzálo Pizarro east of the Andes in hope of discovering El Dorado, and when the expedition ran urgently short of provisions, he volunteered to lead a party downriver in search of food. Prevented from returning by the overpowering current, he and his beleaguered men continued downstream all the way to the river's mouth—a horrific nine-month journey of starvation and warfare with the native peoples. When the survivors finally reached the Atlantic, Orellana sailed for Spain, where he was charged with desertion but ultimately exonerated.

After Orellana's harrowing experience, it was more than a century before anyone tried to emulate his achievement. Then in 1637–38, Spain's rival Portugal dispatched Pedro Teixeira up the Amazon, starting from the mouth on the Brazilian coast. In just under a year, Teixeira managed to reach the river's source in the Andes, becoming the second man to travel the length of the Amazon and the first to make the journey upstream. It was on his return passage, downstream to the river's mouth, that Teixeira was accompanied by Cristóbal de Acuña, whose account of the Amazon added enormously to Europe's knowledge of the river.

La Condamine was the next man to travel the length of the Amazon, in 1743, on his way home to France after his geodetic mission in the Andes. Covering some twenty-six hundred miles in not quite two months, he made a brief survey of the region's geography, plant life, and human inhabitants and brought back to Europe quinine, curare, and rubber, along with tales of the Casiquiare Canal. Though his journey was cursory, La Condamine was the first scientist to explore the river, and his published accounts captured the imagination of Enlightenment Europe. He became celebrated—not, ironically, for his backbreaking seven-year effort to measure an arc of latitude at the equator, but for his seven-week sail down the Amazon. In fact, until he was eclipsed by Humboldt nearly seventy years further on, it was La Condamine, more than any other man, who was associated with the rain forest in the public imagination. (Several years later, two of La Condamine's countrymen also journeyed down the Amazon on their way home to Europe—Jean Godin, one of La Condamine's chain bearers, from 1749 to 1750, and Godin's wife, Isabella, the first woman to complete the journey, from 1769 to 1770—but, like Orellana, their purpose was transportation, not exploration.)

Though Humboldt would spend more time in the rain forest than La Condamine and would explore its plants, animals, and peoples in far greater depth, he had no intention (at least on this leg of his journey) of traveling as far as the Amazon itself, preferring to focus instead on the issue of the Casiquiare Canal. Entering the Amazon Basin at the Río Negro, the great river's principal tributary, he would ascend the Negro to its junction with the Casiquiare, then—assuming La Condamine was right—sail up that river to the Orinoco, then downstream on the Orinoco to the town of Angostura, in the delta near the coast. Fourteen hundred miles long (slightly shorter than Europe's Danube), the Río Negro rises in eastern Colombia, where it is known as the Río Guiana, then flows east to the Venezuelan border, then south along the Colombia-Venezuelan frontier, thence into Brazil, where it joins the Amazon at the famous rubber city of Manaus.

As the name implies, the Río Negro is the ultimate blackwater river. In 1854, William Herndon, charged with surveying the Amazon for the United States government, became the first American to descend that river from Peru to the Atlantic. As he passed the mouth of the Río Negro, he wrote, "There has been no exaggeration in the description of travelers regarding the blackness of its water. It well deserves the name of Río Negro. When taken up in a tumbler, the water is a light-red color like a pale juniper water, and I should think it colored by some such berry. An object immersed in it has the color, though wanting the brilliancy, of red Bohemian glass. It may have been fancy," he admitted, "but I thought that the light cumuli that hung over the river were darker here than elsewhere. These dark, though peaceful-looking clouds, the setting sun, the glitter of the rising moon upon the sparkling ripple of the black water with its noble expanse, gave us one of the fairest scenes that I have ever looked upon."

Having passed the small cataract at the junction of the Pimichín and the Río Negro, on May 5, Humboldt's lancha pulled into the thriving mission of Maroa, where he added some live toucans to his floating menagerie. Two hours later the canoe arrived at the mission of San Miguel de Davipe, where the explorers took on provisions, including some fowls and a pig. It had been a good while since the men had had any meat, and spurred on by the prospect of roast pork that evening, they cut short their stay at Davipe. Just above the mission was a branch of the Casiquiare called the Conorichite, which as late as the mid-1700s had been frequented by Portuguese slavers trading beads, knives, fishhooks, mirrors, and other trinkets with the Caribs for human captives. "Thus," Humboldt wrote, "the unhappy natives before they came into immediate contact with the Europeans, suffered from their proximity. The same causes produce everywhere the same effects. The barbarous trade which civilized nations have carried on, and still partially continue, on the coast of Africa, extends its fatal influence even to regions where the existence of white men is unknown."

At sunset, the travelers landed at the island of Dapa, picturesquely situated in the middle of the river, where they found a hut occupied by more than a dozen Indians eating their evening meal of smoked ants mixed into cassava cakes. Father Zea, "whose fever seemed rather to sharpen than to enfeeble his appetite," according to Humboldt, encouraged his companions to try the concoction. "It somewhat resembled rancid butter mixed with crumb of bread" was Humboldt's assessment. But "some remains of European prejudices prevented our joining in the praises bestowed by the good missionary on what he called 'an excellent ant paste.' "

That night there was a violent rainstorm, and instead of sleeping out of doors, the explorers were invited to share the Indians' hut. Rest was elusive, though, since their hosts, like most tribes in the area, habitually slept from about eight in the evening to two in the morning, after which they would lie in their hammocks and chat, throw wood on the fire, and prepare a bitter, caffeinated beverage from the seeds of the soapberry plant. Knowing the customs of the native peoples, the Spanish and Portuguese used to launch their slaving raids between the hours of nine and twelve, when the Indians were in their first, deepest slumber of the night.

Leaving the mission long before daybreak, the explorers

paddled for twelve hours to the Fort of San Carlos, at the fork of the Río Negro and Casiquiare. The upper story of the fort afforded a fine view of the latter river, which, although notoriously tortuous over most of its course, here ran north to south in a line as straight "as if its bed had been dug by the hand of man." A garrison of seventeen soldiers was assigned to the fort, though ten were detached to outlying missions. Owing to the extreme humidity, only four muskets were in working order.

This intersection of the Río Negro and the Casiquiare was the closest that Humboldt would come to the Amazon on this portion of his journey (it was about four hundred miles away), before turning north to complete his circuit back to the Venezuelan coast. In fact, it wouldn't have taken much more time to sail down the Río Negro to the Amazon and all the way to the Atlantic. To be within striking distance of the fabled river, to be traveling on its principal tributary, Humboldt must have been sorely tempted to continue in that direction. But the Río Negro and the Amazon had already been explored, and he knew he could make a greater contribution to knowledge by charting the controversial Casiquiare. Besides, the other route would have meant crossing from Spanish territory into Portuguese, and given the tense relations between those countries, that would have been very difficult if not impossible.

Though not one of the longest rivers in South America, the Río Negro had always had strategic importance disproportionate to its length, because the Spanish controlled the upper reaches, in present-day Colombia and Venezuela, while the Portuguese claimed the more populated lower course, in present-day Brazil. For the better part of three centuries, the Iberian rivals had been unwilling neighbors in the New World, sometimes arguing their boundary disputes before the pope and other times simply seizing whatever land they could—on which occasions the river provided a ready-made invasion route into the other's territory. To make matters worse, Portugal remained the closest trading partner and military ally of Spain's perennial nemesis, Great Britain. At the time of Humboldt's journey, tensions were running particularly high between the two nations, as Britain and Portugal were still fighting against Napoleon's France, whereas Spain had sued for peace four years before. As a result of these continuing hostilities, security was especially tight on the border between their American colonies. Commanders of frontier outposts were on constant alert, and even the Indians in neighboring Spanish and Portuguese villages were on dangerous terms. Though they were ignorant of the raging geopolitical storm, the Indians could see that their missionaries wore cloaks of different colors, and thus were the lines of mutual antipathy drawn.

Against this background, it's not surprising that outsiders were objects of suspicion, especially when poking around the frontier seeking geographical knowledge that could conceivably be valuable to one's enemies. By venturing into the borderlands between the hostile colonies, Humboldt realized he was incurring a new element of risk. Yet the courses of the Orinoco and the Río Negro were fixed by Nature, not by the courts of Lisbon and Madrid. If he were to verify the existence of the Casiquiare Canal, Humboldt had no choice but to follow where the rivers led. And that meant accepting this new political threat, along with the by-now-familiar perils of river rapids, cannibal tribes, savage animals, and tropical diseases.

Indeed, had Humboldt journeyed south on the Río Negro instead of north, he would in all probability have ended up in a Portuguese prison. For, though he didn't know it at the time, a Brazilian newspaper had published an account of his activities based on reports from the locals and, fearing that the foreigner would incite the native peoples against Portugal, had implored the governor to take action: "A certain Baron von Humboldt, a native of Berlin, has been traveling in the interior of America making geographic observations for the correction of certain errors in existing maps, collecting plants . . . a foreigner who, under pretext of this kind, might possibly conceal plans wherewith to spread new ideas and dangerous principles among the faithful subjects of this realm. Your Excellency should investigate at once . . . as it would be extremely injurious to the political interest of the Crown of Portugal if such were the case. . . . " In response, orders had been issued for the outsider's arrest and for the seizure of all his instruments and notes. If captured, Humboldt would have had his voyage down the Amazon after all—in leg irons—and thence across the Atlantic to Lisbon. Though he would have been released eventually, it would, of course, have meant the end of his South American adventure.

CONTINUING UPSTREAM, the party encountered a group of Indians who collected green minerals called Amazon stones. Carved into cylinders and inscribed with sacred symbols, the amulets were worn around the neck for protection against fevers, nervous disorders, poisonous snakes, and other hazards. In fact, a few years before, Amazon stones were believed to be a powerful febrifuge even in Europe. "After this appeal to the credulity of Europeans," Humboldt scoffed, "we cannot be surprised to learn that the Spanish planters share the predilection of the Indians for these amulets, and that they are sold at a very considerable price." Though the exact source of the stones was unknown, they were thought, by long tradition, to originate to the north, in the country of the 'women without husbands,' " i.e., the Amazon warriors of legend.

The original Amazons were a people of Asia Minor who were renowned for their horsemanship. Their name is thought to derive, through the Greek, from the Iranian *ha-mazan*, meaning "fighting together." As incorporated into Greek mythology, the Amazons were a tribe whose women served as warriors and statesmen while the men took care of the domestic chores. Fighting on horseback like their real-life namesakes, the mythological Amazons cut off one of their breasts to improve their bowmanship, and they supposedly conquered a large swath of Asia Minor and did battle with many Greek heroes, including Hercules.

The Amazons of the New World were originally described by Francisco de Orellana in 1541, the first European to travel the length of the river that now bears their name. Among the many native peoples that his expedition battled en route, Orellana told of a ferocious race of tall women warriors, whom he named after the Amazons of Greek mythology. His account, perhaps an exaggeration of a tribe in which the men and women took up arms together, was perpetuated by the Europeans who followed, including Sir Walter Raleigh. La Condamine accepted the stories and even reported that the Amazons had migrated up the Río Negro. In Humboldt's time, the legend was still current. "Since my return from the Orinoco and the river Amazon," he wrote in the *Personal Narrative*, "I have often been asked, at Paris, whether I embraced the opinion of that learned man [La Condamine], or believed, like several of his contemporaries, that he undertook the defense of the Amazons, merely to fix, in a public sitting of the Academy, the attention of an audience somewhat eager for novelties." Though the perpetrators of the legend had a vested interest in exaggerating the wonders of the New World, Humboldt concluded that, considering the number of similar independent reports, these questionable motives may not be sufficient to reject the stories outright. Could it have been, he wondered, that a group of women, growing tired of mistreatment by the men of their tribe, had struck out into the forest to live independently, learning the martial arts and, to perpetuate their race, periodically admitting the company of the opposite sex? Yet no one ever substantiated the legend of the South American Amazons.

Humboldt's party passed three nights at the fort of San Carlos, not far below the fork of the Río Negro and the Casiquiare. Each night, Humboldt sat up in the darkness, instruments at the ready, waiting for the skies to clear long enough to permit an astronomical sighting that would fix the junction of the two rivers. But his sleeplessness went for naught. "What a contrast," he exclaimed in frustration, "between the sky of Cumaná, where the air is constantly pure as in Persia and Arabia, and the sky of the Río Negro, veiled like that of the Faroe Islands, without sun, or moon, or stars!"

Before sunrise on May 10, the explorers finally resumed their journey upriver. The morning broke clear, but as the temperature rose, clouds gathered and eventually obscured the sun. Humboldt was worried that the overcast skies would prevent him from charting the Casiquiare, particularly the crucial points where it branched from the Río Negro and entered the Orinoco. As if that weren't bad enough, their priceless plant collection, each one painstakingly selected, dried, and identified, was rotting. "We grieve almost to tears when we open our plant-boxes," Humboldt wrote. "The extreme humidity . . . has caused more than one third of our collection to be destroyed. Daily we discover new insects destructive to paper and plants. Camphor, turpentine, tar, pitched boards, and other preservatives so effective in a European climate prove quite useless here." For a naturalist, it was the ultimate disaster—the same one, in fact, that had reportedly caused La Condamine's botanist, de Jussieu, to go mad several decades before.

Faced with the prospect that they would have nothing but

insect bites to show for their grueling journey—no latitude and longitude readings, no botanical specimens, no tangible validation of any kind—Humboldt questioned anew the wisdom of their itinerary. It would be far shorter and less strenuous to retrace their route to the coast rather than to venture another thousand miles—nearly twice as far again—into an unsettled, unknown wilderness. Should he discard his original ambition to prove the existence of the Casiquiare Canal? Worn out from a month of misery on the river, Humboldt, uncharacteristically, hesitated before the challenge.

Ultimately, he determined to press on. Though the route was long, he told himself, they had only ten days' paddling upstream; after that, the entire way would be with the current. Having come so far, he felt they must continue: "It would have been blamable, to have suffered ourselves to be discouraged by the fear of a cloudy sky, and by the mosquitoes of the Casiquiare." Besides, the Indian pilot assured them that as soon as they left the black waters of the Río Negro, they would find clear skies and "those great stars that eat the clouds." Still, it's an indication of the party's wretchedness that the doggedly confident and ferociously determined Humboldt would even consider the alternative to forging ahead.

Eight miles from San Carlos, above the rapids near Piedra de Uinamane, the expedition entered the Casiquiare. Until now the rain forest had presented more opportunities for botanizing than for geologizing. However, as they'd traveled, Humboldt had been noting the location and composition of the various mountains and outcroppings they'd passed. Always toward the front of his mind was the great controversy still raging between the neptunists and the vulcanists. With every geological observation he made, Humboldt asked himself what new light it might shed on this central dispute.

Here at Uinamane, Humboldt noticed that the granite was streaked with numerous veins several inches wide. Some of these blazes were of white quartz mixed with feldspar, while others consisted of mica combined with black tourmaline. Moreover, wherever the two types of veins crossed, the black always obscured the white. Werner and the neptunists held that such veins were created when clefts in existing rock were filled by sediments deposited from above. The fact that the pattern here was consistent—that black always covered white—suggested that the former had been laid down more recently. It also seemed to support Werner's explanation, since one wouldn't have expected to find such a consistent pattern in Hutton's more violent vulcanist model. Indeed, this precise configuration of veins was often seen in Europe, and its discovery here in the New World seemed to lend new credence to Werner's ideas. "Being a disciple of the school of Freiberg," still casting his lot with the neptunists, Humboldt found, "I could not but pause with satisfaction at the rock of Uinamane, to observe the same phenomena near the equator, which I had so often seen in the mountains of my own country."

At its mouth, the Casiquiare was a majestic sight, rivaling the Río Negro in width. In fact, the rivers greatly resembled each other, with thick forest growing down to their banks—except that the Negro ran black and the Casiquiare white. That night Humboldt sat up several hours waiting for the clear skies that the Indian pilot had promised, so that he could fix the mouth of the Casiquiare on the world's maps. But "the air was misty," he was forced to report, "notwithstanding the *aguas blancas*, which were to lead us beneath an ever-starry sky."

The first European settlements on the Casiquiare and elsewhere south of the great rapids of the Orinoco had begun as crude blockhouses boasting a pair of small-caliber cannon and a detachment of two soldiers. Only in the past fifteen years had these military outposts been converted to missions, under the governance of the Franciscan monks. One such mission was San Francisco Solano, near the mouth of the Casiquiare, and it was here that Humboldt's party stopped for the night.

One of the Indians of the mission had two fine birds, a young toucan and a purple macaw, which Humboldt bought and added to his zoological collection. Not counting these most recent acquisitions, the cramped, unstable canoe was now carrying—besides the dozen human crew and passengers, scientific instruments, provisions, and thousands of mineralogical and botanical specimens—a dog, seven parrots, seven other assorted birds, and nine monkeys (including three species previously unknown to science). Although some of the animals were confined to cages, others roamed over the boat. Father Zea, for one, was not enthusiastic about sharing the tight quarters with yet more creatures, and he "whispered some complaints at the daily augmentation of this ambulatory collection." Already, at the threat of rain, two titis (small monkeys) would run to take shelter in the ample sleeves of the priest's cassock. And for the remainder of the journey, the toucan took delight in teasing the two *cusicusis* (small, nocturnal monkeys). The toucan also amused the crew by seizing food in its beak, throwing it up in the air, and catching it before swallowing it. In drinking, it would make extraordinary gestures over the water (owing to the shape of its beak), which the Indians interpreted as the sign of the cross, the reason the bird was known locally as *diostede*, or "God give it to thee." Watching the antics of the animals, the travelers were occasionally able to forget the constant misery of the insects. For though the weather hadn't cleared as expected, the white water of the Casiquiarie had brought the predicted swarms of crocodiles and mosquitoes.

But on May 11, a light east wind began to blow, heralding a

**Humboldt's sketch of a cacajao that he purchased
from Indians along the Casiquiare.**
Courtesy of Staatsbibliothek, Berlin.

general clearing. Humboldt happily reckoned that he'd be able to get an astronomical reading to fix their position that night, and the party delayed their departure so as not to stray too far from the mouth of the Casiquiare before those observations could be made. With some difficulty the crew ascended the rapids of Cunanivacari that afternoon, paddling against a current of more than six miles an hour. Four miles farther upriver, they spied a strange formation jutting from the plain, a perpendicular rock wall eighty feet high, with two turrets so symmetrical that they bore an eerie resemblance to the ruins of an ancient fort. Humboldt wondered, Was the formation the remnant of an island that had once risen out of an inland sea, as Werner would have postulated, or had it been heaved up by volcanic forces deep within the earth, as Hutton would have argued?

At five in the evening, they beached the lancha for the night at a large rock known as the Piedra de Culimacari. The evening was clear, and taking a sighting on the Southern Cross, Humboldt was able to fix the latitude of the Casiquiare's junction with the Río Negro at 2 degrees, 0 minutes, 42 seconds; using his chronometer and observing the stars in the feet of the Centaur, he fixed its longitude at 69 degrees, 33 minutes, 50 seconds. Thus, with mathematical precision, Humboldt established one terminus of the natural canal that supposedly didn't exist.

Desperate to escape the misery of the mosquitoes, the travelers broke camp after these observations, at half past one in the morning. The current was running at nearly eight miles an hour against them, and fourteen hours of constant paddling yielded a gain of only ten miles. It was afternoon before the lancha finally arrived at the mission of Mandacava. Like the other missions on the Casiquiare, Mandacava was sparsely populated, with only about sixty Indians in residence. In fact, since the coming of the padres, the native population had actually declined along the river, as the native people had retreated into the forest to preserve their independence and traditional way of life.

The missionary, a kindly, aging man, had, in his words, spent "twenty years of mosquitoes in the forest of the Casiquiare," and his legs were so mottled by insect bites that the original color of the skin could scarcely be discerned. That night the priest spoke of his solitude—and his regret that he could not dissuade some of his charges from the ancient custom of cannibalism. Usually, the priest explained, the practice was confined to enemies killed in battle, when the body would be cut into pieces and carried as a trophy to the huts of the victors. However, there were exceptions. Just a few years before, an Indian had held his wife prisoner in their *conuco*, and when she was sufficiently fattened, he had butchered and eaten her as one would a farm animal. Another case involved a man who had previously lived peacefully in the mission. While accompanying an injured Indian through the forest one day, the man grew impatient with the other's slowness, killed him, and hid the body in a copse of trees. Returning home, he began to make preparations for a feast, until his children convinced him to give up his grisly plan.

Today it's generally believed that accounts of cannibalism among indigenous peoples were often overstated by early European visitors, due to misunderstanding, fear, cultural chauvinism, and the desire to justify the forced spread of their own "civilizing influences." That said, the Caribs, Aztecs, and other groups in South America undoubtedly practiced the custom, ritualistically consuming parts of enemies killed in battle, as a way of consolidating their victory, or of deceased loved ones, as a means of honoring the dead.

Humboldt related that one evening, as the explorers were eating roasted monkey, a young Indian member of their crew commented that the meat, though darker, reminded him of the taste of human flesh. Among his people, he explained matter-of-factly, the meat from the inside of the hands was particularly prized. Asked if he was still tempted sometimes to indulge in cannibalism, the man answered that now that he was living among the missionaries he would eat only what they ate—not out of any moral repugnance, he seemed to imply, but out of a sense of propriety, just as Indians accustomed to going naked in their own village would cover themselves when visiting the mission. "Reproaches addressed to the native on the abominable practice which we here discuss, produce no effect," Humboldt found; "it is as if a Brahmin, traveling in Europe, were to reproach us with the habit of feeding on the flesh of animals. In the eyes of the Indian of the Guaisiia [tribe], the Cheruvichahena [tribe] was a being entirely different from himself; and one whom he thought it was no more unjust to kill than the jaguars of the forest."

On the night of May 13, Humboldt took some astronomical

sightings, then the travelers left the mission at two o'clock in the morning. The following day, the mosquitoes and biting ants again drove the travelers from their camp in the middle of the night. And as the expedition continued upriver, the insects grew only worse. The Europeans' hands and faces swelled to new dimensions of misery, and even Father Zea, who had been bragging that he would pit his mosquitoes at Maipures against any in the forest, admitted that the sting of the mosquitoes along the Casiquiare was the most painful he had ever experienced.

As they paddled upstream, the explorers felt they were entering a barbarous, forbidding land. There was nothing to attract men here, and the Casiquiare didn't see five Indian boats a year. In fact, except at the scattered missions, the travelers hadn't seen a soul along the river since they'd left the rapids at Maipures, a month before. The area was so wild and so little frequented that, except for a few rivers, even the native guides couldn't name any of the passing landmarks. The skies clouded over again, making astronomical observations impossible, and for the next eight days the expedition had no firm idea where they were. The river was cluttered with half-submerged, nearly invisible trees capable of crushing a light canoe, and "the vegetation along the banks became ever thicker," Humboldt wrote, "in a manner of which it is difficult even for those acquainted with the aspect of the forests between the tropics, to form an idea. There is no longer a bank: a palisade of tufted trees forms the margin of the river. You see a canal two hundred toises [about thirteen hundred feet] broad, bordered by two enormous walls, clothed with lianas and foliage. We often tried to land, but without success. . . . We sailed along . . . seeking to discover, not an opening (since none exists), but a spot less wooded," where the Indians could hack out a campsite.

Though they were in the midst of a vast forest, the travelers had trouble finding wood dry enough for their campfires, which were necessary to keep the jaguars at bay. The humidity had also dampened their powder, making their firearms useless for hunting. They had been able to buy scant provisions at the past few missions, and food was growing scarce. To take the edge off their appetites, they resorted to gathering cocoa, grinding it, mixing the bitter powder with river water, and drinking it.

Hungry, wet, exhausted, tormented by insects, the men were

clearly approaching their physical and mental limits during these long days on the Casiquiare. Since childhood, Humboldt had longed to make a journey such as this, and when opportunity had finally shone, he'd left every known thing to pursue that dream. Yet now, faced with the particular miseries of the Casiquiare, he may well have begun to regret that ambition. For it was on the Casiquiare, especially, that youthful ideal met unfeeling reality, and there that the expedition began to resemble an endless slog through hostile territory, recalling Orellana's nightmarish journey down the Amazon two and a half centuries before. On the Casiquiare, even Humboldt had trouble summoning his usual rapture at the wonders of nature. One night, when the party camped in a grove of magnificent palm trees to escape a violent rain, Humboldt found that, however glorious the tropical foliage and however thrilling the storm, "to have enjoyed it fully we should have breathed an air clear of insects. . . . I advise those who are not very desirous of seeing the great bifurcation of the Orinoco," he appended with typical understatement, "to take the way of the Atabapo in preference to that of the Casiquiare." Meanwhile, the precious botanical specimens continued to molder in their cases.

On the night of May 20, a meteor shower was visible through the high, thin clouds (the Indians called meteors "the urine of the stars," and referred to the dew as "the spittle of the stars"). But the overcast was too thick to permit astronomical measurements. During the night, the cries of jaguars were heard very near camp, and the mastiff began to bark. When the big cats grew too close, the dog scampered under the hammocks and started to howl. "How great was our grief," Humboldt wrote, "when in the morning, at the moment of reembarking, the Indians informed us that the dog had disappeared!" The men waited in camp, hoping the mastiff had only strayed, but when he failed to return, they were forced to conclude that he had been carried off in the night by the great cats. "The dog," Humboldt grieved, "who had followed us from Caracas, and had so often in swimming escaped the pursuit of the crocodiles, had been devoured in the forest." The loss of their mascot only deepened the travelers' sense of malaise.

The very next day, the explorers were rewarded for their perseverance. They again entered the Orinoco—proving once and for all the existence of the Casiquiare Canal. Humboldt is character-istically terse in his description of the event, writing only, "On the 21st May, we again entered the bed of the Orinoco three leagues below the mission of Esmeralda." But it must have been a moment of great triumph for him, the achievement of his single most important goal in the rain forest and compensation for weeks of misery on the river. Persevering over hunger, mosquitoes, and doubt, he had vindicated La Condamine and added an important new waterway to the map of South America.

In fact, Humboldt saw the confirmation of the canal's reality as far more than the resolution of an esoteric academic contro-versy. He strongly believed that the discovery would affect the future of the continent and bring real benefits to the people liv-ing there, and he foresaw a day when "the Casiquiare, as broad as the Rhine, and the course of which is one hundred and eighty miles in length, will no longer form uselessly a navigable canal between two basins of rivers which have a surface of one hundred and ninety thousand square leagues. The grain of New Grenada will be carried to the banks of the Río Negro, boats will descend from the sources of the Napo and the Ucuyabe, from the Andes of Quito and of Upper Peru, to the mouths of the Orinoco, a dis-tance which equals that from Timbuktu to Marseilles. A country nine or ten times larger than Spain, and enriched with the most varied productions, is navigable in every direction by the me-dium of the natural canal of the Casiquiare." Alas, it wasn't to be. Despite a brief role in the rubber trade in the early nineteen hundreds and again during World War II, the Casiquiare never became the bustling waterway that Humboldt prophesied; today, as in ages past, the principal inhabitants are still the *moscas* and the *zancudos*.

Though Humboldt had solved the geographic puzzle of the Casiquiare, it was a bit early for celebration. It had been a month since the explorers left the Orinoco near the mouth of the Gua-viare. Seven hundred fifty miles remained to their destination of Angostura.

TEN MILES UPSTREAM on the Orinoco was the mosquito-plagued village of Esmeralda, so named because European prospectors, in a fit of wishful thinking, had mistaken the worth-less rock crystals in the nearby mountains for diamonds and emeralds. The town was situated on an open plain graced with picturesque streams and hillocks. But the place was so notorious for its isolation and misery that the very name had become a threat, with the clerical authorities terrorizing obstreperous monks with the promise of banishment there. Though the village had only eighty inhabitants, three different Indian peoples were represented, plus a number of mixed-bloods (including many malefactors who had been sent there as a form of internal exile).

The missionary from Santa Barbara, some 150 miles away, made only five or six visits to Esmeralda every year. The rest of the time, the denizens' temporal and spiritual welfare was left to an aged churchwarden, who taught the children the rosary, rang the mission bell when the spirit moved, and occasionally whacked his charges with a chorister's wand to keep them in line. Mistaking his visitors for Spanish merchants, the warden smiled condescendingly over their packages of paper used for drying plant specimens. "You come to a country where this type of merchandise has no sale," he explained; "we write little here; and the dried leaves of maize, the *plátano* [banana], and the *vijaho* [heliconia], serve us, like paper in Europe, to wrap up needles, fish hooks, and other little articles of which we must be careful."

Esmeralda's principal export was a particularly fine form of curare, which commanded a high price. On Humboldt's arrival, most of the Indians were just returning from an expedition to collect plants used in producing the poison. Their homecoming was marked by a great festival among the men, with two days of feasting on roasted monkey and dancing to the music of crude reed pipes. While his neighbors drank themselves into a stupor, the old Indian charged with producing the poison went about his deadly business, allowing Humboldt to bring back to Europe the first detailed recipe for the drug.

First, the poison master took the bark of the vines, which had already been stripped and pounded into fibers. To this he added water, which slowly filtered through the bark in a cone fashioned from plantain and palm leaves. The resulting yellow liquid was then boiled in large, shallow pots, with the poison master occasionally tasting the liquid, which became progressively more bitter as it boiled down. (Humboldt also took a taste of the poison, which was nontoxic as long as it didn't come into direct contact with the blood; in fact, it was drunk as a stomach palliative, which was perfectly safe—as long as one had no open cuts or sores in the mouth.) When the liquid reached the desired concentration, the poison master strained it through rolled-up plantain leaves to remove the fibrous matter. Even in this concentrated form, the poison was still too thin to adhere to an arrow tip, so it was next mixed with the glutinous juice of another plant to give it body; this also imparted curare's characteristic tarry color. The finished preparation was then poured into small calabashes, in which it was sold.

As he worked, the poison master lectured his visitors. "I know," he said, "that the whites have the secret of making soap," whose mysteries he seemed to find second only to those of curare, "and manufacturing that black powder which has the defect of making a noise when used in killing animals. The curare, which we prepare from father to son, is superior to anything you can make down yonder. It is the juice of an herb which kills silently, without any one knowing whence the stroke comes." Applied to the tip of an arrow and delivered through a long blowpipe, the curare maximized hunters' yields, since several monkeys could drop noiselessly to earth before the rest of the troop became suspicious; with a gun, at most one animal could be taken at a time, because the others would scatter at the first shot. The curare would kill a bird in two to three minutes and a pig in ten to twelve. According to the missionaries, meat killed any other way simply didn't taste as good.

Producing its characteristic symptoms of dizziness, nausea, extreme thirst, and spreading numbness, curare was also quite capable of killing human beings, as the conquistadors had discovered. Humboldt himself soon received a lesson on the care with which the poison must be handled. On leaving Esmeralda, he packed a calabash full of curare beside his clothes, and in the warm, humid air the poison liquefied and leaked onto a stocking. As he was about to slip the stocking on, he happened to feel the gelatinous liquid in time: Since his feet were covered with bleeding insect bites, the curare would surely have entered his bloodstream, with fatal effect.

At Esmeralda Humboldt toyed with the idea of traveling eastward, or upriver, in an attempt to discover the source of the Orinoco. However, not long before, the Indians of the region had driven off a force of Spanish soldiers. The resulting carnage had left the native people more hostile than ever toward whites, and there was no hope that Humboldt's small party could accomplish what the Spanish army had not. But even under the best of circumstances, it's unlikely that such a mission could have succeeded, since the headwaters of the Orinoco are now known to be unnavigable. In fact, it wasn't until 1956 that the river's source was finally located—by aerial survey.

By this point in the expedition, the oppressive heat and humidity, the constant misery from biting insects, the irregular food, the cramped conditions in the canoe, had already taken their toll on the travelers. In Europe, once past his sickly youth, Humboldt had enjoyed tremendous physical stamina and near miraculous health, undimmed by long days working in German mines or climbing Alpine peaks. But now, after weeks of slogging through tropical rain forests, his ferocious constitution seemed to have met its match. Both Humboldt and Bonpland had begun to feel a "languor and weakness" that seemed to be growing worse.

On May 23, at three o'clock in the afternoon, Humboldt ordered the pilot to steer the canoe westward on the Orinoco, retracing their route toward the coast. Four hours later the explorers reached the sandy bank where the ship's dog had vanished a few days before. The men made a thorough search of the area but found no sign of the mastiff, whom Humboldt still considered "the most affectionate and faithful companion of our wanderings." All through the night, they heard jaguars prowling very near camp, just as they had on their initial stay at the beach. Leaving before sunrise, the party made good time through a landscape devoid of any human presence. There were no rapids in this stretch of the Orinoco, and some nights the travelers didn't even bother to disembark. To make better time, the pilot would steer the boat downstream as the other men slept, with the river propelling it effortlessly on its course.

On May 25, the canoe arrived at the mission of Santa Barbara, on the banks of the Río Ventuari, one of the Orinoco's principal tributaries. Two days later, the party regained San Fernando de Atabapo. The Orinoco was running fast and deep here, and the canoe was making good progress through familiar territory. The worst appeared to be behind them. "We seemed to be traveling as through a country which we had long inhabited,"

Humboldt wrote. "We were reduced to the same abstinence; we were stung by the same mosquitoes; but the certainty of reaching in a few weeks the term of our physical sufferings kept up our spirits." Even so, Father Zea was taken sick with one of his periodic fevers, and Humboldt and Bonpland were still experiencing their ominous symptoms of ill health.

The party rested for two days at the mission of Maipures while the canoe was portaged back over the great rapids. Maipures was the home of Zerepe, the young Indian who had been whipped into joining the expedition at Pararuma and who had since become Humboldt's interpreter. Entering the village, the young man eagerly sought out his betrothed, whom he'd been forced to leave behind, only to discover that she was no longer there. It seemed the fiancée hadn't taken to life in the mission, and having been told that the white men would continue into Brazil instead of honoring their promise to return with Zerepe, she had lost heart and, with another girl her age, had stolen a canoe and fled into the forest. When Humboldt's party arrived, the village was still buzzing over the girls' adventure. Disconsolate at first, Zerepe soon recovered. "Born among the Christians," Humboldt explained, not without condescension, "having traveled as far as the foot of the Río Negro, understanding Spanish and the language of the Macos [his own people], he thought himself superior to the people of his tribe, and he no doubt soon forgot his forest love." The young man remained at the mission, while the four Europeans continued on with the rest of their Indian crew.

In the late afternoon of May 31, the party landed at el Puerto de la Expedición, to inspect the cavern of Ataruipe, an Indian burial ground. "The grand and melancholy character of the scenery around fits it for the burying-place of a deceased nation," Humboldt found. To reach the cave, it was necessary to climb a steep granite face affording only a few precarious hand- and footholds. But the view from the top was breathtaking, if somber. "On reaching the summit, the traveler beholds a wide, diversified, and striking prospect," he discovered. "From the foaming river-bed arise wood-crowned hills, while beyond the western shore of the Orinoco the eye rests on the boundless grassy plain of the Meta, uninterrupted save where at one part of the horizon the mountain of Uniama rises like a threatening cloud. . . . All is motionless, save where the vulture and the hoarse goat-sucker [a nocturnal bird also known as the nightjar] hover solitary in mid-air, or, as they wing their flight through the deep-sunk ravine, their silent shadows are seen gliding along the face of the bare, rocky precipice until they vanish from the eye."

A narrow ridge led to a neighboring summit, on which rested huge natural granite spheres fifty feet in diameter, seemingly poised to roll into the abyss at the slightest nudge. In a shady, solitary spot on the slope of an adjoining peak lay the cavern of Ataruipe. The entrance to the cave was festooned with Bignonia and aromatic vanilla plants, and the summit above was crowned with murmuring palms. Inside reposed nearly six hundred carefully preserved skeletons, the oldest perhaps a century old, which had been doubled over and placed in square baskets woven from palm fronds. Ranging from tiny infants to mature adults, the skeletons were all that remained of an extinct people, the Atures, for whom the great rapids had been named.

To prepare a skeleton, the Indians would bury the corpse in damp ground to facilitate decomposition. Then, several months later, the body would be disinterred and the remaining flesh scraped from the bones, which would then be bleached or painted red or covered with an aromatic resin. (Stripping the bones in this way was still common at the time of Humboldt's journey; in fact, some Indians wrapped their dead in nets and lowered them into piranha-infested rivers to let the voracious fish strip the bones clean. In areas where there were no caverns to serve as natural mausoleums, the Indians would generally wrap the deceased in his or her hammock and bury the remains in the floor of the house.) The bones were then arranged in their basket and deposited in this cavern. Also in the cave were handsome earthenware urns, some as large as five feet high and three feet long. Oval in shape and greenish in color, the urns had handles shaped like serpents or crocodiles, and edges painted with geometric patterns. Intermingled inside were the bones of an entire family.

Though there was nothing of obvious monetary value there, a persistent rumor claimed that on their expulsion from the Spanish territories three decades before, the Jesuits had concealed a fortune in the cavern. In fact, Father Zea had been summoned to the Audencia, or High Court, in Caracas—some 450 miles away—to defend himself on charges of appropriating the fictitious treasure. Unpersuaded by his testimony, the Audencia had named a commission to inspect the cavern, but Father Zea wasn't expecting a visit in the near future. "We shall wait long for these commissioners," he boasted. "When they have gone up the Orinoco as far as San Borja, the fear of the mosquitoes will prevent them from going farther. The cloud of flies which envelops us in the *raudales* [rapids] is a good defense."

To the consternation of the guides, Humboldt insisted on opening some of the baskets and urns to examine the shape of the skulls. Though most were Indian, some were unmistakably Caucasian, belonging to Portuguese traders, he supposed, who had intermarried with the Atures. Worse, over the guides' objections, Humboldt removed several skulls from the cave, along with the complete skeletons of a small child and two full-grown men. Horrified by this desecration, the Indians failed to mention two similar caves located nearby, which Humboldt heard about only much later. Knowing what he considered the "superstitions" of the Indians concerning the dead, he concealed the bones in mats of woven palm fronds, but thereafter whenever the travelers stopped and villagers gathered round their luggage to admire their monkeys, the local Indians would invariably detect the odor of death. Though the explorers would try to convince them that the mats held only skeletons of crocodiles and manatees, the villagers would insist that the men were carrying the bones of their ancestors.

In some ways, this visit to the caverns of Ataruipe constitutes an uncomfortable, contradictory episode in Humboldt's journey. A man so surprisingly modern in outlook, so sympathetic to Native American culture, was able to rob Indian graves with no more apparent unease than if they were middens of animal bones (which, indeed, he tried to pass them off as when questioned). Was it the graves' antiquity that allowed him to rationalize such behavior? Their anonymity? Their racial provenance? One wonders if Humboldt would have dug up unmarked, century-old European graves with so little obvious concern. Perhaps it was his unbridled curiosity, along with his lack of religious feeling, that caused him to view human remains as just so many natural-history specimens. Or perhaps it was unfettered scientific acquisitiveness—the same drive that would fill dozens of cases with botanical, geological, and zoological samples—that rendered everything else secondary, including reverence for the dead.

Whatever Humboldt's self-justification, the visit to the cata-
comb put him in a meditative state of mind. "We turned our steps
in a thoughtful and melancholy mood from this burying-place of
a race deceased," he wrote. "It was one of those clear and cool
nights so frequent in the tropics. The moon, encircled with colored rings, stood high in the zenith, illuminating the margin of
the mist, which lay with well-defined, cloud-like outlines on the
surface of the foamy river. . . . Thus perish the generations of
men!" he mused. "Thus do the name and the traces of nations
fade and disappear! Yet when each blossom of man's intellect
withers—when in the storms of time the memorials of his art
moulder and decay—an ever new life springs forth from the bosom of the earth; maternal Nature unfolds unceasingly her germs,
her flowers, and her fruits; regardless though man with his pas-
sions and his crimes treads under foot her ripening harvest."

THE BOTTOM OF THE CANOE had worn so thin that the explorers had to take great care with it during the portage at Ature. At
the mission, Humboldt, Bonpland, and Soto also bid farewell to
Father Zea, who, despite his unreconstructed attitude toward the
native peoples, had been of enormous service to the expedition
and who, even with his recurring fevers, had gamely shared their
privations over the past two months. During the party's second
stay at the village, the majority of the inhabitants were stricken
by a fever so severe that they were confined to their hammocks,
unable to even prepare their own food. Though Humboldt didn't
know the cause of the ailment, he worried that it was contagious,
and the party left promptly.

Below Ature, the travelers encountered an unnavigable stretch
of river. They unloaded all the equipment and specimens onto an
island, where Humboldt, Bonpland, and Soto ventured into an
underwater cave to collect algae specimens. While the naturalists were exploring, the crew was to haul the lancha along the island, then stop to retrieve the passengers and gear. "This spot
displayed one of the most extraordinary scenes of nature, that
we had contemplated on the banks of the Orinoco," Humboldt
wrote. "The river rolled its waters turbulently over our heads. It
seemed like the sea dashing against reefs of rocks; but at the entrance of the cavern we could remain dry beneath a large sheet
of water that precipitated itself in an arch from above the bar-

rier. In other cavities, deeper, but less spacious, the rock was pierced by the effect of successive filtrations. We saw columns of water, eight or nine inches broad, descending from the top of the vault, and finding an issue by clefts, that seemed to communicate at great distances with each other."

The Europeans had more time to examine the cave than they had counted on. It should have been a simple procedure to maneuver the lancha upstream, but after an hour and a half the boat had not appeared. Then, as the brief tropical dusk was falling, a tremendous storm broke, with the travelers trapped in the middle of the river. Had the crew been set upon by crocodiles or piranha? Had the frail canoe been wrecked, and the Indians returned to the mission? Drenched, the three men had no interest in spending the night on the unprotected island. Bonpland proposed swimming through the rapids to shore and enlisting the help of Father Zea at the mission. Even in daylight this would have been a hazardous proposition, but in the dusk, it could have been disastrous; at length, Humboldt and Soto managed to dissuade him. Cold and wet in their cages, the monkeys began to cry plaintively, and while the men were debating what to do, two huge crocodiles, apparently attracted by the whimpering, crawled ashore and began to eye the animals. Just before dark, the Indians finally arrived with the canoe. It seemed the river had proved impassible along the island, and the crew had had to pick their way through the maze of rocky shoals. But the boat had sustained no damage in the process, and instruments, provisions, animals, and passengers were soon reembarked— though the experience had done nothing to alleviate Humboldt's and Bonpland's vague feelings of unwellness.

Stopping at San Borja, where they had stayed on their up-river journey, the explorers were surprised to discover that their earlier visit had been the inadvertent cause of the settlement's dissolution. After the Europeans' departure, some dissident Guahíbos, who wanted to resume their traditional nomadic way of life, had spread the rumor that the white men intended to return to the village to enslave the Indians and carry them off to Angostura. Accordingly, when the villagers received word that Humboldt and his party were approaching again, they had fled en masse to the savannahs to the west. Thus Humboldt saw firsthand how easy it was to disrupt the fragile missions. Perhaps because of his skepticism surrounding the missionary enterprise, he betrays no pangs of conscience over the incident, but instead ascribes the desertion to the inhabitants' inherent wanderlust. After all, he writes, "No tribe is more difficult to fix to the soil than the Guahíbos. They would rather feed on stale fish, scolopendras, and worms, than cultivate a little spot of ground."

The travelers stopped next at the mission of Carichana. By this time, Humboldt's and Bonpland's lethargy had deepened. The men needed time to recuperate from their exertions on the river. But the area around the mission was rich with plant life, and Bonpland couldn't resist making extensive forays into the countryside in search of unusual specimens. As a result, he got very little rest, and with the frequent showers, his clothes became drenched several times a day.

The party pressed on. More than two miles wide here, the river was flowing at about a mile and a half per hour. Humboldt took astronomical measurements wherever he could, but the steady overcast prevented him from fixing many locations. In another two days the expedition arrived at the mission of Uruana, set at the foot of a high granite mountain and graced with a spectacular view of the river, which at that point ran wide and straight, divided into channels by two narrow islands. The missionary, Ramón Bueno, seemed genuinely concerned with the welfare of his charges, the Ottomacs, who made a marginal living from hunting and fishing. Savage, vindictive, ill kempt, overly fond of homemade liquor and intoxicating snuff, the Ottomacs were despised by the other Indians, who had a saying, "Nothing is so loathsome but that an Ottomac will not eat it."

Indeed, for two or three months of the year, the Ottomacs reportedly subsisted on nothing but clay, which they called *poya*, and apparently lost no weight and suffered no ill effects from this strange diet. In the Indians' huts, Humboldt found balls of fine yellowish-gray earth, five or six inches in diameter, piled into pyramids three or four feet high. To prepare the clay for eating, the Ottomacs would bake the balls in a fire until they formed a hard, reddish crust (owing to the iron oxide in it), then would moisten them with water. Having acquired a taste for the *poya*, in fact, the Indians would even mix it with other foods during times of plenty. Though earth eating had been documented on every continent (including Europe, where German miners would spread on their bread a fine clay called *Steinbutter*, or "stone butter"), Humboldt was at a loss to explain the Ottomacs' apparent health and vigor despite "so extraordinary a regimen" that "furnishes nothing, or probably nothing, to the composition of the organs of man." Perhaps, he suggests, the Ottomacs had been able to adjust their digestive organs to the practice over the course of many years.

On June 7, the travelers resumed their downriver journey. The mosquitoes seemed to become less thick with each passing day, and there were other indications that they were reentering more settled territory. The next day, they reached the point where the Orinoco made its great turn from north to east. The forest on the right bank thinned, while vast steppes appeared on the left. The next morning, the lancha encountered several merchant canoes sailing upriver toward the Apure, which was heavily trafficked between Angostura and the port of Varinas. Nicolás Soto took his leave here, boarding one of the merchant boats to return to his family. The population along the river grew more dense as Humboldt and Bonpland, alone again except for their Indian crew, neared Angostura. But the settlements of Alta Gracia, Piedra, Real Corona, and Borbón were miserable-looking places, peopled by whites, blacks, and mestizos, with few Indians to be seen.

On June 13, the lancha pulled into Angostura (namesake of the bitters still used by bartenders). It was the welcome end of a long and exhausting journey. "It would be difficult for me to express the satisfaction we felt on landing . . . ," Humboldt wrote. "The inconveniences endured at sea in small vessels are trivial in comparison with those that are suffered under a burning sky, surrounded by swarms of mosquitoes, and lying stretched in a canoe, without the possibility of taking the least bodily exercise."

In seventy-five grueling days, the explorers had journeyed some fifteen hundred miles on two of the great river systems of South America. They had ascended the great rapids at Ature and Maipures to penetrate a wilderness seen by only a handful of Europeans. Straining the limits of their endurance, they had survived insects, hunger, and treacherous waters in some of the most dangerous, desolate country in the world. Moreover, they arrived in Angostura with an incredible bounty of scientific data and specimens, which would add hugely to Europe's trove of knowledge about South America and its rain forests. Humboldt had proved the existence of the Casiquiare Canal, connecting the vast river basins of the Orinoco and the Amazon. Despite the overcast that dogged them, he had fixed the latitude and longitude of more than fifty places—missions, mountains, rivers, streams— on the world's maps. None of his predecessors had examined the rain forest with such an eclectic, penetrating eye.

These accomplishments had not been without cost, however. For both Humboldt and Bonpland still suffered from the lethargy that they'd been experiencing for the past three weeks. In fact, with their arrival at Angostura, their symptoms seemed to be growing only worse.

EIGHT

Cuba

WHEN IT WAS FOUNDED IN 1764, THE PRINCIPAL CITY OF the Orinoco Delta was christened Santo Tomé de la Nueva Guayana. But the official designation proved cumbersome, and the town soon became known simply as Angostura, the Spanish word for "strait." Strategically situated between the rain forest and the coast, the port developed an active trade with Spain and the West Indies, and by the time of Humboldt's arrival, thirty-seven years after its founding, the city had grown to nearly seven thousand souls. Angostura's streets were laid out parallel to the river and lined with high, pleasant houses built of stone. Above the town rose a barren slate outcropping, and behind, to the southeast, lay fetid marshes. Though Humboldt found little in the monotonous landscape to recommend it, Angostura did afford an undeniably majestic view of the Orinoco, which here formed a vast canal running southwest to northeast. During the rainy season, when the quays flooded, crocodiles were a hazard even in the center of town. While Humboldt was there, an Indian man was killed by one of the creatures while mooring his canoe in less than three feet of water.

Though their clothes were in tatters from the long river journey, Humboldt and Bonpland presented themselves to Governor Felipe de Ynciarte, who received them cordially and offered them accommodations. After the wild, desolate country they'd grown accustomed to, Angostura seemed a metropolis and the townfolk urbane citizens of the world. "We admired the conveniences which industry and commerce furnish to civilized man," Humboldt wrote. "Humble dwellings appeared to us magnificent; and every person with whom we conversed, seemed to be endowed with superior intelligence. Long privations give a value to the smallest enjoyments; and I cannot express the pleasure we felt, when we saw for the first time wheaten bread on the governor's table."

In the days after their arrival, Bonpland went to work organizing the surviving two thirds of their herbarium, while Humboldt took measurements to fix Angostura's latitude and longitude. But the travelers continued to experience the fatigue and weakness they'd felt on the river, and it wasn't long before both men, plus a servant who'd been with them since Cumaná, contracted a severe fever. Humboldt wasn't surprised by the delayed onset. "It is common enough for travelers to feel no effects from miasma till, on arriving in a purer atmosphere, they begin to enjoy repose," he explained. "A certain excitement of the mental powers may suspend for some time the action of pathogenic causes." He suspected typhoid, which was long confused with typhus (the disease that had swept through the *Pizarro*) but is now known to be a distinct illness.

It would be another eighty years before the cause of typhoid was discovered, a bacterium of the *Salmonella* genus that enters the body through contaminated food or water. After an incubation period ranging from ten days to four weeks, the victim feels the early symptoms of lethargy, headache, body aches, fever, and restlessness, which may be followed by a cough, nosebleeds, red spots on the trunk, and digestive symptoms such as loss of appetite, constipation, or diarrhea. The fever gradually rises over the course of about a week, till it plateaus around 104 degrees, where it holds steady for another week or more. In the next week, the fever gradually declines and the long recuperation begins—unless the sufferer has already succumbed to complications, which could include heart failure, a perforated intestine, pneumonia, encephalitis, or meningitis.

The servant's symptoms advanced alarmingly, and the man was soon prostrate. Humboldt's fever also rose precipitously, and he was advised to take a concoction of honey and angostura bark, a bitter, aromatic substance similar to cinchona, the source of quinine. That night his fever worsened, but it broke the following day. Meanwhile, Bonpland, fearing he wouldn't be able to tolerate the potent angostura bark, prescribed for himself other herbal remedies that he considered better suited to his constitution. But his fever continued unchecked, now accompanied by dysentery.

Through the course of his illness, Bonpland never lost "that courage and mildness of character, which never forsook him in the most trying circumstances," Humboldt explains, while Humboldt himself brooded obsessively over his companion's condition. "I can barely describe to you the worry I suffered during his illness," he wrote his brother, Wilhelm. "Never could I have hoped to meet again with a friend as loyal and devoted as he. I shall never forget how he saved my life in a storm that overtook us on the Orinoco . . . when he offered to swim ashore from the boat with me on his back." We can forgive Humboldt's exaggeration under the circumstances, when he was tortured by self-recrimination. "It was I who had chosen the path of the rivers," he confessed in the *Personal Narrative*. "Instead of going up the Orinoco, we might have sojourned some months in the temperate and salubrious climate of the Sierra Nevada de Mérida [in western Venezuela]." The missionaries had warned of the perils waiting beyond the rapids of Ature and Maipures, but he had obstinately pressed ahead, and now "the danger of my fellow-traveler presented itself to my mind as the fatal consequence of this imprudent choice."

On the ninth day, the servant's death was announced. Rushing to his bedside, Humboldt discovered that the man wasn't dead, however, only comatose. After several hours, the servant regained his senses, and soon afterward his temperature peaked, then started to subside. Bonpland never experienced a similar crisis, and as the days wore on, his fever and dysentery gradually abated. To Humboldt's immeasurable relief, his companion's condition began to improve, and before long he was asking from his sickbed to see the branches and flowers of the tree that had cured Humboldt, which the naturalists realized constituted a previously unknown genus. (When Humboldt later shipped the specimen to the famed German botanist Karl Ludwig Willdenow for classification, he asked that the plant be named in his friend's honor. Willdenow complied, and the new genus was called *Bonplandia*.)

Though Bonpland continued to recover, it would be nearly a month before he'd be well enough to travel. It was not until July 10, with their mule train packed with instruments, botanical and geological specimens, Indian skeletons, and caged monkeys and birds, that the explorers would set out once more for the Venezuelan coast.

After the misery and illness of the past twelve months, it wouldn't have been surprising if Humboldt had decided to return to Europe at this point. And indeed, with their magnificent herbarium and the map of the Casiquiare, the expedition would have been acclaimed a resounding success even if it had concluded here. But having finally realized his lifelong dream of scientific exploration, Humboldt had no intention of terminating it so soon. As he wrote his brother, "I could not possibly have been placed in circumstances more highly favorable for study and exploration than those which I now enjoy. I am free from the distractions constantly arising in civilized life from social claims. Nature offers unceasingly the most novel and fascinating objects for learning. The only drawbacks to this solitude are the want of information on the progress of scientific discovery in Europe and the lack of all the advantages arising from an interchange of ideas." Humboldt had lost himself in the jungles and mountains of the New World, just as in childhood he had immersed himself in the fields and forests of the family estate at Tegel.

From Venezuela Humboldt elected to visit Havana. The city was Spain's principal port in the Americas, and having missed it on the outbound journey due to the typhus outbreak aboard the *Pizarro*, he was eager to see it now. Then from Cuba he planned to resume their original itinerary, through North America and the Philippines, before sailing across the Pacific and back to Europe. But first they had to recross the Llanos.

STIFLING EVEN DURING THE RAINY SEASON, the eastern Llanos were as wild and forbidding as those farther west, which the travelers had traversed en route to the Orinoco. But with the coming of the rains, the vegetation had emerged from its annual dormancy. The grass now formed a thick turf, and the stubby Mauritia palms bore enormous clusters of red fruits, of which Humboldt's monkeys were exceedingly fond. Always seeking to tease out the tangled skein of cause and effect, Humboldt was struck by the enormous influence a lone palm tree could exert on the landscape: "We observed with astonishment how many things are connected with the existence of a single plant," he wrote. "The winds, losing their velocity when in contact with the foliage and branches, accumulate sand around the trunk. The smell of the fruit, and the uprightness of the verdure, attract from afar the birds of passage, which love to perch on the slender, arrow-like branches of the palm-tree. . . . If we examine the soil on the side opposite to the wind, we find it remains humid long after the rainy season. Insects and worms, everywhere else so rare in the Llanos, here assemble and multiply." Indeed, "this one solitary and often stunted tree, which would not claim the notice of the traveler amid the forests of the Orinoco, spreads life around it in the desert."

On their third day out of Angostura, the travelers rested at the mission of Cari. As they prepared to depart again, there was some trouble with the Indian muleteers. Reloading the animals, the teamsters discovered the skeletons that Humboldt had taken from the cavern of Ataruipe. The men refused to load "the bodies of their old relations," for fear the mules would be struck dead on the road, and nothing Humboldt said could persuade them otherwise. Finally, the missionary came to the rescue (as had so many other missionaries throughout their journey), mustering his ecclesiastical authority and ordering the Indians to pack the skeletons and get under way.

After a dangerous ford through the high waters and quicksand of the apparently misnamed Río de Agua Clara, the party reached the little town of Pao, where the small reed huts were roofed with leather and the *peones llaneros* guarded herds of cattle, horses, and mules. Even at this time of year, the heat was unbearable. The travelers would have preferred to ride at night, as they had done on the southbound crossing, but bandits had been reported in the area, robbing and killing anyone who fell into their hands. So the group pressed on during the day, their skin burned by the blazing sun and chafed by the windblown sand. As they went, Humboldt kept a nervous eye on Bonpland, to see how his friend was bearing the rigors of the journey.

Despite the heat, Humboldt greeted the open spaces of the Llanos with a sense of euphoria. "After having passed several months in the thick forests of the Orinoco," he wrote, "in places where one is accustomed, when at any distance from the river, to see the stars only in the zenith, as through the mouth of a well, a journey in the Llanos is peculiarly agreeable and attractive. The traveler experiences new sensations; and, like the Llanero, he enjoys the happiness 'of seeing well around him.' " But this enjoyment was short lived. "There is doubtless something solemn and imposing in the aspect of a boundless horizon, whether viewed from the summits of the Andes or the highest Alps, amid the expanse of the ocean, or in the vast plains of Venezuela. . . . Infinity of space, as poets in every language say, is reflected within ourselves; it is associated with ideas of a superior order; it elevates the mind, which delights in the claim of solitary meditation. . . ." Nevertheless, "the dusty and creviced Llano, throughout a great part of the year, has a depressing influence on the mind, by its unchanging monotony. . . . After eight or ten days' journey, the traveler . . . loves again to behold the great tropical trees, the wild rush of torrents, or hills and valleys cultivated by the hand of the laborer."

The travelers trudged on toward the coast. On the night of July 16—the first anniversary of their arrival in the New World—they slept at the mission of Santa Cruz de Cachipo. Soon after, they were delighted to see the mountains of Cumaná rising in the distance, at first hazy and broken like a bank of fog, then gradually growing more solid and real. Finally, on the twenty-third, the party arrived at the coastal city of Nueva Barcelona, where they settled into the home of a wealthy merchant of French extraction named Pedro Lavié.

Founded in 1637 by Spanish conquistador Juan Urpín, Nueva Barcelona had grown over the past decade from ten thousand inhabitants to more than sixteen thousand. The town was celebrated for a miraculous image of the Virgin Mary, which had been discovered inside the trunk of an old *tutumo* (calabash tree) at the nearby Indian village of Cumanagoto. The Virgen del Tutumo had been carried in a solemn procession to Nueva Barcelona, but whenever the residents there had incurred the displeasure of the padres, the image would miraculously return by night to the tree where it had been found. This phenomenon continued for years, until a fine monastery was built for the Franciscans and the nocturnal peregrinations abruptly ceased.

Despite Humboldt's worries, Bonpland had borne the journey across the Llanos remarkably well. Unfortunately, the same couldn't be said of Humboldt himself. Nueva Barcelona was extremely damp, and after getting caught in a tropical rainstorm, Humboldt came down with a fever even more severe than the one he'd recently suffered. Typically, the *Personal Narrative* doesn't linger over the severity of his symptoms or the depth of his concern over his health. But from the fact that he and Bonpland rested in Nueva Barcelona for nearly a month, we can infer that Humboldt was seriously ill, his constitution perhaps still weakened by the weeks of privation in the rain forest and by his earlier siege in Angostura.

During their forced stay, the Europeans became friendly with a young Franciscan monk named Juan Gonzáles, whom Humboldt found "cheerful, intelligent, and obliging." Interested in natural history, Gonzáles had earlier made his own journey into the Upper Orinoco, and he now examined the explorers' animal and plant specimens with a relish tinged with nostalgia. Since Fray Juan was preparing to return to Europe, he arranged to accompany his new friends as far as Cuba, where he could find passage across the Atlantic. Altogether, the three men would spend the next seven months in each other's company.

When Humboldt was feeling well enough, he and Bonpland borrowed some of the finest saddle horses belonging to their host, Don Pedro, and, with local guides, made an excursion to the hot springs about five miles southeast of town, where they examined the rock formations and investigated the mineral content of the water. They had been warned not to ford the Río Narigual, but on the return trip, finding the crude bridge impassible, they rode their horses into the river. As the animals swam across, Humboldt's mount suddenly sank from under him, struggled underwater for some time, then vanished. Stunned, the men searched for the horse but could find neither the animal nor an explanation of its disappearance. Crocodiles were numerous in the area, and the guides suggested that the horse had been seized by one of the reptiles and dragged down to its death. Whatever the cause, the accident left Humboldt less concerned about his own near escape than about his awkward position vis-à-vis his host: Having done exactly what he'd been warned not to do, he had lost one of Don Pedro's prize horses. Worse, decorum would not permit him to even offer reimbursement. The perfect gentleman, Don Pedro tried to allay his guest's distress by exaggerating the ease with which fine horses were captured on the Llanos and broken to the saddle, but Humboldt was mortified by the whole affair.

The packet boat from La Coruña was three months overdue in Nueva Barcelona, leading to speculation that it had been caught in the English blockade. With no certainty that the ship would ever arrive, Humboldt searched for a vessel that he could charter as far as the larger port of Cumaná, where he hoped to locate a ship bound for Cuba. What he managed to find was a smuggler, a lancha carrying on a contraband trade with the British island of Trinidad, just to the north. Because the English were actively encouraging such activity to undermine the Spanish mercantile monopoly, and since he held a passport from the governor of Trinidad, the lancha's captain confidently explained that he had no reason to fear the blockade. So Humboldt, Bonpland, and Fray Juan stacked the animal cages, instruments, specimen boxes, and other gear beside the smuggler's cargo of cacao, and under a fine sky, the lancha sailed out of the harbor—and straight into hostile gunfire.

Heaving to, the lancha was boarded by privateers from Halifax, Nova Scotia. In common use at the time, privateers were freelancers (or "pirates," depending on one's point of view) authorized by a government to prey on enemy shipping in exchange for a share of the spoils. Having captured a bona fide Spanish vessel, even a modest one like the lancha, the mercenary captain had every intention of towing it to Nova Scotia to claim his reward. One of the boarding privateers happened to be Prussian, and Humboldt interceded in his native language, but to no avail. The passengers were ushered aboard the gunship, still arguing, but the captain was uninterested in political niceties and insisted that the lancha was a legal prize. Humboldt was in the skipper's cabin, demanding that the passengers and their effects be rowed ashore, when a crewman entered and whispered something to the captain, who hurried topside.

As it happened, the British sloop *Hawk* had witnessed the capture and signaled the privateer to heave to. When the order was ignored, the *Hawk* halted the ship with a cannonball across the bow, then sent a midshipman aboard to investigate. Humboldt and his friends were taken aboard the English vessel to be interviewed. Although all three—a Prussian, a Frenchman, and a Spaniard—were citizens of Britain's wartime enemies, they were received kindly by the commander, Captain Garnier, who had read of Humboldt's exploits in the English newspapers. Having served on voyages of exploration himself, including a cruise to the American Northwest with George Vancouver, the captain was eager to hear about the expedition. As a result, the trio spent a festive evening swapping yarns and news with the British officers. In fact, Humboldt "had not, during the space of a year, enjoyed the society of so many well-informed persons. . . . Coming from the forests of the Casiquiare, and having been confined during whole months to the narrow circle of missionary life," he wrote, "we felt a high gratification at meeting for the first time with men who had sailed around the world, and whose ideas were enlarged by so extensive and varied a course." The captain lent Humboldt his own stateroom for the night and even presented him with updated astronomical tables, which would permit more accurate geodetic measurements for the remainder of the journey. The next day, to prevent additional interference, the sloop saw the lancha safely to its destination.

The explorers' arrival at Cumaná seemed like a homecoming. "We gazed with interest at the shore," Humboldt wrote, "where we first gathered plants in America. . . . Every part of the landscape was familiar to us; the forest of cactus, the scattered huts and that enormous ceiba, beneath which we loved to bathe at the approach of night. Our friends at Cumaná came out to meet us: men of all castes, whom our frequent herborizations had brought into contact with us, expressed the greater joy at the sight of us, as a report that we had perished on the banks of the Orinoco had been current for several months," due to stories either of their illness in Angostura or their earlier near capsize.

Their old friend Governor Vicente Emparán found Humboldt, Bonpland, and Gonzáles a spacious house in the center of town, with lovely views of the sea and with terraces ideal for setting up their instruments. Pleasant though the accommodations were, Humboldt was eager to continue on. But with Cumaná tightly blockaded, there was little traffic entering or leaving the harbor. His impatience growing, he considered making a dash for a group of Danish islands nearby, but in the end decided it wasn't worth the risk: Given the uncertain political climate, he couldn't be assured of reentry into Spain's possessions, despite his royal passport, in which case, he would have no alternative but to return to Europe.

During this second stay at Cumaná, Humboldt and Bonpland occupied themselves with more geological investigations on the Araya peninsula, checked and rechecked their measurements of the town's latitude and longitude, and conducted experiments on refraction, evaporation, and atmospheric electricity. They also spent countless hours organizing their huge botanical collection, which, having now grown to some twelve thousand specimens, was by far the largest New World herbarium ever made. Even so, Humboldt was frustrated by its incompleteness. "We were barely able to collect a tenth of the specimens met with," he complained in a letter to the botanist Willdenow. "I am now perfectly convinced of a fact that I would never admit while visiting with botanists in England . . . that we do not know three fifths of all the existing plants on earth!"

As during their first visit, the naturalists were often interrupted in their work by curious townspeople eager to get a glimpse of their menagerie. Especially popular was the humanlike capuchin monkey (so named because its black head, sitting atop a black-and-white body, was reminiscent of the cowl worn by the Capuchin monks), as was the sleeping monkey, a variety never before seen on the coast. Humboldt intended to donate the monkeys and birds to the Jardin des Plantes in Paris, and when a French squadron arrived in Cumaná for refitting after a failed attack on nearby Curaçao, the commander agreed to transport the creatures to the French island of Guadeloupe, whence they could be conveyed to France. Humboldt was later disappointed to learn that the animals, which the travelers had nursed through some of the most challenging terrain on earth, died on Guadeloupe, perhaps after failing to receive there the cosseting to which they'd become accustomed. However, some of the skins did eventually make the journey to Paris and were installed in the Jardin de Plantes.

The arrival of their French allies created a sensation at Cumaná, with the populace gawking at the elaborate uniforms and peppering the officers with questions. Alarmed by the outpouring of interest in the foreign visitors—and especially by the pointed inquiries about the extent of self-government permitted on nearby Guadeloupe—the Spanish authorities rushed to reprovision the squadron and see it on its way. "These contrasts," Humboldt wrote, "between the restless desires of the colonists [against Madrid's restrictive rule], and the distrustful apathy of the government, throw some light on the great political events which, after long preparation, have separated Spain from her colonies." There had been sporadic uprisings in the Spanish colonies before Humboldt's arrival, and Venezuela would be engulfed in open rebellion just seven years later.

Losing hope of finding a Spanish ship to break the blockade, Humboldt managed to secure passage on a neutral United States vessel laden with an odiferous cargo of salted meat. On November 16, the travelers finally sailed from Cumaná—not to Havana, but back to Nueva Barcelona, where the trader was being loaded. "The night was cool and delicious," Humboldt reports. "It was not without emotion that we beheld for the last time the disc of the moon illuminating the summit of the cocoa-trees that surround the banks of the Manzanares. The breeze was strong, and in less than six hours we anchored near the Morro of Nueva Barcelona, where the vessel which was to take us to the Havannah was ready to sail." The travelers were back where they had started ten exasperating weeks before.

Eight days later, Humboldt, Bonpland, and Juan Gonzáles departed Venezuela at last. Planning to continue to Mexico and then on to the Philippines, Humboldt believed he was leaving South America for good.

The first five days at sea passed uneventfully, with the naturalists charting their location, observing the passing islands, measuring the temperature of air and water, and recording anything else that seemed noteworthy. But by noon on November 29, all signs indicated a change of weather. The temperature dropped precipitously, bands of black clouds appeared to the north, and though the breeze remained light, the seas grew high. The next day, the wind veered to the northeast, the surge grew heavier, and their small vessel began to roll violently. That night the cook managed to set the deck on fire, and though the flames were soon extinguished, the incident heightened the sense of foreboding on board. But on the morning of December 1, the sea gradually calmed and the wind subsided.

Two days later, the sentries called out a pirate ship on the horizon, and passengers and crew rushed to the rail. Eventually, they were able to read the name on the hull of the approaching vessel: *Balandra del Frayle* (*Sloop of the Monk*). Not a pirate ship at all, the sloop was operated by a Franciscan missionary who had grown rich carrying on an illegal, lucrative trade with the Danish islands nearby. The smuggler monk was harmless, but the false alarm unsettled the captain, who now determined to take the most direct, but also potentially most hazardous, route to Cuba—over the Pedro Shoals, a maze of reefs occupying an area almost as large as Puerto Rico.

On December 4, as the trader picked its way through the shoals, it was buffeted by heavy rain, thunder, and increasingly violent gusts from the north-northwest. Then on the night of the sixth, with the storm still raging, the passengers heard the unmistakable roar of breakers. The gale had driven the ship dangerously out of position, and now it was speeding straight for the reef. Peering through the rain at the looming, phosphorescent surf, Humboldt had a flashback to all the rapids they had negotiated on the Orinoco. Their little lancha had survived those encounters, but the outcome of this mismatch was anything but certain. As the shoals grew closer, the terrified passengers waited for the violent shudder of keel on reef. But the captain managed to bring the helm around in time and guide the ship through an opening in the barrier. A quarter of an hour later, the trader was in open water once more. The next day the storm abated, and as the ship neared Cuba, a "delicious aromatic odor" announced the proximity of the island. Finally, on December 19, after an interminable passage of twenty-five days in bad weather and heavy seas, the trader dropped anchor in Havana harbor.

CUBA WAS BOOMING at the time of Humboldt's arrival, but the economy hadn't always been so vital. The big island, the largest in the Antilles (slightly smaller than England), had been "discovered" by Columbus on October 28, 1492, on his first voyage, after Indians had alerted him to its presence. Sailing along the northeast coast, he first believed that he'd found Japan, then decided the region must be part of China. Columbus averred that he "never saw a lovelier sight: trees everywhere, lining the river, green and beautiful. They are not like our own, and each has its own flowers and fruit. Numerous birds, large and small, singing away sweetly. . . . It is a joy to see all the woods and greenery, and it is difficult to give up watching all the birds and come away. It is the most beautiful island ever seen, full of fine harbors and deep rivers. . . ." Then he added prophetically, "The Indians tell me that there are gold mines and pearls on this island. . . ." He christened the land Juana, after the daughter of his sponsors, Ferdinand and Isabella of Spain. On his second voyage, in 1494, Columbus investigated Cuba's southern shore, concluded that the landmass was too vast to be an island, and, threatening dissenters with a huge fine and the removal of their tongue, compelled all his men to sign an oath to that effect. Then the explorer sailed on to Hispañola, never to return to the island (at least in life; remains that some believe to be Columbus's were later reinterred on Cuba, but the body has never been positively identified).

In 1508, the Spanish explorer Sebastian Ocampo sailed the entire coastline, proving Cuba to be an island. Then, just three years later, responding to rumors of gold there, King Ferdinand directed Diego Columbus, Christopher's son and governor general of Hispañola, to conduct the conquest of Cuba. Diego Columbus dispatched Diego Velásquez with a force of three hundred men (including Hernán Cortés, the future conqueror of Mexico), and despite an inspired resistance under the cacique Hatuey, Velásquez succeeded in subduing the island. As related by the sympathetic Dominican friar Bartolomé de Las Casas, when Hatuey was captured he was offered the choice of converting to Catholicism before he was burned at the stake, so that his soul might be conveyed to heaven.

"And to heaven the Christians also go?" the chief supposedly asked.

"Yes," the priest responded, "they go to heaven if they are good and die in the grace of God."

"If the Christians go to heaven, I do not want to go to heaven," the cacique told his executioners. "I do not wish ever again to meet such cruel and wicked people as Christians who kill and make slaves of the Indians."

Settlement began immediately, and over the next few decades, tens of thousands of Native Americans followed Hatuey into an untimely grave, slaughtered by Spanish soldiers, stricken by European diseases, or worked to death in the gold mines. At the time of Columbus's landing, the Indian population of Cuba was estimated at one hundred thousand; by 1550, less than sixty years later, it had dwindled to just three thousand. The gold soon petered out, and over the next two centuries Cuba languished under a weak economy dominated by small-scale tobacco farming. Then in the early 1700s came the increased importation of African slaves and the establishment of large plantations of coffee and, especially, sugar (still the mainstay of the Cuban economy three hundred years later), and the island finally began to come into its own. The British occupation of 1762–63 has been called a watershed because the English introduced a taste for free trade and a liberal approach to government, both of which would help to enflame the Cuban independence movement. And in 1796, just a few years before Humboldt's arrival, Cuba was able to seize a greater share of the world sugar market after the slave rebellion on Hispañola disrupted production there; the colony also received a surge in population and an infusion of expertise from the thirty thousand white refugees who fled from Hispañola to Cuba at that time.

But Havana had been bustling even during the years when the rest of the island had been struggling. Thanks to its strategic location, the harbor was the principal Spanish naval base in the Western Hemisphere. It was also the busiest commercial port in the Americas, "the Gateway to the New World," which for centuries had been a port of call for ships arriving from Spain (just as Humboldt's ship, the *Pizarro*, was supposed to have landed there before the typhus outbreak). Similarly, it was the customary port of departure for vessels returning to Spain, with ships reprovisioning there and collecting into convoys for the dangerous voyage across an Atlantic awash with British privateers. With a steady influx of transients filling its streets, Havana boasted shipyards, chandlers, slaughterhouses, boardinghouses, bars, brothels, and every other establishment catering to a ship's captain and crew. On Humboldt's arrival, the city was a colorful, expensive, freewheeling port.

In 1800, Havana proper had a population of forty-four thousand, including twenty-six thousand blacks and mixed-bloods, and another forty thousand or so inhabitants in the two outlying *arrabales* (suburbs) of Jesú-Maria and La Salud (Health). With a narrow neck of harbor guarded by two fortresses, El Morro and San Salvador de la Punta, the port of Havana was, Humboldt found, "one of the gayest and most picturesque on the shore of equinoctial America, north of the equator. . . . It boasts not the luxuriant vegetation that adorns the banks of the river Guayaquil, nor the wild majesty of the rocky coast of Rio de Janeiro; but the grace which in those climates embellishes the scenes of cultivated nature, is at the Havana mingled with the majesty of vegetable forms, and the organic vigor that characterizes the torrid zone."

Inside the city things were different. Crammed onto a slim promontory with less than a square mile inside its walls, Havana was cramped, noisy, dirty. The streets were narrow and for the most part unpaved, owing to a lack of stone. Not long before, the city had experimented with mahogany logs for paving, and now great trunks jutted up through the mud, vestiges of the aborted effort. In some places, the mire reached the pedestrians' knees, which—along with the many carts loaded with barrels of sugar, the jostling porters, and the ever-present odor of tasajo (jerked beef)—made walking through the streets a disagreeable and even hazardous affair. "At the time of my sojourn there," Humboldt decided, "few towns of Spanish America presented, owing to the want of a good police, a more unpleasant aspect."

Still, there were the spectacular royal palms towering over the skyline and two pleasant public walks where one could escape the bustle of the metropolis. And some elegant houses, mail-ordered from the United States, had been built along the shore. "In the coolness of the night, when the boats cross the bay, and, owing to the phosphorescence of the water, leave behind them long tracks of light, these romantic scenes afford charming and peaceful retreats for those who wish to withdraw from the tumult of a populous city." But the main public buildings—the cathedral, the government house, the post office, the tobacco factory—were "less remarkable for their beauty than for solidity of structure." There was also, regrettably, a bustling slave market, a "place fitted to excite at once pity and indignation."

Humboldt and Bonpland settled into the house of a Señor Cuesta, wealthy partner in one of the great commercial companies of Spanish America. (The house still stands, now converted into a Humboldt museum.) They stored their instruments and collections at the home of Count O'Reilly, where the terraces were well suited for astronomical observations. For the next three months, Humboldt's restless curiosity would range over Havana and the Güines Valley south of the city.

Despite the port's crucial importance, there was still some disagreement concerning Havana's longitude, with earlier measurements varying by a fifth of a degree (ranging from 5 degrees, 38 minutes, 11 seconds, to 5 degrees, 39 minutes, 1 second), or more than ten miles. Latitude (distance north or south of the equator) could be fixed by measuring the height above the horizon of a pole star or the sun. Determining longitude (east-west position) was more complex, because it must also take account of the earth's rotation. After John Harrison's invention of the chronometer in 1735, the standard procedure for fixing longitude was to mark noon, or the sun's highest point in the sky, at the traveler's location, then to refer to a precision timepiece that had previously been set to local time at another point, such as Greenwich or Paris. By comparing the time discrepancy between the two places, the navigator could fix his distance relative to the other city, as each hour of difference represented 15 degrees of longitude, or approximately 1,035 miles at the equator.

An older method of fixing longitude, originated by Galileo in the early seventeenth century, used the eclipses of the moons of Jupiter as a kind of universal timepiece. These eclipses, caused when the moons pass into Jupiter's shadow, are visible at the same moment from every point on earth and can be predicted with mathematical precision. Therefore, a traveler could note the local time of an eclipse, compare that to published tables giving the time that the eclipse was expected to occur at a given point on earth, then use the time discrepancy to calculate the difference in longitude between the reference point and the traveler's own location, as with a chronometer.

Perhaps eager to try out the new navigation tables given to him by the English captain Garnier, this latter method was the one Humboldt chose on this occasion. Working with Dionisio Galeano of the Spanish navy, he netted a mean result of 5 degrees, 83 minutes, 50 seconds, of longitude—which later experts confirmed to within a single second. Finally, three centuries after its founding, the principal port of the Western Hemisphere could be accurately fixed on the world's maps.

A highly developed colony, Cuba was critically important to Spain owing to its size, its naval base, and its strategic location atop the Caribbean Sea, all of which made it a timely subject and an area of widespread interest throughout Europe. "I wanted to throw light on facts," Humboldt explained, "and give precision to ideas by comparing the island's condition with that of South America." To accomplish this, he produced a Herculean feat of what today we would call physical and human geography. Surveying the island from north to south and synthesizing a wealth of secondary information besides, he made extensive geological, hydrological, botanical, and meteorological studies, including Cuba's precise shape and area, its topography, the composition of the soil, the location of mineral deposits, the height and placement of mountain ranges, the availability of water, the distribution of plants, and the state of agriculture (especially the key exports of sugar, coffee, and tobacco). He also examined the island's population, its commerce, and the state of its finances. Ultimately, these myriad observations—the most extensive, systematic study ever made of the island—were collected into a two-volume compendium, complete with statistical tables and map, entitled the *Political Essay on the Island of Cuba*, which Humboldt would publish in 1828.

The work was important for the way it interrelated, in true Humboldtian fashion, physical characteristics (such as weather and soil) to human issues (such as population and economic development). It is also remembered as Humboldt's most comprehensive condemnation of slavery—excised, to Humboldt's fury, from the first British translation of the work—which he called "the greatest evil that afflicts human nature. . . . It is for the traveler who has been an eyewitness of the suffering and the degradation of human nature," he wrote, "to make the complaints of the unfortunate reach the ear of those by whom they can be relieved." Since blacks comprised more than eighty percent of the population of the West Indies, Humboldt argued, it was foolish to think they could be subjugated forever by the white minority. The Caucasians would either voluntarily grant the slaves their freedom, or the oppressed masses would eventually take it by force: "If the legislation of the West Indies and the state of the men of color do not shortly undergo a salutary change," he warned, "if the legislature continues to employ itself in discussion instead of action, the political preponderance will pass into the hands of those who have strength to labor, will to be free, and courage to endure long privations."

Cuba was in a unique position to avoid such a racial catastrophe, since an unusually high proportion of the island's population, about seventy-five percent, were already free, including substantial numbers of blacks. In addition, Cuba had a strong tradition of manumission, owing (among other reasons) to more liberal Spanish laws, more intense religious sentiment, and greater opportunities for slaves to earn money with which to purchase their freedom. But time was running out, Humboldt warned, and it could not be left to the planters and slave owners to regulate themselves. Direct legislative action was needed to ameliorate the misery of slavery and gradually to eliminate it. First, the illegal importation of slaves must be halted, as more slaves only exacerbated the problem. Also, various laws could be enacted that would, say, free slaves after a prescribed period of service, or set them free with the condition that they work a certain number of days for their previous master, or give them a certain percentage of a plantation's profit, or set aside a fixed amount from the public budget for buying slaves' freedom. Humboldt suggested that by these or other incremental measures, the manumission of the slaves could proceed at a measured pace, avoiding the sort of bloodshed seen on Hispañola—until all enchained men, women, and children were liberated.

Such warnings were lost on Madrid, and as the nineteenth century wore on, the issue of slavery became intrinsically linked with the struggle for independence on the part of Cuba's Creoles, who saw in the practice a stark reminder of the unsavory prerogatives of the wealthy as well as a symbol of colonial oppression. A bloody revolution (the Ten Years' War) would erupt from 1868 to 1878, in which more than fifty thousand Cubans and two hundred thousand Spaniards gave their lives. Though the rebellion was unsuccessful, Spain did ultimately outlaw slavery on the island in 1886 (leaving Brazil as the last country in the New World to sanction slaveholding, until 1888). Nine years later another revolt was led by José Martí, and, after the intervention of the United States, the island would finally win its independence in 1898.

AT THE END OF APRIL, Humboldt was preparing to sail to Veracruz, on the eastern coast of Mexico, when he saw a newspaper report that would profoundly affect his journey—not to mention the course of his future life and even the history of the sciences. The French government had finally funded Captain Baudin's scientific voyage around the world, the same expedition that Humboldt and Bonpland had been recruited for the previous year. Baudin's two ships, the *Géographe* and the *Naturaliste*, had already left for Cape Horn, on the tip of South America, and intended to sail up the Pacific Coast before continuing to Australia. The year before, as he was leaving Europe, Humboldt had written Baudin promising to join the admiral's voyage en route. Now he was excited by the prospect of being able to realize at last all the plans that he and Bonpland had made back in Paris. Instantly, Humboldt decided to forego the journey to Mexico and to return to South America, to intercept Baudin's expedition at Peru the following year.

There was no guarantee, of course, that the ships would have room for two more naturalists on board. If that were the case, Humboldt now assured Baudin by letter, he would understand. "I shall never grumble over events that lie beyond our control," he wrote. "Such frankness in your decision will be to me the most cherished proof of your friendship. In this event I would simply continue my journey from Lima to Acapulco and Mexico, thence to the Philippines, Persia, and Marseilles. I would much prefer, however, to be a member of your party."

Before they could depart Havana, Humboldt and Bonpland had to finish organizing their huge botanical collection and other specimens and preparing them for shipment to Europe. Considering the vicissitudes of war and weather, Humboldt decided not to risk the priceless artifacts to a single ship, but to divide them into three roughly comparable lots, keeping only a small herbarium for their own reference. "In this journey around the world," he wrote Willdenow from Havana, "when the seas are infested with pirates, and when passports of neutral nations are as little respected as ships of non-warring countries, nothing makes me more anxious than the safety of manuscripts and herbariums. It is really quite uncertain, almost unlikely, that both of us,

Bonpland and myself, will ever return alive. . . . How sad if by such misfortune the fruits of all our labors should be lost forever!"

Humboldt determined that one part of the collection would remain at Havana for safekeeping, one would be sent with their friend Juan Gonzáles to Spain and thence to Paris, and the third would travel to London with Scottish botanist John Fraser, who had been shipwrecked off the Cuban coast and stranded in Havana without passport or money until Humboldt had come to his rescue; from London, this latter portion would be forwarded to Willdenow in Germany as soon as military and political conditions allowed. So much for the specimens, but what about the copious notes and data that the expedition had generated thus far? Humboldt considered entrusting his only copy to Gonzáles but in the end decided to hold on to it himself. It was a crucial decision, for the manuscripts were to be the irreplaceable basis of Humboldt's extensive program of publications, which he had already begun to lay out in his mind.

In the event, Gonzáles's ship was wrecked off the coast of Africa. Their young friend was lost, along with a third of the botanical specimens, the entire insect collection, and the troublesome skeletons from the cavern of Ataruipe. Gonzáles was also carrying letters to family and colleagues detailing Humboldt and Bonpland's new itinerary. In consequence, the pair didn't receive a single letter from Europe over the next two years, and they would have no idea of the fate of the other precious specimens they had dispatched.

While Bonpland labored around the clock to put the collections in order, Humboldt tried to secure transportation back to South America. There were no vessels in port traveling in that direction, and the men he asked "seemed to take pleasure in exaggerating the difficulties of the passage of the isthmus, and the dangerous voyage [around the tip of South America] from Panama to Guayaquil, and from Guayaquil to Lima and Valparaiso." Finally, unable to find a neutral vessel, Humboldt made arrangements with the captain of a small Spanish freighter docked at Batabano, some forty miles below Havana on Cuba's southern coast, and on March 6, the travelers set out across the island. The road from Havana to Batabano ran through uncultivated, partially forested country, and as they walked, the naturalists stopped often to add to their now empty specimen boxes. Among their discoveries was a species of palm with fan-shaped fronds; though it was common along the island's southern shore, the tree had never been recorded by science.

Batabano was a poor village surrounded by bleak, crocodile-infested marshes. The *ciénaga*, or swamp, reminded Humboldt of the Llanos at flood stage. "Nothing can be more gloomy than the aspect of these marshes around Batabano," he found. "Not a shrub breaks the monotony of the prospect: a few stunted trunks of palm-trees rise like broken masts. . . ." He was keen to investigate the two types of crocodiles living in the marshes, which he suspected were different species from the ones they had seen in the Orinoco, but their one night in Batabano left no time to test this hypothesis.

On March 9, their vessel, a trader of forty tons, left Batabano bound for either Cartagena (in present-day Colombia) or Portobelo (in present-day Panama), depending on the weather they would meet. The ship had no cabin, only a hold barely large enough for the passengers' instruments and provisions. The only sleeping accommodations were on deck. "Luckily," Humboldt assures us, "these inconveniences lasted only twenty days. But those twenty days and nights would be spent in all types of weather! Our several voyages in the canoes of the Orinoco, and a passage in an American vessel laden with several thousand arrobas [one arroba equals twenty-five pounds] of salt meat dried in the sun had rendered us not very fastidious."

Sailing southeast, the sloop entered the greenish-brown water of the Gulf of Batabano, bordered by a low and marshy coast. At the gulf's southern entrance, like a stopper in a basin, lay the large island known as Isla de Pinos, for its forested mountains. (It had been christened El Evangelista by Columbus and would later be called Isla de Santa María and Isla de la Juventud.) The gulf was generally plied by smugglers, euphemistically known as "traders" in these parts, but theirs was the only vessel in sight. For the next three days, the ship negotiated the maze of islands known as los Jardines y Jardinillos ("Gardens and Bowers"), which Columbus had named in 1494, on his second voyage, when he found them *"verdes, llenos de arboledas y graciosos"* ("green, pleasant, and filled with trees"). But, prepossessing though it was, to a mariner the archipelago was a nightmare. It had taken Columbus fifty-eight days, battling contrary winds and currents, to sail from El Evangelista to Cuba's eastern tip, a distance of less than five hundred miles. And in 1518, Hernán Cortés had fared even worse: His navy had been wrecked on one of these bars, causing the invasion of Mexico to be delayed till the following year.

As the sloop weaved through the archipelago, Humboldt occupied himself with measuring the latitude and longitude of the various islands and with studying the relationship between bottom composition and the color and temperature of the water. The surface of the gulf was like glass, and he found, "The most absolute solitude prevails in this spot, which, in the time of Columbus, was inhabited and frequented by great numbers of fishermen." There was no village between Batabano and the town of La Trinidad, some 150 miles to the east.

Pausing at the tiny islets for geologizing and botanizing, the boat made slow progress. At one stop, the crew fished for lobsters and, disappointed in their luck, took revenge on a flock of pelicans roosting in a mangrove tree. "The young birds defended themselves valiantly with their enormous beaks, which are six or seven inches long," Humboldt reported with disgust; "the old ones hovered over our heads making hoarse and plaintive cries. Blood streamed from the tops of the trees, for the sailors were armed with great sticks and cutlasses. In vain we reproved them for this cruelty. Condemned to long obedience in the solitude of the seas, this class of men feel pleasure in exercising a cruel tyranny over animals, when occasion offers. The ground was covered in wounded birds struggling in death. At our arrival a profound calm prevailed in this secluded spot; now, everything seemed to say: 'Man has passed this way.'"

On March 12, the sea turned indigo and the breakers disappeared, indicating that the trader had finally found open water. Running before favorable winds, the ship continued eastward toward La Trinidad in order to gain a more beneficial tack for the remainder of the voyage. La Trinidad had been founded in 1514 by Spanish conquistador Diego Velásquez, following the discovery of gold in the valley nearby. The day of their arrival, Humboldt stayed up most of the night making astronomical readings to fix the town's location. The next evening, the lieutenant governor, a nephew of the noted Spanish astronomer Antonio Ulloa, hosted a gala dinner in the visitors' honor, where Humboldt met some French immigrants from Hispañola, a few of the thousands of refugees who had fled the bloody slave revolt there in 1791.

Another night, Humboldt and Bonpland were the guests of Antonio Padrón, one of the city's wealthiest residents, who introduced them to the cream of local society. On this occasion, Humboldt and Bonpland "were again struck with the gaiety and vivacity that distinguish the women of Cuba. These are happy gifts of nature, to which the refinements of European civilization might lend additional charms, but which, nevertheless, please in their primitive simplicity." The travelers departed La Trinidad on the night of March 15, amid a strange spectacle. A fine carriage with damask seats was hired to conduct them to their boat, and on the dock, a priest, "the poet of the place," dressed in a velvet suit despite the tropical heat, recited an original sonnet immortalizing the naturalists' journey up the Orinoco. The mortified Humboldt was delighted to escape back to the little trader.

Navigating south-southeast, the travelers gradually "lost sight of the palm-covered shore, the hills rising above the island of Cuba." A natural-born traveler, Humboldt developed quick affinities to places and generally left them wistfully. Cuba was no exception. "There is something solemn," he found as they sailed away from the island, "in the aspect of land from which the voyager is departing, and which he sees sinking by degrees below the horizon of the sea."

Humboldt's stay in Cuba had been brief and had included only a narrow corridor from Havana due south, plus a few southern coastal towns and islands. However, well into the twentieth century he would still be remembered with affection and gratitude by the Cuban people for his thoughtful study of the island— and for his insistent, courageous condemnation of slavery. In 1939, the University of Havana would present a statue of Humboldt to the city of Berlin, citing him as "the second discoverer of Cuba." In 1969, the two hundredth anniversary of his birth, Cuba would issue a commemorative Humboldt stamp. In 1989, a one-peso coin would be minted in his honor. And in 2001, the Alexander von Humboldt National Park on the eastern end of the island would be designated a World Heritage Site. By any

standard it was a remarkable string of honors for a Prussian
aristocrat who had visited the island for only three months,
nearly two centuries before.

WITH THE WIND FRESHENING from the northeast, the sloop
was driven relentlessly off course, toward the Cayman Islands.
On the morning of March 17, it came within sight of Cayman
Brac, the easternmost island of the archipelago. On account of
the dangerous reefs, the captain kept the sloop a half mile from
shore, but Humboldt could see that the south and southeast
coastline formed a bare, rocky wall, while the north and northwest were low and sandy, with scrubby vegetation. Huge sea
turtles followed the boat, and the sailors were about to dive in after the animals, before they noticed an accompanying school of
sharks.

The wind turned to the southeast and continued to strengthen,
dipping the trader's gunwale underwater. Soon after they passed
the shoals of La Víbora, the weather became fine, the wind
shifted to the north-northwest, and the water turned cobalt. On
March 24, the ship entered the Gulf of Santa Marta, on the
Colombian coast, known for its violent gales. The next day the
wind rose out of the northeast with increasing force. The sea became exceptionally rough during the night, and in the morning
the captain sought shelter in the Río Sinú, west of Cartagena,
just as a violent rain began to fall on the unsheltered passengers.
After a miserable passage of nearly three weeks, Humboldt and
Bonpland had again reached the shores of South America, but by
all indications they had landed in a wild area rarely visited by
outsiders.

At the tiny village of Zapote, the residents regarded the European visitors with suspicion, asking pointed questions concerning their itinerary, their books, and their instruments, and trying
to intimidate them with tales of jaguars and poisonous snakes.
But their long weeks in the rain forest had inured the travelers
to such exaggerations, which, Humboldt believed, "arise less from
the credulity of the natives, than from the pleasure they take in
tormenting the whites." To find some peace, the naturalists retreated to the forest for some botanizing.

After about an hour's walk, they came upon a clearing where
a group of workers were making palm wine by cutting the trees

and allowing the sap to collect, then ferment, in the hollowed trunk. Among the dark-skinned people was a slight man incongruous for his fair hair and pale complexion. At first Humboldt suspected the stranger was a deserter from a North American ship, but he soon discovered that the man was a fellow German who had served in the Danish navy before coming to the Río Sinú, he explained, *"Para ver tierras, y pasear, no más."* ("To see other lands and to walk about, nothing more.") During his five years in the New World, Humboldt met exactly two persons with whom he could speak his native German; the first had been a hand on the privateer that had wanted to tow their trader to Nova Scotia as a prize of war, and the second was this man on the nearly uninhabited coast of Colombia. Humboldt quizzed the stranger on the surrounding territory, but the man only smiled enigmatically and replied "that the country was hot and humid; that the houses in the town of Pomerania [in Germany] were finer than those of Santa Cruz de Lorica; and that, if we remained in the forest, we should have the tertian [every-other-day] fever from which he had long suffered." When Humboldt tried to give him some money, the man refused with hauteur, explaining that he could never accept anything "in the presence of those vile colored people."

Faced with thick forest and oppressive humidity, the naturalists returned to the boat before sunset, laden with mosses, lichens, and other botanical specimens. The wind was still blowing furiously when they weighed anchor the next morning, and it continued to rage all day. That evening, waves began to wash over the deck, and the captain was again forced to seek shelter along the coast. Unable to find a likely cove, he dropped anchor in shallow water, but, discovering the area to be a coral reef, decided to take his chances in the open sea. At length, it began to rain, the wind abruptly diminished, and the boat was able to pull up to a small island for the night.

The next morning a dangerous gale blew up. "The sea was fearfully rough," Humboldt wrote. "Our tiny craft could not master the waves, and suddenly was thrown on her beam-ends. A tremendous wave broke over us and threatened to engulf the ship. The man at the helm remained undismayed at his post. All of a sudden he called: *'No gobierna el timón!'* ['The rudder won't steer!']. We all gave ourselves up for lost. In this, as it seemed to us, our utmost danger, we cut away a sail that had been flapping loosely, when the ship righted herself on top of another wave, enabling us to find refuge behind the promontory. . . ." It was Palm Sunday, and, as they waited out the storm, Humboldt's servant reminded him that it was one year to the day since their lancha had nearly capsized in that other dangerous gale, on the Orinoco.

A lunar eclipse was predicted for that evening, and the next night there was to be an eclipse of one of the moons of Jupiter, which would be very useful in determining the longitude of nearby Cartagena. Humboldt urged the captain to allow one of the four crew members to make the five-mile overland trek with him to the foot of Boca Chica (Little Mouth), the smaller of the two openings to Cartagena harbor, so he could fix the port's longitude. But to Humboldt's annoyance, the captain refused, citing the lack of a village or even a path through the jungle. So instead Humboldt and Bonpland paddled the ship's canoe to shore to collect plants in the moonlight.

As they landed, the Europeans were approached by a young, powerfully built black man, who emerged from the forest wearing only chains and carrying a machete. The young man directed them to land at a protected mangrove thicket, where he offered to guide them inland in exchange for some clothing. But, alarmed by the man's "cunning and wild appearance, the often-repeated question whether we were Spaniards, and certain unintelligible words which he addressed to some of his companions who were concealed amidst the trees," the Europeans leaped into the canoe and fled back to the ship. Humboldt concluded that the men were escaped slaves hoping to steal their boat; though he pitied them, he also knew enough to fear them, for "they ha[d] the courage of despair, and a desire of vengeance excited by the severity of the whites."

Still, Humboldt was deeply moved by the encounter. "The aspect of a naked man," he wrote, "wandering on an uninhabited beach, unable to free himself from the chains fastened round his neck and the upper part of his arm, was an object calculated to excite the most painful impressions." The captain and crew didn't share his sympathy. "Our sailors wished to return to the shore for the purpose of seizing the fugitives," Humboldt continued, "to sell them secretly at Cartagena. In countries where slavery exists, the mind is familiarized with suffering, and that instinct of pity which characterizes and ennobles our nature, is blunted." But the captain had clearly been right to forbid the excursion to Boca Chica, which, Humboldt had to admit, could have had fatal consequences. For the rest of the evening, he satisfied himself with observing the lunar eclipse from the safety of the boat. The next morning, the ship sailed to Boca Chica, where she took on a *práctico*, or pilot, to conduct her the seven or eight miles to the anchorage near town.

Humboldt and Bonpland stayed six days at Cartagena, confirming earlier determinations of the town's latitude and longitude and exploring the surrounding mountains and coast. The city was one of the principal ports in the Spanish colonies, and the authorities had long worried over the difficulty of defending a harbor with two openings. In 1741, the British had invaded the port and destroyed three forts along the coast before being forced out. As a result, a dike had been erected across Boca Grande in 1795, in order to seal off the wider entrance. But the sea continued its assault, and in the few years since, the water had already started to erode the dike, while at the same time filling Boca Chica with silt.

Cartagena was bordered on the north and east by marshes, which were separated from the town by a chain of low hills. Crowning the hills were a fort and monastery, the latter built on such unstable land that its continued existence was taken as a literal miracle, attributed to the image of the Virgin housed there. The hills were covered with cactus, and while herborizing in the area, Humboldt and Bonpland were shown a large, infamous acacia tree covered with huge biting ants as well as sharp thorns two inches long. There, the guides related, "a woman, annoyed by the jealousy and well founded reproaches of her husband, conceived a project of the most barbarous vengeance. With the assistance of her lover she bound her husband with cords, and threw him, at night, into [the tree]. The more violently he struggled, the more the sharp woody thorns of the tree tore his skin. His cries were heard by persons who were passing, and he was found after several hours of suffering, covered with blood, and dreadfully stung by the ants. This crime is perhaps without example in the history of human turpitude," Humboldt concluded:

"It indicates a violence of passion less assignable to the climate than to the barbarism of manners prevailing among the lower class of people."

Under his revised itinerary, Humboldt planned to sail from Cartagena to the eastern coast of Panama. Then he intended to cross the isthmus, taking topographical measurements and charting a route for a possible canal, a superhuman undertaking that had been suggested as far back as the sixteenth century and had been advocated by Benjamin Franklin and other luminaries. From Panama's Pacific coast, Humboldt proposed to sail south to Guayaquil, in present-day Ecuador, where he hoped to intercept the Baudin expedition.

However, once in Cartagena, he discovered that this plan had two serious flaws. First, though the journey across Panama would be a relatively simple matter, he would probably have a long wait for a ship traveling south at this time of year. And, second, even if he were lucky enough to find a vessel, he would be in for an exceptionally long, uncomfortable journey down the Peruvian coast, against strong currents and seasonally unfavorable winds. Rather than risk missing the rendezvous with Baudin due to forces outside his control, Humboldt decided to travel overland. Not only did his sources at Cartagena assure him that this would be more practicable, he realized that the route through the Andes would provide an unparalleled opportunity for scientific exploration. In Bogotá, he would even be able to call on José Celestino Mutis, a former student of Linnaeus and the world's leading authority on South American plants, who could help him and Bonpland make taxonomic sense of the thousands of species they had collected thus far. (The change in plans meant that Humboldt never did make it to Panama. But he did eventually obtain detailed topographical data of the isthmus and propose various courses for a canal to be built there; one of his suggested routes was later chosen by the French engineers who began the project in 1881. An American team would finally complete the massive undertaking in 1914.)

Thus, for the third time since leaving Europe, serendipity had swept Humboldt's journey in a novel direction—just as it had when the typhus outbreak on the *Pizarro* had redirected him, against all expectation, to the coast of Venezuela and thence to the Orinoco and Amazon; and again when news of Baudin's expected arrival in Peru had turned Humboldt away from his anticipated voyage to Mexico and the Philippines and back to South America. Had Humboldt realized what would be entailed in such an overland expedition, he might have reconsidered yet again. But, fortunately for science, he was still ignorant of the time and incredible effort required for a trek through the Andes, some of the highest, most rugged mountains in the world. And he was right about one thing: The cordillera did indeed provide an "immense field for exploration." In fact, his travels through this scientifically virginal region would revolutionize many disciplines, including geology, geography, and anthropology—and even influence the fine arts. Through a potent combination of happenstance and daring, Humboldt was about to embark on a segment of his journey that, though totally unanticipated, would prove one of the most original and influential of his entire five-year odyssey. It would also make him world famous.

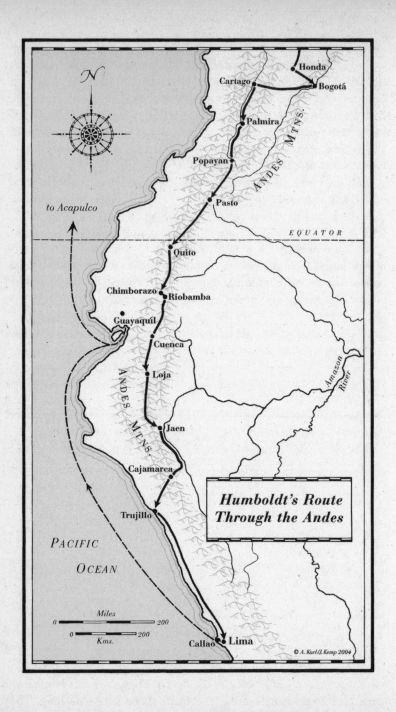

Humboldt's Route Through the Andes

to Acapulco

N

Honda
Bogotá
Cartago
Palmira
Popayan
Pasto

ANDES MTNS.

EQUATOR

Quito
Chimborazo — Riobamba
Guayaquíl
Cuenca
Loja
Jaen

ANDES MTNS.

Amazon River

Cajamarca
Trujillo

PACIFIC OCEAN

Miles
0   200
Kms.
0   200

Callao — Lima

© A. Karl / J. Kemp 2004

NINE

Chimborazo

ON APRIL 21, 1801, HUMBOLDT AND BONPLAND BOARDED an Indian canoe on the Río Magdalena to begin the long journey from Cartagena to the foot of the Andes. The rainy season had begun. The tropical storms were torrential, the insects voracious. Struggling against the swift current, the Indian paddlers were able to average only a little over ten miles a day.

For Humboldt and Bonpland, conditions must have been depressingly reminiscent of their excruciating expedition up the Orinoco and Río Negro. As on those other rivers, the rain forest here hung in a dense curtain along the unpopulated banks, crocodiles dozed in the mud, and long-legged water birds stalked their prey. Humboldt passed the days recording meteorological data and, wielding his compass, making the first chart of the river, while Bonpland went ashore at every opportunity to add to their herbarium. After six agonizing weeks, their canoe finally landed at the tiny river port of Honda, some four hundred miles inland. Towering above was the Cordillera Oriental, the eastern spur of the mighty Andes.

No one knows the origin of the name *Andes*, but it is thought to derive from either of two Indian words—*anti*, for "east," or *anta*, for "copper." Whatever the source of their name, the mountains, like the Amazon, which rises among their peaks, constitute one of the geographical wonders of earth. Snaking in a multibranched spine along the western coast of the continent, the chain runs for 5,500 miles, from Tierra del Fuego in the south all the way to the Caribbean in the north. Though Asia's Himalayas are taller, the Andes are more than three times as long. They comprise the highest peaks in the Western Hemisphere and some of the greatest mountains in the world. Aconcagua in Argentina, at 22,834 feet the tallest point outside Asia, is just one of many Andean peaks over 20,000 feet high. (By contrast, the tallest mountain in North America, Alaska's Denali, or Mount McKinley, is 20,320 feet.)

The Andes were formed over millions of years, as the South American Tectonic Plate shifted westward and forced itself into the adjoining Pacific Plate, buckling the earth's surface along the leading edge. The theory of plate tectonics—the idea that the planet's landmasses were once united in a huge supercontinent, which gradually broke up as the continents drifted into their current locations—was suggested around 1920 by the German geologist Alfred Wegner. Though the theory was at first considered absurd, in the latter half of the century, it revolutionized our understanding of the history of the earth. It's not known for certain why the twenty or so tectonic plates located on the planet's rocky outermost layer, each up to sixty miles thick, roam over the more pliant stratum beneath, but the process is thought to be due to differences in temperature between the earth's core and the cooler mantle above. As the plates shift and collide, they trigger earthquakes, open ocean trenches, raise mountain ranges, and otherwise reshape the surface of the planet.

Geological violence on such a grand scale has produced an extremely complex mountain system in South America. Just as the Amazon can be seen not as one river but as a network of interconnected waterways, the geologically young Andes can be viewed as a labyrinth of ranges that converge and diverge as they wend their way up the long Pacific Coast. Like the various segments of the Amazon—the Manañon and Solimões rivers, for instance—these ranges are called by different names in various regions, and each boasts its own characteristics of height, width, vegetation, snow cover, and so on. In their southernmost zone, the Andes form a single, relatively narrow range that, while not among the highest in South America (averaging only about 6,500 to 8,500 feet), are rugged and covered with glaciers and ice fields. As the range travels up the coast, it widens, becomes more arid and taller, then diverges into eastern, central, and western spurs, with high plateaus, called altiplano, in between. In the eastern, older region, the mountains are lower, having had more time to erode. The newer, geologically active western region includes the tallest peaks, as well as the greatest concentration of volcanoes. It was to this area that Humboldt naturally turned.

In Humboldt's day, before the Himalayas had been surveyed, the Andes were believed to be the highest mountains in the world. Always sensitive to the grand spectacle of nature, Humboldt was deeply moved by the range's majesty—it would become the yardstick by which he gauged every other geographical feature he encountered in South America, from the Llanos to the Orinoco. Though he found several phenomena that approached the mountains for splendor, he would never find any to surpass them.

Humboldt's first destination en route to Quito and Lima was Santa Fe de Bogotá, located on a high plain southwest of Honda. Though the city (today's Bogotá, capital of Colombia) was only fifty miles from the Río Magdalena, it lay nearly nine thousand feet higher, over incredibly rugged roads that hadn't been improved since the time of the Incas. In fact, in places the route wasn't a proper road at all, just crude steps carved between the steep rock faces, so narrow that a laden mule could barely slip through. The explorers hadn't been at such altitude since their climb up el Pico del Teide, in the Canaries, two years before. As they made the arduous ascent now, Bonpland experienced severe headache and nausea, along with a fever possibly from malaria contracted on the coast.

When the travelers finally reached the sunny, settled Sabana de Bogotá, the savannah surrounding the city, Humboldt sent a messenger ahead with news of their arrival. The next day, as the party approached the city, they were met by a colorful procession of dozens of horsemen and hundreds of citizens on foot, most of whom had never seen anyone who did not count himself a subject of the Spanish King. Humboldt was swept into the city in the archbishop's fine six-horse carriage, while Bonpland rode in another vehicle behind. Though astonished and somewhat embarrassed by the spectacle, Humboldt was also flattered. "There hadn't been such bustle and tumult in this dead town for years and years," he wrote in his journal.

The world-famous botanist José Celestino Mutis had hastily assembled a distinguished reception committee, who were waiting on the steps of Mutis's house. As Humboldt climbed out of the carriage, clutching his barometer (which he refused to entrust to anyone else), he was embraced by the aging botanist, who showed such deference and modesty that the guest became uncomfortable. Humboldt launched immediately into work, then realized his faux pas. "We started talking straight away about scientific matters," he related. "I began to tell of the plants I had seen during the day," but Mutis "very cleverly steered the conversation round to more general topics so that the others standing nearby could understand." A marvelous banquet had been laid inside, and Humboldt was again astonished when the renowned scientist Salvador Rizo materialized to serve them at table. Mutis had temporarily evicted his brother's widow from the house next door, and the visitors were to lodge there.

José Celestino Mutis was born in 1732 in Cádiz. One of Linnaeus's earliest disciples in Spain, he emigrated in 1760 to Colombia, where, as head of the Royal Botanical Expedition of New Grenada, he pioneered the study of quinine and became the world's foremost authority on the plants of South America. Mutis's botanical library was second only to that of Joseph Banks in London, and arranged on shelves in an adjoining room was a collection of more than twenty thousand dried plant specimens, which a team of thirty artists had been painting for the past fifteen years, under Mutis's personal supervision. The illustrations were intended for the mammoth *Flora de Bogotá o de Nueva Granada*, which Mutis didn't live to finish (in fact, the illustrations wouldn't see publication for another century and a half). A man of wide-ranging interests, Mutis also founded in Bogotá the first astronomical observatory in South America and transformed the city into an international center of learning, which Humboldt considered the "Athens of America."

The mountains around Bogotá had for centuries been home to the Chibcha Indians, one of the most technologically sophisticated indigenous peoples of Colombia. Among their many industries—agriculture, weaving, pottery, copper smelting—the Chibchas practiced gold making. In fact, the legend of El Dorado, the inspiration for much of the fevered, disastrous early exploration of the South American interior, is thought to have originated with this group. Each year, according to the legend, the Chibchas chose a new leader, who was covered in gold dust and ceremonially washed in a nearby lake, into which golden ceremonial items were also thrown. (One of these chiefs, Bacatá, is immortalized in the name Bogotá.) First conquered by the Inca, the Chibchas were subdued only a few decades later, in 1538, by lawyer-and-writer-turned-conquistador Jiménez de Quesada, who enslaved them, destroyed their temples, and stole their gold. In 1740, Santa Fe de Bogotá was named capital of New Grenada (which also included present-day Colombia, Venezuela, Ecuador, and Panama), continuing the city's long tradition as a governmental center despite its remote location.

While Bonpland recovered from his malaria attack, Humboldt made scientific excursions to the surrounding countryside, where he inspected deposits of rock salt and coal. He also measured the heights of the nearby mountains, some of which rose to sixteen thousand feet above sea level, and he visited Lake Guatavita, thought to be the body of water in the legend of El Dorado. On one foray, Humboldt traveled to el Campo de Gigantes (the Field of Giants) to examine the mastodon bones there. The forebears of modern elephants, the forest-dwelling mastodons evolved in Africa in the Oligocene epoch (thirty-eight to twenty-four million years ago) then, during the Miocene epoch (twenty-five to five million years ago) migrated from Africa to Europe and thence to Asia and across the Bering Land Bridge to the Americas, only to perish as the Ice Age of the Pleistocene era (two million to eleven thousand years ago) swept over the world.

In 1833, during his voyage to South America aboard the *Beagle*, Charles Darwin also investigated a deposit of mastodon bones (in Argentina). Projecting from the face of a cliff, the skeletons were badly decomposed but identifiable. When he asked his guides how they supposed the bones had gotten into this unlikely location, Darwin was amused by their response: Clearly, the huge mastodon had been a burrowing animal! From the creatures' wide range over both the Old and the New worlds, Darwin concluded that the climate in those regions must once have been similar. But if that were the case, why were so many species of plants and animals now living in North America—such as the wolf, the bear, and the elm tree—not found in South America? And why were so few species that were native to South America—such as the jaguar, the spider monkey, and the ceiba tree—found farther north? To resolve this dilemma, Darwin pointed to the mountainous barrier that Humboldt had reported running east to west across southern Mexico (the Trans-Mexican

Volcanic Belt). The formation of this range, he suggested, had created a barrier impassible by plants and animals. Then, over millions of years, the isolated populations to the north and south had evolved in divergent directions according to the law of natural selection, until the products were recognized as different species.

AFTER TWO MONTHS OF RECUPERATION from his fever, Bonpland was finally well enough to travel. On September 8, the explorers set out toward Quito, nearly five hundred miles away. From Bogotá, they trekked westward, down the Cordillera Oriental and across the Magdalena Valley. Navigating narrow, twisting trails with solid rock on either side, deep mud underfoot, and thick vegetation obliterating the light above, they descended through uninhabited woods and passed the towns of Pandi, Espinal, Contreras, and Ibagué. Just as he had earlier eschewed the easiest route through the rain forest, here Humboldt forsook the relatively level way through interconnecting river valleys and elected to climb over the Cordillera Central via the treacherous Quindío Pass, which, at nearly twelve thousand feet, was one of the most demanding trails in all the Andes. As they ascended through stands of bamboo, wax palms, and tree ferns, punctuated with orchids, passion flowers, and fuchsias, the party was struck by driving rains. The animals sank deep in the mud, and the men's boots were destroyed by bamboo spikes jutting from the swampy ground.

Just beyond the pass, the travelers were met by a group of *cargueros*, Indian porters who eked out a living by strapping a chair to their back and, walking doubled over and supporting themselves with a cane, conveyed Spanish mining officials over the mountain trails. Humboldt, the self-proclaimed republican, was infuriated to see such a degrading practice and to "hear the qualities of a human being described in the terms that would be employed in speaking of a horse or a mule," such as surefootedness and an easy gait. Rather than mount the human beasts of burden, he and Bonpland elected to walk down the mountain to the town of Cartago, though their feet were bare and bleeding. To Humboldt, this was undoubtedly a noble, democratic deed, but the gesture didn't impress the *cargueros;* to them, it was just an act of stinginess that deprived them of much-needed income. In fact, the porters were vociferous in their objection to a new road being built through the mountains, on the grounds that it would rob them of their livelihood.

The travelers were now about 150 miles west of Bogotá. From Cartago, they continued some 175 miles southward through the fertile Cauca Valley, which joins the valley of the Magdalena. For the month of November they remained at the town of Popayán, which had been founded in 1537 and, thanks to its position on the road over which gold traveled between Quito and Cartagena, had quickly grown into an important administrative center. On one of their excursions to explore the local geological formations and flora, Humboldt and Bonpland climbed the 15,604-foot-tall active volcano Puracé, which at the time was giving off noisy jets of steam.

Though not one of the tallest or best-known peaks on the continent, Puracé is notable as the first of the dozens of active volcanoes that Humboldt would climb in South America. During the ascent, he would have been mulling the host of questions that had been occupying him since his climb of el Pico del Teide, and even before: How were volcanoes formed? What caused their subterranean fire? Did they communicate underground, or was each volcano an independent entity? At the time, scientists' scant knowledge of volcanoes had been based almost exclusively on their observations of Europe's two active peaks—Etna and Vesuvius. But it was a mistake to rely too heavily on these models, Humboldt believed, since they might not be representative of all volcanoes. In fact, at about 11,000 and 4,300 feet, respectively, they were both relatively low. Taking them as archetypes of the world's volcanoes, Humboldt wrote, one might be led into a fallacy like the one made by "Virgil's shepherd, who thought he beheld in his humble cottage the type of the Eternal City, Rome."

On the contrary, "to become completely acquainted with the important phenomena of the composition, the relative age, and mode of origin of rocks, we must compare together observations from the most varied and remote regions"—such as the great peaks of South America. Not that Humboldt necessarily had expected to find differences of kind between American and European volcanoes; indeed, he believed that the study of both would illuminate the universal laws governing them all. "If . . . new zones do not necessarily present to us new kinds of rock . . . ," he wrote, "they, on the other hand, teach us to discern the great and everywhere equally prevailing laws, according to which the strata of the crust of the earth are superposed upon each other, penetrate each other as veins or dykes, or are upheaved or elevated by elastic forces."

Humboldt's mentor Werner had encouraged his students to travel so as to observe the earth's whole range of geological phenomenon firsthand. Now Humboldt was taking him at his word, intent on testing his teacher's theories against this mass of new data. Were the earth's mountains created in a one-time act of creation, as Werner believed? Were they really formed by matter settling out of a great primordial ocean? Humboldt would not resolve any of these questions on the slopes of Puracé, but he would return to them repeatedly throughout his exploration of this magnificent landscape.

As the party left Popayán, the sky was shrouded in perpetual gray, and the traveling conditions remained miserable. "Thick woods intersperse with swamps," Humboldt wrote, "where the mules sank up to their girths, and narrow paths winding through such rocky clefts that one could almost fancy one was entering a mine, and the road paved with bones of mules that had perished from cold or fatigue." At about eight thousand feet, the travelers entered one of the barren, frigid plateaus called *páramos* (from the Spanish word for "wasteland"), where only scrubby alpine plants could survive. On the Páramo de Pasto, the clouds mingled with volcanic gases, and the rain turned to sleet and snow. At night, the Indian guides erected tents made from wooden poles covered with the large, waxy leaves of the heliconia (a member of the banana family); then in the morning they would carefully roll them up and repack them for next time.

The travelers passed Christmas in the town of Pasto, in the southwestern tip of present-day Colombia. Then, crossing the equator, they finally entered Quito on January 6, 1802. It had been a harsh journey of more than eight months and nearly eight hundred miles over the mountains, far longer than Humboldt had planned—and the party was still some eight hundred miles from Lima, where they hoped to rendezvous with Captain Baudin. The French expedition was expected in less than six months, and Humboldt was anxious to reach the coast in time.

Though Quito is situated only about fifteen miles south of the equator, due to its high elevation, it is graced with a pleasant, springlike climate. The city's name commemorates the Quitus Indians, who made the area their home for at least a thousand years. Before the arrival of the Spanish, the Quitus merged with the Cara, Shyri, and Puruhás peoples into the Kingdom of Quitu, whose capital was located at present-day Quito. Around 1500, Quitu was conquered by the Inca, who established a capital of their own there, only to be conquered by the Spanish three decades later. Perched at 9,300 feet above sea level, Quito spread out on picturesque, rolling hills surrounded by volcanoes. In 1660, an eruption of its closest volcanic neighbor, Pichincha, destroyed the city, and in 1797 it was severely damaged by an earthquake at Riobamba, which killed forty thousand people over a wide area. Five years later, when Humboldt arrived, the city was still experiencing aftershocks. If he wanted to study volcanic activity, he couldn't have chosen a better place.

Despite the earthquake damage, Humboldt found Quito a handsome, lively city, with impressive monasteries and churches. After the 1797 quake the sky had become perpetually cloudy, and the mean temperature had dropped noticeably. Yet, despite the geologic disasters they were forced to endure—or perhaps because of them—the city's forty thousand inhabitants exhibited a nonchalant, carpe diem attitude toward life. Humboldt found the Quiteños charming, sophisticated, and determined to enjoy the moment, since they never knew when the ground might tremble again. "The town," he wrote, "breathed an atmosphere of luxury and voluptuousness, and perhaps nowhere is there a population so entirely given up to the pursuit of pleasure. Thus can man accustom himself to sleep in peace on the brink of a precipice."

Humboldt and Bonpland were taken in by the Marqués de Aguirre y Montúfar, the provincial governor and one of the wealthiest men in the city. Montúfar's young son Carlos, whom Humboldt found "an estimable youth," became the visitor's fast companion for the remainder of the journey—and for nearly a decade afterward. As with Humboldt's other close friendships, it's tempting to speculate on the nature of his relationship with Montúfar. Accentuating the sense of mystery is the fact that Humboldt wrote an entire fourth volume of the *Personal Narrative*, then destroyed it on the eve of publication. Some have speculated that he wanted to shield his relationship with Montúfar from inquisitive eyes and to protect the reputation of his socially prominent friend. (However, most historians today believe that, attracted though he may have been to young, handsome men, Humboldt did not engage in sexual relations with them. One is reminded of his earlier admonition that "serious themes, and especially the study of nature, become barriers against sexuality.")

Whatever the nature of his friendship with Montúfar, it seems doubtful that it was the motivation for the destruction of Volume Four. Humboldt divulged virtually no personal details in the first three volumes. If he felt the need to obscure the nature of his relationship with Montúfar, or with anyone else, it would have been a simple matter for him to skip over it and to concentrate instead on the more factual aspects of the journey. Indeed, though Humboldt is prolix in his scientific ruminations and his physical descriptions in the three existing volumes, despite the *Personal Narrative*'s title he includes very little that could be described as truly personal. Various other ad hoc companions— Carlos del Pino, Nicolás Soto, Juan González—are little more than names who pass in and out of the story. Even Bonpland, Humboldt's dear companion, barely emerges from the book as a flesh-and-blood character. So even if Montúfar had been included in Volume Four, how would that have provided grist for the scandalmongers?

The alternative motivation put forward for Humboldt's destruction of the fourth volume is that the material had, by the time of publication, become politically embarrassing. After his return to Europe, Humboldt was named chamberlain to the king of Prussia. Perhaps Volume Four was too generous in its praise of the developing independence movements in Spanish America, to a degree that was unfitting for a member of a European royal court. Yet Humboldt was certainly outspoken concerning his republican views in his other writings from those years, including the rest of the *Personal Narrative*.

We will never know the exact nature of Humboldt's friendship with Montúfar, any more than we will ever know the precise character of his relationship with any of the other men that he was drawn to over the course of his long life. The destruction of Volume Four will most likely remain a mystery, along with much else about the personal life of this intensely private man.

IN QUITO, Humboldt received a stunning setback to his plans: He learned that the Baudin expedition would not be calling on the South American coast after all. Instead of sailing across the Atlantic and around Cape Horn, Baudin had chosen a route around the tip of Africa. The hoped-for rendezvous—which Humboldt had traveled more than eight months and some eight hundred backbreaking miles to achieve—would never happen.

Yet Humboldt preferred to focus on the positive. "We have come to feel that man ought not to count upon anything which he cannot obtain by his own enterprise," he wrote French astronomer and mathematician Jean-Baptiste-Joseph Delambre. "Accustomed to disappointments, we consoled ourselves in the thought that we had been prompted by a good purpose in all the sacrifices we made. In going over our herbariums, our barometric and trigonometric observations, our drawings and our experiments on the atmosphere of the Cordilleras, we see no reason for regretting our visit to countries that have remained largely unexplored by scientists."

Thrown back on his own resources, Humboldt immersed himself in his study of the Andes with a ferocity that left little time for anything else. The marqués de Montúfar's beautiful daughter, Doña Rosa, found Humboldt charming, but elusive. "At table," she recalled many years later, "he never remained longer than was absolutely necessary to still his hunger and pay the customary courtesies to the ladies. He seemed always glad to be outdoors again, examining rocks and collecting plants. At night, long after we all had retired, he would observe the stars. To us young women this manner was more difficult to understand than for my father, the marqués." Her description is reminiscent of Humboldt's early preference of the woods and fields of Tegel, as well as his sister-in-law Caroline's futile efforts to domesticate him in Paris. Clearly, two and a half years in the wilderness had done nothing to whet Humboldt's appetite for polite society or feminine companionship.

Mounts were always on call to take Humboldt into the countryside, and he often availed himself of the opportunity. Over the coming months, he systematically climbed every volcano on the plateau outside Quito—Pichincha, Cotopaxi,

Antisana, Tunguragua, Illiniza, Chimborazo, and others besides—
sometimes with Bonpland and sometimes with only his inexperi-
enced Indian guides. As he climbed, he was constantly observing,
measuring, collecting, seeking to glean the mountains' secrets.
What types of rock were they made of? What gases spewed from
their craters? What geologic process had formed them? How
were they different from and similar to the volcanoes of Europe
and each other? Did they fit into any natural patterns that might
elucidate their structure or history?

Typical of Humboldt's climbing efforts during this time were
his three attempts on the active volcano Pichincha, which, rising
above Quito at nearly 15,700 feet above sea level, had been previously climbed only by La Condamine. But the Frenchman had
gone up without his instruments, so he was unable to perform
any meaningful scientific analysis, and he had lingered over the
crater only fifteen minutes before being driven down by the biting cold. Humboldt resolved to give the volcano a more thorough
look. His first effort was a complete failure. High up on the side
of the volcano, he was stricken with dizziness, passed out, and
had to be carried down; if he had gone up alone, he might well
have frozen to death.

Not long after, on May 26, he tried again, accompanied by an
Indian porter named Aldas. "Since Condamine had approached
the crater from the snow-covered side of the rim," Humboldt
wrote his brother, "I followed . . . in his footsteps. That was
nearly the end of us." When the porter sank up to his chest in a
crevasse above the crater, they realized to their horror that they
were standing on an unstable snow bridge. "Frightened but undeterred," as Humboldt wrote, they climbed a rocky outcropping
where the snow had been melted by steam from below. On top of
this tiny perch, he discovered a cantilevered stone that formed a
kind of balcony. Though the stone was rocked by tremors every
twenty to thirty seconds, Humboldt and the porter erected the
instruments there. "We lay down flat to get a better look at the
bottom of the crater," Humboldt recorded; "nobody can imagine
anything more sinister, mournful, and deathly than what we saw
there. The gorge of the volcano forms a circular hole of about one
mile in circumference; the broken edge is covered with snow and
the interior a dull black in color. But the depth is so enormous

that it contains several mountains whose peaks can be just discovered. They seemed to be perhaps 1,900 feet below us. How far down can it be to the bottom?" he wondered. "I believe that the bottom of the crater lies at the same height as the town of Quito. La Condamine found the crater extinct and covered with snow. We were forced to tell the people the sad news that their nearby volcano is on fire. The unmistakable signs did not leave any doubts. We were nearly suffocated by sulfurous vapors at the approach to the opening. We saw bluish flames flicker in the depths and felt violent tremors every two or three minutes. They shook the rim of the crater. . . . It seems that the catastrophe of 1797 has relit the fires of Pichincha."

From the top of the mountain, Humboldt eagerly searched in the west for the Pacific, the fabled ocean discovered by Balboa and once sailed by Cook, Bougainville, and the other heroes of Humboldt's boyhood. But however much he strained, he found that "no sea horizon can be clearly distinguished, by reason of the too great distance of the coast and height of the station: it is like looking down from an air-balloon into vacancy. One divines, but one does not distinguish."

Humboldt didn't return to Quito till well beyond midnight, after more than eighteen hours of climbing. But, just two days later, on May 28, he climbed the mountain again, this time lugging an even greater armamentarium of scientific instruments. As he struggled up the peak, he recorded fifteen seismic shocks within thirty-six minutes—an observation which, when reported in the town below, would give rise to rumors that he was hurling gunpowder into the crater in order to set off the mountain.

Less than two weeks after the third climb up Pichincha, Humboldt turned his sights to Chimborazo, the tallest peak in the northern Andes and at the time thought to be the tallest mountain in the world. "After the long rains of winter," he wrote in *Researches Concerning the Institutions and Monuments of the Ancient Inhabitants of America*, "when the transparency of the air has suddenly increased we see Chimborazo appear like a cloud at the horizon; it detaches itself from the neighboring summits, and towers over the whole chain of the Andes, like that majestic dome, produced by the genius of Michael Angelo, over the antique monuments, which surround the Capitol." Chimborazo's height and remoteness proved an irresistible challenge, and on June 9, the travelers left Quito to begin the eighty-mile trek south, following the road that would later become the Pan-American Highway.

On the morning of June 23, 1802, Humboldt, Bonpland, and the young Montúfar set out from the house of the alcade, or mayor, of Calpi. They had been preparing for this ascent for months, refining their climbing skills and testing their stamina on the smaller peaks nearby. "Fortunately, the attempt to reach the summit of Chimborazo had been reserved for our last enterprise among the mountains of South America," Humboldt wrote, "for we had by then gained some experience and knew how far we could rely on our own strength." Indeed, by this point in his journey, Humboldt was one of the most experienced climbers in the world.

From the base of the mountain, Humboldt squinted up the slope before him, hoping for a glimpse of the great Chimborazo, or *Urcorazo* ("Snow Mountain") in the Quichua Indian language. But he saw only clouds. How ironic that the tallest peak in the world should lie hidden in mist, massive yet invisible. The going was relatively easy for the first six thousand feet, but after that the terrain got progressively rougher. Leaving the mules behind, the men continued on foot.

Keeping close so as not to lose sight of each other, the party made slow progress in the reduced visibility. The ridge snaked up the eastern side of the mountain and narrowed in some places to a width of only eight or ten inches. To the left was a snow-covered slope of about thirty degrees. On the right lay an abyss a thousand feet deep, with huge rock formations projecting from the bottom. They had no climbing equipment, and at some places the ridge rose so steeply that they had to pull themselves up with their bare hands, which bled on the sharp rocks. Everyone was gasping in the thin air.

One by one the climbers began to feel the nausea and giddiness of soroche, the altitude sickness that had first been reported by Pizarro's men in Peru. Eventually their noses were streaming blood as the vessels began to rupture. The temperature had been dropping by one degree for every three hundred feet of altitude, and they were all shivering in their lightweight clothing. Still they struggled up the narrow ledge, stopping to take temperature and barometric measurements as they went. At the snow line, all but one of the Indian porters deserted them.

Abruptly, the mist cleared and the dome-shaped summit loomed very close in front of them. "What a grand and solemn spectacle!" Humboldt declared. "The very sight of it renewed our strength." The rocky, snow-covered ridge became somewhat wider. They hurried on.

Not long after, they came to an abrupt halt before a ravine some four hundred feet deep and fifty feet wide. The snow that filled the crevasse was soft and flaky, certainly too soft to climb. Unable to scale the walls, freezing, weakened from loss of blood, they had no choice but to turn back. They had climbed as high on Chimborazo as they would be able. Humboldt took his watch out of his pocket. It was one o'clock in the afternoon.

The naturalists unpacked their instruments. The thermometer read three degrees below freezing, but it felt colder. Setting up the barometer with great care, Humboldt computed their altitude at 19,286 feet. The summit was a tantalizing 1,300 feet above.

Standing in the clear, cold air above the clouds, Humboldt "felt as isolated as in a balloon." He was exhilarated, for though they hadn't reached the summit, they had tasted the mountain's majesty and isolation. Indeed, the little party had achieved what few in their age had even believed possible. They had ventured higher into the atmosphere than anyone in the history of mankind. They had climbed to the putative roof of the world. Humboldt bent over and picked up a few stones; he knew his colleagues in Paris would be wanting to see them.

Reluctant but grateful, the party began their descent. On the way down, they fought through violent hail and a snowstorm that concealed the trail. Then the weather passed just as abruptly, and they could feel themselves reentering the familiar terrestrial realm. At 16,920 feet, they encountered their first sign of life, a lichen growing on a rock above the snow line. At 16,600 feet, they saw a fly, and at 15,000 feet Bonpland captured a yellow butterfly—apparently the first time insects had even been reported on a snowfield. At a few minutes past two, they reached the spot where they had left the mules.

HUMBOLDT WOULD BE PROVEN WRONG in believing Chimborazo to be the highest mountain in the world. Indeed, at 20,600 feet, it isn't even the highest mountain in the Andes, where nine peaks surpass 22,000 feet. But all those mountains, like the Himalayas themselves, wouldn't be surveyed for decades to come.

And in some ways, Humboldt wasn't so far wrong in his estimation of Chimborazo. For one thing, the mountain used to be much taller, before its main crater was blown away in an eruption. For another, there is no higher peak north of Chimborazo in the New World, as its Andean superiors are all found in the southern part of the chain. And the mountain is the world's "tallest" in the sense that its summit is farthest from the center of the earth—because it is near the equator, where, as La Condamine demonstrated, the planet is at its widest. As a result, one weighs less on top of Chimborazo than anywhere else on earth, and one rotates through space faster than anywhere else on the globe. More to the point, no one would climb higher than Humboldt's band did that day for decades to come. Chimborazo itself wouldn't be conquered until 1880, by the renowned British mountaineer Edward Whymper.

Humboldt always took great pride in his altitude record. "All my life," he later wrote, "I prided myself on the fact that of all mortals I had reached the highest point on earth, I mean on the slopes of Chimborazo!" When, three decades after his climb, European surveyors ventured into Tibet to take the first accurate measurements of the mountains there, Humboldt expressed good-natured envy that he couldn't go along. But he didn't begrudge his young, largely English successors their Himalayan adventures. Indeed, Humboldt would manage to secure Prussian state support for three of his own protégés—the brothers Emil, Hermann, and Adolph Schlagintweit—to explore the region. Comforting himself with the Andes' preeminence in the New World, he also felt a sense of pride that his own exploits in South America had helped to inspire Whymper and the others first to conquer Europe's Alps, then to turn their attention to Asia's great mountains. "I have consoled myself over the achievements in the Himalayas in the justified assumption that my labors in America gave the English a first impulse to pay more attention to these snow mountains than had been accorded them over the last century and a half," he wrote.

The birth of mountain climbing is generally fixed at 1760, the year that Swiss scientist Horace Bénédict de Saussure offered a reward to the first person to scale Mont Blanc, at 15,771 feet the tallest peak in Europe. The prize wasn't claimed for twenty-six years, when physician Michel-Gabriel Paccard and his guide, Jacques Balmat, summitted the peak. The Jungfrau was successfully climbed in 1811, and the Finsteraarhorn (the highest peak in Switzerland's Bernese Oberland) in 1829. Dominated by the Swiss, these early climbs were motivated, like Humboldt's own ascent of Chimborazo, primarily by scientific interest. But by the mid-nineteenth century, the dawn of the so-called Golden Age of Alpine Climbing, the torch had passed to the British, who were inspired instead by the pure challenge of the pursuit, and mountaineering evolved into more of a sport and less of a scientific undertaking. Within the next decade virtually all the major Alpine peaks were conquered. (Whymper successfully climbed the Matterhorn in 1865.) It would be the 1920s before the first assault was made on Everest, by the British again. And it wouldn't be for three decades after that, until 1953, that Sir Edmund Hillary and Sherpa Tenzing Norgay, would finally stand atop the true roof of the world. However, in straddling Everest, Hillary was in a very real sense standing on Humboldt's shoulders. For, by being the first to penetrate the world's highest places, Humboldt and his party had, on that June day in 1802, helped set into motion a chain of events that would stretch all the way to Hillary, and beyond.

It was also Humboldt's ascent of Chimborazo, reported months afterward in the French scientific journals, that completed Humboldt's transformation into an international celebrity. An audacious, unprecedented achievement, the climb proved that man could attain altitudes previously thought unsurvivable, and it captured the public imagination in the same way that Hillary's conquest of Everest would a century and a half later. Yet Humboldt's interest in climbing didn't stem so much from a yearning for adventure as from a restless curiosity. And though it was his daring on Chimborazo that attracted the admiration of the world, it was the quality of his mind that held and deepened that veneration over the course of his long life and after his death.

Humboldt's scientific achievements in the Andes are among the most important and far reaching of his entire South American odyssey. In the Alps and on Tenerife, he had already seen how plant life changes with altitude. But by compressing many different climate zones into a tight geographical area, the much taller Andes provided a vertical laboratory ideal for studying the influence on plants of a host of physical factors, including altitude, atmosphere, rainfall, and soil type. Humboldt, who pioneered the study of such factors, named this new science "plant geography."

After his climb of Chimborazo, Humboldt sat down literally at the foot of the mountain and started to draft his *Essay on the Geography of Plants*, the work that established the discipline. Always searching for connections between physical phenomena, Humboldt was the first to systematically study why different plants grew where they did—palms on the tropical coast, broadleaf trees on the cooler hills, conifers in the mountains, and so on. In fact, Humboldt considered the *Essay* so fundamental to his researches in South America that he planned to use it as the introduction to the entire corpus of scientific writings growing out of the journey.

Included in the *Essay*'s pages was a complex fold-out engraving entitled "The Physical Table of the Andes," which, depicting an east-to-west cross-section of the entire range at the latitude of Chimborazo, condensed much of the information that Humboldt had gathered over his months in the mountains— geological formations, temperatures, locations of plant and animal species, crops grown. Thus, in one data-rich, visually

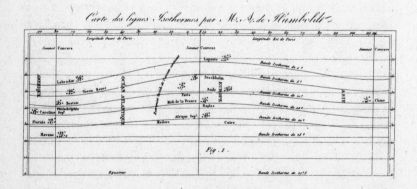

Isothermal lines.

From Humboldt's *Annals de chemie et physique*.

stunning illustration, Humboldt captured for a specific region both the surface complexity and the deep interconnectedness of the natural world. Recognizing the tremendous power of showing such correlations graphically, Humboldt became a pioneer in the visual presentation of scientific data. Thus, in addition to revolutionizing meteorology with the invention of isotherms (lines connecting places of the same average mean temperature), he transformed geography with the introduction of these geographic profiles depicting the relative elevation of neighboring landmasses.

These new techniques proved invaluable to other researchers (the isotherm remains an essential instrument of meteorology to this day), but to Humboldt they were far more than technical tools. It was one thing to be told that vegetation varies predictably with altitude; it was another entirely to be able to see those changes for oneself, via a compelling representation. In all these graphic techniques, the intent was to present apparently discordant data with such visual force that their fundamental interrelations were inescapable. That is, Humboldt's breakthroughs in the visual presentation of data were meant as literal portraits of the underlying unity of nature.

But Humboldt's contributions in the Andes went far beyond new ways of depicting scientific data. During his ascents of the mountains around Quito, he revolutionized the science of geology, especially vulcanology. As he trekked the various peaks, Humboldt was forced to conclude that the harsh, patently volcanic landscape could not be adequately explained by the neptunist theory of his mentor at Freiberg, Abraham Gottlob Werner. From the abundant lava and pumice, it was clear that the Andes had been created by heat, not sedimentation. Moreover, the landforms were obviously recent—and still in the forming, as witnessed by the frequent earthquakes and dozens of active volcanoes—not the product of a one-time, long-ago process of creation. In fact, everywhere he looked, the landscape seemed to support the vulcanist model and to contradict the School of Freiberg, of which, until now, Humboldt had considered himself a member.

In addition, volcanoes could not be the purely local phenomena that the neptunists believed—the result of subterranean coal fires—but must result from far more extensive, interconnected phenomena. Why else would volcanic peaks congregate together as in the Andes? "It seems probable," Humboldt wrote his brother, "that the whole of the more elevated portion of the province [of Quito] is but one huge volcano, of which the peaks of Cotopaxi and Pichincha rise as giant summits whose craters are only vents for the subterranean lava." In *Aspects of Nature*, he elaborated this idea: "These assemblages of volcanoes, whether in rounded groups or in double lines, show in the most conclusive manner that the volcanic agencies do not depend on small or restricted causes, in their proximity to the surface of the earth, but that they are great phenomena of deep-seated origin . . . ," he began. "The subterranean fire breaks forth sometimes through one and sometimes through another of these openings, which it has been customary to regard [incorrectly] as separate and distinct volcanoes. . . . Even the earthquakes which occasion such dreadful ravages in this part of the world afford remarkable proofs of the existence of subterranean communications." He concludes, "All volcanic phenomena are probably the result of a communication either transient or permanent between the interior and exterior of the globe." Isolated coal fires clearly had nothing to do with it.

Not only do volcanoes tend to congregate together, Humboldt was the first person to notice that they also are apt to run in more or less straight lines, as up the coast of South America. In recognition of this phenomenon, he named the majestic row of peaks extending south from Quito the "Avenue of the Volcanoes," a term so perfectly descriptive it is still used today. The reason for this linearity, Humboldt realized, is that volcanoes are positioned along fissures in the earth's crust, through which the lava finds an outlet to the surface.

By the time he left the Andes, Humboldt had more direct experience of active volcanoes than anyone else who had ever lived. Thanks to his perspicacity and scientific rigor, he also understood the volcanic processes more clearly than any of his predecessors. True, there were many more questions to resolve. And Humboldt by no means was correct on every point. For instance, to his dying day he believed, along with his friend the great geologist Leopold von Buch, in the theory of "craters of elevation," which held that mountains were formed by the subterranean upwelling of gases; if the earth's crust were weak at a particular point, the lava forced its way through and created a volcano, but if the lava were unable to break the surface, the crust swelled like an unpopped blister. Though wrong, the idea was one of the most important geological theories of the nineteenth century, defended by Darwin and many others. Similarly, Humboldt, along with other geologists of his day (including the greatest, Charles Lyell) believed that the lava expelled from volcanoes originated deep within the earth's core, whereas today we realize that it is formed much closer to the surface, as the result of powerful tectonic forces.

"The philosophical study of nature endeavors, in the vicissitudes of phenomena, to connect the present with the past," Humboldt wrote in *Aspects of Nature*. And there is no doubt that by the time Humboldt quit the Andes, Hutton's principle of uniformitarianism—the crucial concept that the processes that shaped the earth are still at work and still visible today—offered a much more plausible connection between past and present than Werner's idea that the world's landforms had been sedimented out of the primordial sea in a one-time act of creation. After seeing the great mountains of the New World, Humboldt would never again consider himself a neptunist. Thus, "problems which long perplexed the geologist in his native land in these northern countries, find their solution near the equator." And after his reports to the scientific societies of Europe and his copious publications, neptunism became an increasingly moribund theory. The old paradigm didn't succumb all at once, but over the next two or three decades became so compromised, as it was perpetually modified to accommodate contradictory data, that it gradually ceased to resemble itself. It would be left to Sir Charles Lyell to deliver the coup de grâce, in 1830.

That was the year that Lyell, a Scottish lawyer-turned-geologist, published the first volume of his *Principles of Geology*, which created an immediate sensation after its release. A landmark of scientific publishing, the three-volume work effectively reconciled vulcanism and neptunism, adopting the best parts of each and merging them into a coherent whole that formed the modern basis of the science. In fact, Lyell's ideas went far beyond geology, and, by helping to redefine science as the study of universal, continually operating laws, influenced a host of other scientific disciplines and made a deep impact on nineteenth-century thought.

Lyell acknowledged his obvious debt to his countryman James Hutton—and to Humboldt. In fact, Lyell visited Humboldt in Paris in 1823, when the latter was world famous and the former a twenty-six-year-old student. The older man kindly showed him around the observatory, prompting Lyell to write his father, "There are few heroes who lose so little by being approached as Humboldt." But Lyell clearly wasn't one to be cowed by Humboldt's celebrity. "He was not a little interested in hearing me detail the critiques which our geologists have made on his last geologic work [*Essai géognostique sur le gisement des roches dans les deux continents*]," Lyell also told his father, "a work that would give him a rank in science if he had never published aught besides."

In the *Principles*, Lyell cited Humboldt no fewer than twenty-six times, on topics ranging from earthquakes to the distribution of plants and animals. He also used Humboldt's meteorological data as well as Humboldt's technique of isotherms to illustrate his own ideas on climate. Indeed, Lyell's theory of climate built directly on the other's: Whereas Humboldt had suggested that differences in climate between the northern and southern hemispheres were the result of differences in their relative sizes, Lyell generalized from this to state that all differences in climate were due to differences in physical geography and that past climatic shifts had been caused by changes in the proportion of land and sea. (Today we understand that other factors, such as the composition of the atmosphere, play important roles in climate change as well.) Lyell accepted Humboldt's ideas on several other points, including that volcanoes tended to be positioned along distinct lines on the earth's crust, that they were connected beneath the surface, and that volcanoes and earthquakes were different manifestations of the same underlying phenomenon.

Not even all Lyell's theories were vindicated, though—for example, he believed that volcanoes were somehow dependent on seawater (since so many volcanoes were found in the ocean or along the coast), and, as mentioned above, that volcanoes' heat came from the earth's primordial core. Still, Lyell's work was a fundamental breakthrough in geology, representing the resolution of the science's principal controversy and, as the book's title implied, the first compelling articulation of the principles that would guide the discipline into a new age.

Just as Lyell had drawn from Humboldt, others were greatly influenced by Lyell, including perhaps the greatest name in nineteenth-century science—Charles Darwin. Along with the Bible and Humboldt's *Personal Narrative*, Darwin shipped on board the *Beagle* the *Principles of Geology*, which he found a revelation. As the young Briton made his round-the-world voyage, Lyell's ideas supplied a crucial framework on which to fix his own thoughts concerning geologic and biological change on earth. For one thing, the sort of incremental modifications in species that Darwin would ultimately theorize required an immensely long period over which to operate—clearly longer than the six thousand years that had, according to contemporary interpretations of the Bible, lapsed since the Creation. However, the enormous geologic modifications that Lyell suggested required an even longer period. If the earth were ancient enough to accommodate those changes, it was clearly old enough to accommodate natural selection.

In addition, as Lyell had adapted to geology Isaac Newton's principle of *vera causa*—the idea that a cause must be independently verifiable—Darwin adapted the principle of uniformitarianism to biology: Just as the earth's landforms had not been produced by a unique act of creation, but had been shaped by age-old processes that were still operating and still observable today, the world's species had not resulted from a one-time act of creation, but had evolved—and were still evolving—according to universal principles. Recognizing the congruence with his own work, Lyell became an early supporter of natural selection after *On the Origin of the Species* was published in 1859. For his part, Darwin acknowledged, "The science of geology is enormously indebted to Lyell—more so, I believe, than to any other man who ever lived."

HUMBOLDT'S UNPLANNED TREK across the Andes would have momentous significance for him as well as for the disciplines he studied. Not only would his partial ascent of Chimborazo catapult him to international celebrity, Humboldt's thinking began taking exciting new turns in the Andes, especially concerning the origin of volcanoes and the formation of mountains.

In the coming years, the ideas he had germinated in this barren landscape would revolutionize geology—and, through their influence on such seminal figures as Charles Darwin and Charles Lyell, would exert a profound effect on the course of science itself.

But Humboldt wasn't yet through with the great mountains of South America. Lima, where he hoped to find passage to Mexico, lay some eight hundred miles to the south.

TEN

Cajamarca

AT MORE THAN NINETEEN THOUSAND FEET, THE MAS-
sive, snow-covered, nearly symmetrical cone of Cotopaxi presents
the picture-perfect image of a volcano. The type geologists call
composite (i.e., with layers of lava alternating with those of
ejected stones and ash), the peak towers over the Ecuadorean altiplano about forty miles south of Quito. But on its southern face,
Cotopaxi's symmetrical perfection is marred near the summit by
a thousand-foot-tall protuberance called la Cabeza del Inca, "the
Head of the Inca," which, according to legend, was created by an
explosive eruption in 1533, the year Francisco Pizarro executed
the inca Atahualpa.

One of the tallest active volcanoes in the world, Cotopaxi has
been the most destructive in all of South America. Since 1534, it
has erupted ten times, including five particularly devastating
outbursts in the eighteenth century, in 1742, 1743, 1744, 1766,
and 1769. In the last of these, the shower of ash was said to be
so dense that residents of the indomitable town of Latacunga
(wiped out by the volcano on three separate occasions) were forced
to use lanterns to navigate in the middle of the afternoon.

But when Humboldt visited Cotopaxi, in September 1802, the
volcano was quiet. Stopping at the nearby hacienda La Ciénega,
where the "Humboldt Suite" has been preserved to this day, he
sketched the mountain, which he considered "one of the most
majestic and awe-inspiring views I ever beheld in either hemisphere." Fresh from their triumph on Chimborazo, the travelers
decided to make an attempt on Cotopaxi. It's not known what
route they chose, but at fourteen thousand feet, just below the

snow line, they were forced back by loose, wet ash. Humboldt pronounced the peak unclimbable. And so it would remain for seventy years, till 1872, when German scientist and explorer Wilhelm Reiss and his Colombian companion Angel Escobar gained the rim of the crater by following a still-warm lava flow up the western side. Eight years later, Edward Whymper, hero of the Matterhorn and Chimborazo, climbed Cotopaxi's north face and spent the night on the summit; today his route is the standard path to the top.

From Cotopaxi, the explorers turned south. Humboldt intended to cross the mountains to the Pacific Coast, where he hoped to find a ship to Mexico. From there, he planned to sail to the Philippines, then continue his voyage around the world.

At Riobamba, the town about a hundred miles south of Quito where the disastrous earthquake of 1797 had been centered, the travelers stayed for a few weeks with Carlos Montúfar's brother, who served as the local magistrate. Here Humboldt had the opportunity to inspect some sixteenth-century manuscripts written in a pre-Inca language called Purugayan, which had been translated into Spanish. In the possession of a cultured man named Leandro Zapla, who claimed to be an Indian prince, the manuscripts related the dramatic story of a volcanic eruption and the religious and political significance that the local shamans had ascribed to it.

To Humboldt, the manuscripts were an epiphany, captivating the romantic side of his nature with their glimpse of bygone glories. In fact, they "revived in me," he wrote his brother, "the wish to study the early history of the aborigines of these countries . . . ," whose "languages suffice to give evidence of a higher civilization before the Spanish conquest in 1492. . . . The priests of those ages," for instance, "possessed sufficient knowledge of astronomy to draw a meridian line and to observe the actual moment of the solstice. They changed the lunar into the solar year by the intercalculation of days, and I have in my possession a stone in the form of a heptagon which was found at Bogotá, and which was employed by them in the calculation of their calendar."

This evidence convinced Humboldt that a sophisticated civilization had once flourished in these mountains. But the Spanish, intent on their agenda of salvation and conquest, had never cared to delve into the astounding achievements of these native cultures. In Venezuela and Colombia, Humboldt had suggested that the Indian peoples he saw were the debased remnants of an earlier, more advanced nation. Now in the Andes, he at last found tangible evidence of a great indigenous civilization. In the coming months, as he trekked through Peru, the Inca's ancient homeland, Humboldt would fix his attention on this magnificent culture with the same determination and perspicacity with which he'd examined the continent's botanical, zoological, and geological treasures.

Traveling south-southeast, the party passed over stormy, ten-thousand-foot-high páramos dotted with alpine vegetation and ravaged by violent hailstorms. Humboldt was still eager for his first glimpse of the Pacific, and the muleteers had assured him that from the Inca ruins on the Páramo de Guamini, between Loja and Guancabamba, he would be able to make out "the sea itself which we so much desired to behold." But when the travelers reached the site, they were dismayed to see that thick mist covered the plain, obscuring whatever view of the Pacific there might have been. Instead of the ocean, they "saw only various shaped masses of rock alternately rise like islands above the waving sea of mist, and again disappear, as had been the case in our view from the Peak of Tenerife."

Leaving New Grenada, the explorers entered the fabled country of Peru, whose name is thought to derive from a Native American nation called either Virú or Birú. "After a residence of an entire year on the crest of the chain of the Andes or Antis, between 4 degrees north and 4 degrees south latitude . . . ," Humboldt wrote, "we rejoiced in descending gradually through the milder climate of the Quina-yielding forests of Loja. . . ." *Quina* (the Quechua Indian word for "bark"), also known as quinine, was the product of the renowned cinchona tree.

A genus of evergreen trees and shrubs found in high, humid areas ranging from Bolivia to Colombia and in some regions of Central America, cinchona had been used by native peoples to control malarial and other fevers for centuries. The plant's name commemorates Francisca Henríquez de Ribera, the countess of Chinchón, wife of the Peruvian viceroy, who, according to an apocryphal story, carried some of the bark to Spain after being successfully treated with it in 1638. More likely, it was a Jesuit priest who introduced the substance to Europe; in fact, it was

Cinchona condaminea.
From *Plantes équinoxiales* by Humboldt and Bonpland.

once known as Jesuit's bark. But La Condamine repeated the
Chinchón story to Linnaeus, who named the plant in the countess's honor. (When Humboldt and Bonpland discovered a new
species of cinchona, they returned the Frenchman's courtesy,
calling it *Chinchona condaminea*. "This beautiful tree is adorned
with leaves five inches long and two broad . . . ," Humboldt wrote,
"and as it spreads its upper branches, the foliage glistens from a
distance with a peculiar reddish tint when moved by the wind.")

Even after the bark was introduced to Europe, its use spread
slowly among Protestant nations, who were suspicious of its
Spanish, Catholic provenance. In 1820, French chemists J. B.
Caventou and P. J. Pelletier isolated the active ingredient, and
over the next 130 years quinine is credited with saving more
than a million lives, before being replaced by synthesized forms
in the 1950s.

Continuing southward, the travelers were forced to cross the

Rio Guancabamba, a tributary of the Amazon, no fewer than twenty-seven times. The Guancabamba was only 120 to 150 feet wide—a creek compared to the miles-wide rivers they had encountered in the rain forest. But it was a "torrent . . . ," Humboldt said, "so strong and rapid that in fording it, our heavily laden mules were often in danger of being swept away by the flood." Worse, the animals "carried our manuscripts, our dried plants, and all that we had been collecting for a year past [since they had divided their collections in Cuba]. Under such circumstances, one watched from the other side of the stream with very anxious suspense until the long train of eighteen or twenty beasts of burden had passed in safety."

Farther downstream, Humboldt discovered that the Guancabamba's swift current was harnessed to deliver mail all the way to the Pacific Coast, via *el correo que nada*, or "the mail that swims." In this unique postal system, a young Indian man would carefully tie outgoing letters into a large cotton handkerchief, which he would wind around his head like a turban. Then he would jump into the river (crocodile-free here owing to the strong current) and ride it downstream for two days, occasionally gripping a balsa log to rest and climbing ashore wherever necessary to avoid the many waterfalls. At night he would stop at one of the huts along the way, where he would be provided with a meal and a place to sleep. Though *el correo que nada* may not conform to conventional notions of postal efficiency, according to the governor of the province, the mail was rarely lost or damaged, and after returning to Paris, Humboldt himself received letters that had been posted this way. Indeed, the river was used to transport more than correspondence. Humboldt also reported seeing groups of thirty to forty Indians—men, women, and children—bobbing sociably downstream as the current swept them to their destination.

As the travelers approached the Amazon Basin, Humboldt was "cheered by the aspect of a beautiful, and occasionally very luxuriant vegetation," including a new species of bougainvillea with bright-red bracts, the finest orange trees he had ever seen, and a remarkable tree named *Porlieria hygrometrica*, which would predict a coming storm by closing its fine leaflets. ("It very rarely deceived us," Humboldt testified.)

At the town of Chayma, the party boarded balsa rafts and

drifted down the Río Chayma to its junction with the Amazon (called the Marañon here). In Venezuela, tensions between Spain and Portugal had prevented the explorers from traveling as far as the Amazon itself, but now they lingered seventeen days in the great river's headwaters, adding to their herbarium and investigating its wild course. Here in the mountains, the Amazon was a rock-filled, rapid-choked torrent that bore little resemblance to its meandering lower reaches. At the Pongo (Rapids) de Manseritche, where the river squeezed through a mountain ravine, Humboldt measured it at less than 160 feet across. "At some points the overhanging rocks and the canopy of foliage forbid more than a very feeble light to penetrate," he described, "and there all the drift-wood, consisting of a countless number of trunks of trees, is broken and dashed in pieces. . . . " In places, the untamed, unpredictable river was known to rise more than twenty-five feet in the course of a single day, casting huge boulders before it and incessantly rearranging its channel.

As they left the Amazon Valley, the explorers began the steep ascent up the eastern slope of the Andes. One evening at sundown they arrived at the famous silver mines of Chota, where the mountains' silhouette was broken by numerous towers and pyramids, creating an effect the mine owner likened to a *castillo encantado*, or "enchanted castle." But for the workers, there was nothing magical about the pits: Crude sheds to house the miners had been thrown up wherever a flat surface presented itself, and the men "had to carry down the ore in baskets, by very steep and dangerous paths," to the smelter below.

Nearby was Micuipampa, a mountain town of three or four thousand inhabitants. Though situated nearly on the equator, the place occupied a barren desert some eleven thousand feet above sea level, and it was so cold at night that pitchers of water froze inside the houses for much of the year. The only crops that could survive were kale and salad greens; every other necessity had to be laboriously carted up from the surrounding valleys. With its boomtown atmosphere of sudden wealth and widespread boredom, gambling was rampant at Micuipampa, reminding Humboldt of "the soldier of Pizarro's troop who, after the pillage of Cuzco [the Inca capital], complained that he had lost at one night at play 'a great piece of the sun,'" that is, a large gold plate looted from the Temple of the Sun.

The path leading from Micuipampa to the ancient city of Cajamarca ("Frost Town" in the Quechua language) was nearly impassible even for mules. For almost six hours Humboldt's party struggled over a succession of barren páramos, at elevations of ten thousand to eleven thousand feet, where they were stung by needles of hail driven before a howling wind. Yet even in these brutal conditions, Humboldt would stop periodically and unload his instruments. It was in this godforsaken terrain that he would make one of his greatest discoveries in South America—the location of the earth's magnetic equator.

The phenomenon of magnetism had been known for centuries (the term itself comes from *Magnesia*, the name of an area of Asia Minor where naturally magnetic iron ore had been mined for centuries). The magnetic compass had been invented in ancient China and used in Europe for navigation since the Middle Ages, and it had long been observed that the needle didn't point precisely toward geographic north. However, by the turn of the nineteenth century, magnetism was still incompletely understood. Today we know that the phenomenon is a form of electricity, and that spinning electrons generate a minute magnetic field. In certain substances, more electrons spin in a particular direction, and these particles tend to line up and reinforce each other. If such a substance is placed within an outside magnetic field, its electrons will align with that field, creating a magnet. Every magnet has a so-called north and a south pole, at the two end points.

At the time of Humboldt's expedition, geomagnetism—the study of the earth's magnetic properties—was a young but promising field of research. After 1600, when English physician and physicist William Gilbert published *De Magnete*, scientists realized that the earth itself was a huge magnet. But it was not known whether the intensity of the planet's magnetic field varied in any systematic way from place to place. In 1769, French scientist Jacques Mallet-Favre made the first attempt to measure geomagnetism at different latitudes, but his instruments weren't sufficiently sensitive, leading to the erroneous conclusion that no such variation existed. Skeptical of Mallet's results, his countryman Jean Borda repeated the experiment on his 1776 journey to West Africa, but his instruments proved deficient as well. After his return to Paris, Borda invented an improved magnetometer, which he presented to Humboldt to take to South America.

Immediately after leaving Europe, Humboldt began taking regular measurements of the earth's magnetic field, using a kind of vertical compass called a dip needle. During the course of his journey, he made 124 magnetic observations ranging over 115 degrees of longitude and 64 degrees of latitude, also recording in every case the geodetic coordinates, height above sea level, and distance from any mountains or prominent rocks that might influence the results. As he traveled south toward the geographic equator, he noted with increasing excitement a steady decrease in the earth's magnetic field.

Even after Humboldt crossed the geographic equator in Ecuador, the magnetic dip—the angle with which the magnetic needle was attracted downward, toward the earth, continued to decline. But now, as the party traversed the Cajamarca Plateau, he finally registered a dip of zero: He had located the magnetic equator—the line where the vertical component of earth's magnetic force is zero—at 7 degrees, 27 minutes south latitude and 81 degrees, 8 minutes west longitude.

To gauge the intensity of the earth's magnetic force, Humboldt invented a unit of measurement that would become the standard for fifty years, until supplanted by the *gauss*, named in honor of Carl Gauss, the great German mathematician and protégé of Humboldt. In 1804, Humboldt published the results of his measurements—a map showing five zones of equal geomagnetic strength in the Northern Hemisphere and one in the Southern. He also invented *isogonics* (lines connecting points with equal magnetic variation from true north), *isoclines* (lines connecting points with equal magnetic dip), and *isodynamics* (lines connecting points with equal magnetic strength), all of which are still in standard use in the field of geomagnetism.

Humboldt's discovery of the magnetic equator on the desolate Cajamarca Plateau was a landmark achievement, not only proving that the earth's magnetic field varies predictably with latitude but pinpointing the exact location where there is no vertical dip at all. "I have considered the law of the decrease of the magnetic forces from the pole to the equator as the most important result of my American journey," Humboldt wrote. But his contributions to the field did not end even after his return to Europe.

From May 1806 to June 1807, in a rented potting shed outside Berlin, he made more than six thousand measurements demonstrating the daily fluctuations in a compass's variation from true north. He also discovered the phenomenon of the magnetic storm (another term he coined), a disturbance in the earth's magnetic field caused by the sudden release of ultraviolet radiation from the sun. Indeed, Humboldt would continue to study and advance the field of geomagnetism for the rest of his life.

FROM THE LAST OF THE BARREN PÁRAMOS, Humboldt "looked down with increased pleasure on the fertile valley of Cajamarca," which, despite its altitude of nearly nine thousand feet, was graced with wheat fields and orchards and etched by avenues of willows and flowering trees. Occupying a high plain reminiscent of the one around Bogotá, the ancient city of Cajamarca, with a population of seven or eight thousand, enjoyed good soil, ample water, and, thanks to the protection of the surrounding mountains, a relatively mild climate. In the distance, Humboldt could see puffs of steam drifting up from los Baños del Inca (the Baths of the Emperor), the hot-water springs where the Inca Atahualpa had built a palace before the coming of the conquistadors. According to local legend, his golden throne had been cast into the hot springs to keep it out of the Spaniards' hands, but after centuries of searching, it had never been discovered. Though their capital was at Cuzco, more than six hundred miles to the southeast, it was here at Cajamarca that the mighty Inca Empire had fallen to the Europeans. Humboldt was captivated by their story.

About A.D. 1400, the Inca had begun a period of rapid expansion, and over the next century, through a deft mix of diplomacy and conquest, they managed to subsume nearly a hundred distinct peoples into their realm, which they called Tahuantinsuyu, or "the Four Quarters of the World." At its high-water mark, after Inca Huayna Capac's conquest of the neighboring Kingdom of Quito about 1515, the empire stretched from Ecuador to Chile, taking in snow-covered peaks, coastal desert, and dense rain forest.

Presiding over this huge territory and everything in it was the all-powerful inca, or emperor. Believed to be descended from the sun itself, the emperor lived in almost unbelievable luxury.

Adjoining his palace at Cuzco was a magnificent walled garden with fantastic plants shaped from silver and gold, according to sixteenth-century chronicler Garcilaso de la Vega, "with their leaves, flowers, and fruit; some just beginning to sprout, others half grown, others having reached maturity. They made fields of maize, with their leaves, *mazorcas* [ears of corn], canes, roots, and flowers, all exactly imitated. The beard of the *mazorca* was gold, and all the rest of silver. . . . They did the same thing with other plants, making the flowers, or any other part that became yellow, of gold, and the rest of silver." Inside the capital's Temple of the Sun, "all four of the walls . . . were covered, from roof to floor, with plates and slabs of gold. . . . On either side of the image of the Sun were the bodies of the dead kings, arranged according to priority as Children of the Sun, and embalmed so as to appear as if they were still alive. They were seated on chairs of gold, placed upon the golden slabs on which they had been used to sit."

To Humboldt, the emperor was a tyrant, and it is true that Inca society was rigidly structured, with laws to govern every aspect of public and private behavior. Unquestioning obedience was expected from all quarters, from the various levels of nobility down through the great mass of common subjects, and transgressions were severely punished. Garcilaso reproduced the inca Pachacuti's uncompromising ethical code in his *Royal Commentaries of the Inca:*

> When subjects, captain, and curacas [lords] cordially obey the king, then the kingdom enjoys perfect peace and quiet.
>
> Envy is a worm that gnaws and consumes the entrails of the envious.
>
> He that envies another, injures himself.
>
> He that kills another without authority of just cause condemns himself to death.
>
> It is very just that he who is a thief should be put to death.
>
> Adulterers, who destroy the peace and happiness of others, ought to be declared thieves, and condemned to death without mercy.
>
> The noble and generous man is known by the patience he shows in adversity.
>
> Judges who secretly receive gifts from suitors ought to be looked upon as thieves and punished with death as such.

The physician herbalist who is ignorant of the virtues of herbs or who, knowing the uses of some, has not attained to a knowledge of all, understands little or nothing. He ought to work until he knows all, as well the useful as the injurious plants, in order to deserve the name to which he pretends.

He who attempts to count the stars, not even knowing how to count the marks and knots of the quipus, ought to be held in derision.

Drunkenness, anger, and madness go together; but the first two are voluntary and to be removed, whereas the last is perpetual.

Every subject, of whatever station, was guaranteed food, shelter, and the other necessities of life, but by the same token private property was severely restricted. Tribute to the inca was carefully delineated, as recorded by the sixteenth-century Spanish monk Blas Valera:

1. Tribute was to consist solely of time, or skill as a workman or artisan. . . .
2. Except for work as a husbandman or as a soldier, for which any *puric* might be called upon, no man was compelled to work at any craft save his own.
3. If tribute took the form of merchandise produced by the payer's labor, only the produce of his own region could be demanded of him, it being held to be unjust to demand from him fruits that his own land did not yield.
4. Every craftsman who labored in the service of the inca or of his curaca must be provided with all the raw materials for his labor, so that his contribution consisted only of his time, work, or dexterity. His employment in this way was not to be more than two or three months in the year.
5. A craftsman was to be supplied with food, clothes, and medicine at need while he was working, and if his wife and children were aiding him, they were to be supplied with those things also. . . .

As in many other cultures of the time, the secular and spiritual realms were inextricably intertwined, and religion played a central role in daily life. The chief deity in the Inca pantheon was

Viracocha, who had created mankind in his own image on the shore of Lake Titicaca, then walked westward over the surface of the Pacific Ocean. Serving Viracocha were a multitude of lesser gods, including planets, stars, thunder and lightning, some rivers and mountains (such as Cotopaxi), and other natural phenomena. Preeminent among these secondary deities was Inti, the sun god, ancestor of the Inca rulers, who was presented with tremendous works of gold, which was called "the sweat of the sun." (Inti's consort, the moon, was honored with silver, or "the tears of the moon.")

Taking pity at the benighted lives led by men, the sun had sent one of his sons, Marco Capoc, and one of his daughters, Mama Oello Huaco, to civilize mankind and to teach them useful arts such as agriculture and weaving. These divine intermediaries had carried with them a great golden wedge, which, it was prophesied, would sink deep into the ground at the spot where they were to establish their city. The miracle occurred at Cuzco, and the Inca capital was duly founded there.

The Inca had no writing; though they did have the quipu, a system of knotted, color-coded cords used to count and perhaps to send simple nonnumerical messages. They also had no wheeled vehicles, no iron, and no money. Yet they skillfully administered a vast empire and conducted an incredibly ambitious program of public works, including land terraces, irrigation systems, roads, palaces, fortresses, and temples. All these projects, some requiring decades of work, were accomplished through an arrangement called *mita* (Quechua for "turn"), under which each community throughout the empire was required to send a quota of young people to work for the state for a limited period. Far from a time of drudgery, some historians suggest, the mita (which survived in modified form through the colonial period) was for many of these participants a high point of their lives, giving them their only taste of the wider world outside their own village.

The Inca were able to undertake such enormous engineering works (and support activities such as gold making, fine weaving, sculpting, painting, oral poetry, and music) because of an efficient agricultural system, which produced huge surpluses of food and freed labor for other purposes. The most important crop was corn, followed by potatoes and other roots, beans, and a wide variety of other vegetables, some of which were dried for future use. The Inca apparently made a scientific study of agriculture, and as nations were brought into the empire, the newcomers' farming methods and other technologies were examined and the most promising methods adapted throughout the realm. The Inca are even believed to have had agricultural research stations, where new crops were tested and improved.

Though the Inca never discovered the arch, they were able to build palaces and other large structures from great blocks of stone. It's not known how workers managed to transport such tremendous weights and set them precisely in place, but Inca stonework was so finely wrought that it required no mortar, and many stunning examples remain today. However, of all the Inca's engineering feats, it was the network of roads constructed for governmental and military use that particularly struck Humboldt (and many other European observers). "The impressions produced on the mind by the natural characteristics of these wildernesses of the Cordilleras," Humboldt found, "are heightened, in a remarkable and unexpected manner, from its being in those very regions that we still see admirable remains of the gigantic work, the artificial road of the Inca, which formed a line of communications through all the provinces of the Empire. . . ."

With a total of perhaps ten thousand miles of paving, this vast highway system consisted of two main north-south arteries, one along the coast and another along the eastern flank of the Andes, each about two thousand miles long and linked by numerous east-west connecting routes. Bridges and aqueducts were also constructed where needed, and every several miles could be found a tambo, or way station, some of which incorporated natural hot springs. Masterfully engineered, the roads were as good, Humboldt decided, as any of the Roman roads he had seen in Italy or France. And just as the Roman roads had all led to their capital, so the Inca's all led to the main square in Cuzco.

It is thought that, using these highways, relays of runners could cover the fifteen hundred miles between the capital and Quito in just five days. Intended solely for pedestrian traffic and llamas, the roads were occasionally broken by long flights of vertical steps, which would later prove a serious impediment to the Spaniards' horses, before they adopted the more surefooted mules.

Francisco Pizarro, the conqueror of Peru, wrote of these thoroughfares, "In the whole of Christendom there are nowhere such fine roads as those which we here admire." Yet by the time of Humboldt's arrival, the roads, like the other monuments to the great Inca culture, were in a state of ruin, having been looted by the Spanish for materials with which to construct their own cities, fortresses, and places of worship.

Some of these scavenged stones had been used to construct the fine churches in Cajamarca, as well as the state prison and the cabildo (town hall), which had been built on the hill that had been the site of the inca's palace. Humboldt was shown steps cut into the rock known as the Inca's Footbath, where the ruler's feet had supposedly been washed, accompanied, as Humboldt phrased it, "by some inconvenient usages of court etiquette." That is, the inca, owing to his majesty, would spit not on the ground, but into the hands of one of the ladies in waiting. In the palace's principal building, Humboldt viewed the room where one of the great perfidies of history had been committed, when the inca Atahualpa had been held prisoner by Francisco Pizarro.

Pizarro had begun life about 1476 in Trujillo, Spain, as the illegitimate son of an aristocrat and a woman of more humble station. He was apparently abandoned by both parents and as a boy was left to scratch out a living as a swineherd. (According to one, mythologized version of his life, he was actually suckled by a sow.) As a young man, Pizarro had fled to Seville, then traveled to Colombia in 1515 with the conquistador Alonso de Ojeda. In 1513, he was in Panama with Vasco Nuñez de Balboa at the Spaniards' discovery of the Pacific Ocean. Inspired by Cortés's conquest of Mexico in 1521 and spurred on by stories of a golden land beyond the equator, Pizarro made three campaigns in South America.

The first, from 1524 to 1525, ventured no farther than the San Juan River in Colombia and produced nothing but debts. But the second, from 1526 to 1527, made contact with the Inca at Tumbez, on the Peruvian coast, and later reached the mouth of the Santa River. Encouraged, Pizarro's partners elected him to return to Spain, where he impressed Charles I with living llamas, samples of gold, and stories of fabulous wealth. The king agreed to advance the funds for a more extensive campaign, and in 1532, Pizarro sailed again. Landing his men at Tumbez, he marched south along the coastal desert plain, then turned inland toward the heart of the empire, where the inca Atahualpa was waiting.

Atahualpa's father, Huanya Capac, had captured the Kingdom of Quito, which had rivaled the Inca Empire itself in wealth and power. Huanya Capac had taken into his harem the daughter of Duchicela, the last king of Quito, and she had given birth to Atahualpa. According to Inca law, the heir to the throne was to be the firstborn male of the *coya*, the Inca's sister and principal wife. However, Atahualpa became his father's favorite, and Huanya Capoc bequeathed to him the Kingdom of Quito, his mother's homeland, while Huáscar, Atahualpa's half-brother and the rightful heir, was to rule the remainder of the empire.

Time would prove that an unwise decision. Though the two kingdoms lived in peace for several years, tensions escalated between Huáscar, who some sources describe as the more mild mannered of the two, and Atahualpa, who is sometimes described as the more aggressive and warlike. Eventually their hostility erupted into civil war, and in the spring of 1532, just months before Pizarro's second landing, Atahualpa routed Huáscar's army, first at Ambato, in the shadow of Chimborazo, then again on the Plains of Quipaypan, outside Cuzco. Imprisoning his half-brother, Atahualpa claimed both the northern and southern kingdoms for himself. However, the civil war had left the empire in disarray and ill prepared to meet the threat that was, literally, just over the horizon.

Marching down the coast, Pizarro heard of a great Inca city in the mountains, and in November he and his men climbed the ridge of the Andes and had their first, awed sight of Cajamarca. The metropolis "looked like a very beautiful city," wrote one of the conquistadors. "So many tents were visible that we were truly filled with apprehension. . . . It filled us Spaniards with fear and confusion. But it was not appropriate to show any fear, far less to turn back. For had they sensed any weakness in us, the very Indians we were bringing with us would have killed us. So, with a show of good spirit, and after having thoroughly observed the town and tents, we descended into the valley and entered the town of Cajamarca."

With fewer than two hundred men against thousands of Inca warriors, Pizarro decided to risk everything on a stealthy, preemptive blow. Drawing from the reservoir of daring and treachery that would characterize his entire career, he ordered his men to slaughter thousands of the stunned, unarmed Indians in the city's main square. Then he seized Atahualpa.

A prisoner in his own palace, the inca soon recognized the Spaniards' inordinate love of gold and presented his captors with a proposition. The story is related in William H. Prescott's nineteenth-century classic, *The Conquest of Peru,* which, though surpassed by more recent historical research, still stands as a masterpiece of vivid storytelling. Atahualpa, writes Prescott, "one day told Pizarro, that, if he would set him free, he would engage to cover the floor of the apartment on which they stood with gold. Those present listened with an incredulous smile; and, as the Inca received no answer, he said, with some emphasis, that 'he would not merely cover the floor, but would fill the room with gold as high as he could reach'; and, standing on tiptoe, he stretched out his hand against the wall. All stared with amazement; while they regarded it as the insane boast of a man too eager to procure his liberty to weigh the meaning of his words." Though skeptical, Pizarro realized that he had nothing to lose and hastily accepted the inca's offer. Meanwhile, even as he negotiated with Pizarro for his own release, Atahualpa, fearing that Huáscar would take advantage of the situation and bribe his way out of prison, ordered his half-brother drowned in the river of Andamarca. Prophetically, Huáscar declared "with his dying breath that the white men would avenge him, and that his rival would not long survive him."

In the coming weeks, huge quantities of gold were stripped from the Temple of the Sun in Cuzco, as well as other palaces, and carried by the wagonload to Cajamarca. Then, while the ransom was still being collected, Spanish reinforcements arrived on the coast and joined their comrades at Cajamarca. When a comet appeared over the city at about the same time, the inca, Prescott tells us, "gazed on it with fixed attention for some minutes, and then exclaimed, with a dejected air, that a similar sign had been seen in the skies a short time before the death of his father, Huayna Capac. From this day a sadness seemed to take possession of him, as he looked with doubt and undefined dread to the future."

Atahualpa had cause for dejection. With his ranks swelled

with reinforcements, Pizarro chose to renew his campaign of conquest. Instead of waiting in Cajamarca for the promised ransom, he decided to go on the march and seize the gold himself, before the inca could raise an army against him. Thus, having received a tremendous ransom in gold, the conquistador failed to release Atahualpa as promised and instead had him tried on a dozen charges, ranging from the assassination of Huáscar to bigamy to idolatry. After a sham trial, the inca was condemned to be burned alive in the great square at Cuzco.

When the sentence was conveyed to Atahualpa, Prescott writes, "the overwhelming conviction of it unmanned him, and he exclaimed, with tears in his eyes, 'What have I done, or my children, that I should meet such a fate? And from your hands, too,' said he, addressing Pizarro; 'you, who have met with friendship and kindness from my people, with whom I have shared my treasures, who have received nothing but benefits from my hands!' " Then "Atahualpa recovered his habitual self-possession and from that moment submitted himself to his fate with the courage of a warrior."

Prescott continues, "The doom of the Inca was proclaimed by sound of trumpets in the great square of Caxamalca; and two hours after sunset, the Spanish soldiery assembled by torch-light in the plaza to witness the execution of the sentence. It was on the twenty-ninth of August, 1533. Atahualpa was led out chained hand and foot. . . ." As he was bound to the stake, a Dominican friar, Father Valverde, "holding up the cross, besought him to embrace it and be baptized," promising that if he did, his execution would be by means of the less painful garroting. Atahualpa agreed, and at the moment of his death forsook the religion of his ancestors for Christianity.

Pizarro permitted Huayna Capac's eldest remaining legitimate son, Tupac Hualpa, to be crowned inca, with the intent of ruling through him, but the young man complicated things by dying soon thereafter. However, as the Spanish marched from Cajamarca toward Cuzco, they were met by Manco Capoc, legitimate half-brother of Huáscar, who now pressed his claim as rightful heir to the throne. "Pizarro listened to his application with singular contentment," Prescott writes. "He received the young man, therefore, with great cordiality, and did not hesitate to assure him that he had been sent into the country by his master, the

Castilian sovereign, in order to vindicate the claims of Huáscar to the crown, and to punish the usurpation of his rival."

Manco Capoc was crowned in 1534, but did not prove the pliable puppet that Pizarro intended. He ultimately escaped in 1536 and, at the head of a huge army, laid siege to Cuzco. The months-long battle terrified the Spanish and virtually destroyed the city. However, the defenders managed to hold out till spring, when the Indians were forced to disband to plant their crops or face certain starvation. Fleeing to the mountains, Manco Capoc conducted a brilliant guerrilla war for another eight years, before finally being slain in 1544 by Spaniards to whom he had offered refuge after they themselves had rebelled against Pizarro. Thus, like Atahualpa before him, Manco Capoc, the last inca, fell victim to betrayal at the hands of white men to whom he had shown mercy.

Meanwhile, Pizarro took as a mistress Atahualpa's fifteen-year-old daughter Quispe Cusi, also known by her Christian name Inés, with whom he had a daughter, Francisca, and a son, Gonzálo. Later, by the Princess Anas, a wife of Atahualpa also known as Angelina, he had two more sons, Francisco and Juan. In 1534, Pizarro proclaimed the Spanish city of Cuzco on the ruins of the Inca capital, and the following year founded two new cities on the coast, Trujillo, named after his own birthplace in Spain, and Lima, his new capital, whose location he'd chosen for its proximity to the sea and its good supply of fresh water. Having grown fabulously wealthy from his conquests, Pizarro was created a marqués by King Charles in 1539. But his riches and his title could not protect him from the enemies he had made among his own countrymen. Having relied on deceit and cruelty so often himself, he was assassinated on June 26, 1541, by Spanish rivals.

The fatal confrontation is described by William Prescott. As the assassins stormed through his palace in Lima crying, "Death to the tyrant!" Pizarro struggled to buckle his armor. When the intruders broke into his chamber, where he had been having his midday dinner, the sixty-five-year-old marqués "threw himself on his invaders, like a lion roused in his lair, and dealt his blows with as much rapidity and force, as if age had no power to stiffen his limbs. 'What ho!' he cried, 'traitors! Have you come to kill me in my own house?' The conspirators drew back for a moment, as two of their body fell under Pizarro's sword; but they quickly rallied. . . ." One of the assassins, Rada, called out, " 'Why are we so long about it? Down with the tyrant!' and taking one of his companions, Narváez, in his arms, he thrust him against the marqués. Pizarro, instantly grappling with his opponent, ran him through with his sword. But at that moment he received a wound in the throat, and reeling, he sank on the floor, while the swords of Rada and several of the conspirators were plunged into his body. 'Jesú!' exclaimed the dying man, and, tracing a cross with his finger on the bloody floor, he bent down his head to kiss it, when a stroke, more friendly than the rest, put an end to his existence."

HUMBOLDT FOUND CAJAMARCA, the scene of Pizarro's first great triumph, evocative and poignant. His guide, a "pleasing and friendly youth of seventeen" named Astorpilco, claimed to be a descendant of the emperor, and he and his family were living "among the melancholy ruins of ancient departed splendor . . . in great poverty and privation; but patient and uncomplaining." Nevertheless, the young man "had filled his imagination with images of buried splendor and golden treasures hidden beneath the masses of rubbish upon which we trod." He told Humboldt how one of his forefathers had blindfolded his wife and taken her through the rocky labyrinths to a fabulous underground garden of the inca, which had "skillfully and elaborately imitated, and formed of the purest gold, artificial trees, with leaves and fruit, and birds sitting on the branches," similar to those found at the Temple of the Sun in Cuzco, as well as the golden sedan chair of Atahualpa himself. "The man commanded his wife not to touch any of these enchanted riches, because the long foretold period of the restoration of the empire had not yet arrived, and that whoever should attempt, before that time, to appropriate aught of them would die that very night." Astorpilco assured Humboldt that beneath their very feet, just a little to the right of where they were standing, was buried a tree made of solid gold, which had once spread its branches over the inca's throne.

Humboldt was impressed "deeply but painfully" by the force of Astorpilco's faith, "for it seemed as if these illusive and

baseless visions were cherished as consolations in present sufferings." He asked him, "Since you and your parents believe so
firmly in the existence of this garden, are not you sometimes
tempted in your necessities to dig in search of treasures so close
at hand?" But "the boy's answer was so simple, and expressed so
fully the quiet resignation characteristic of the aboriginal inhabitants of the country, that [he] noted it . . . in [his] journal. 'Such
a desire does not come to us; father says it would be sinful. If we
had the golden branches, with all their golden fruits, our white
neighbors would hate and injure us. We have a small field and
good wheat' " with which they must content themselves, though
they believed themselves descendants of one of the greatest
rulers in the history of the world.

Humboldt's party stayed in Cajamarca longer than they had
planned, due to a lack of reliable guides and a shortage of mules
to carry their growing collections. After five days, they headed
southwest toward the coast and climbed down more than six
thousand feet on a precipitous switchback trail, till they reached
once again the Valley of the Magdalena, one of the deepest in all
the Andes. It was a desolate area, where even a "few wretched
huts . . . were called an Indian village."

As the travelers struggled up from the valley over the Pass of
Guangamarca, Humboldt's hopes of seeing the Pacific rose again.
"As we toiled up the mighty mountainside," he wrote, "with our
expectations continually on the stretch, our guides, who were not
perfectly acquainted with the road, repeatedly promised us that
at the end of the hour's march which was nearly concluded, our
hopes would be realized. The stratum of mist which enveloped
us appeared occasionally to be about to disperse, but at such
moments our field of view was again restricted by intervening
heights."

Finally, as they gained the top of the ridge, a sharp wind rose
in the west and dispersed the fog. A cobalt sky appeared, and the
whole of the western slope of the Andes lay before them "in astonishing apparent proximity. We now saw for the first time the
Pacific Ocean itself; and we saw it clearly; forming along the line
of the shore a large mass from which the light shone reflected,
and rising in its immensity to the well-defined, no longer merely
conjectured, horizon. The joy it inspired, and which was vividly

shared by my companions Bonpland and Carlos Montúfar, made
us forget to open the barometer" to measure the altitude.

For Humboldt, not only was the sight of the ocean the crowning moment of a long overland journey, it was the fulfillment of a
childhood dream. "The view of the Pacific was peculiarly impressive to one who like myself owed a part of the formation of his
mind and character, and many of the directions which his wishes
had assumed, to intercourse with one of the companions of Cook.
My schemes of travel were early made known, in their leading
outlines at least, to George Forster, when I enjoyed the advantage of making my first visit to England under his guidance . . . ,"
Humboldt wrote. "Forster's charming descriptions of Otaheite
[Tahiti] had awakened throughout Northern Europe a general
interest (mixed, I might almost say, with romantic longings) for
the Islands of the Pacific, which had at that time been seen by
very few Europeans." In addition, Humboldt saw in this view of
the great ocean a glimpse into his own future, as he planned to
continue his expedition to the Philippines before sailing on to
Europe around the tip of Africa.

From the western Cordillera, the travelers descended to Trujillo, whence they trekked 250 miles southward through the coastal
desert. On October 23, they entered Lima. Pizarro had christened his new capital la Ciudad de los Reyes, "the City of Kings,"
in honor of the Magi, around whose feast day it was founded, but
by the end of the sixteenth century the city became known simply as Lima, an adaptation of the Indian name for the place. In
the seventeenth and eighteenth centuries, the city lived up to its
earlier, royal appellation, when as capital of the Viceroyalty of
Peru and thus the de facto capital of Spain's New World empire,
the place had been synonymous with wealth and luxury. As a cultural center only Bogotá could compete in those palmy days, and
in physical splendor and political might it was approached only
by Mexico City, the capital of New Spain.

However, the intervening years had not been kind to Lima.
Laid out on the typical Spanish grid pattern, the city in 1800
had a population of about fifty thousand, of whom about thirty
percent were slaves and another fifteen percent were Indians.
The heart of the metropolis was the wide and graceful Plaza
Mayor, where a great brass fountain dominated the center and

the sumptuous cathedral and elegant palaces of the viceroy and the archbishop stretched along the perimeter. But the city had been destroyed by three earthquakes, in 1630, 1687, and 1746, had seen a disastrous Indian revolt in 1780, and now was clearly in a long period of decline.

In truth, the city had always been hampered by its distance from Spain. Situated on the far side of South America, it could be approached by water only after a long and perilous journey through the Straits of Magellan at the southern tip of the continent. Havana, Mexico City, and Buenos Aires were far better situated for communication with the mother country, and by 1800 the latter port had surpassed Lima as the busiest in South America. In 1776, it had suffered another setback, when silver-rich Upper Peru (present-day Bolivia) had been incorporated into the new Viceroyalty of Río de La Plata, depriving Peru of much-needed revenue. With its agriculture struggling, its roads in notoriously poor condition, its continuing Indian troubles, the loss of Bolivia, and steadily decreasing production from the rest of its silver mines, Peru and its capital were suffering. No wonder Herman Melville, some years later, called Lima "the saddest city on earth."

Cultural life had declined along with the economy. By the time of Humboldt's visit, there was only one theater still open, and the bullring and cockpits provided the dominant pastimes. Yet, despite the failing economy, the Church remained powerful and rich, and ties to the mother country were unusually strong. (Though Peru would win its independence in 1821, Lima's port, Callao, would stubbornly hold out against the revolutionaries until 1826.) With government sinecures the principal source of income for the city's elite, political intrigue and sheer idleness had elevated character assassination to a high art. Every visitor to the city seemed to remark on the prevalence of slander and backbiting at the time.

Humboldt, who had seen many of Spain's New World capitals, wasn't favorably disposed toward the city either. "Lima has declined greatly . . . ," he wrote in a letter. "Here I never saw well-furnished homes or well-dressed women. . . . At night it is impossible to travel the streets by carriage, obstructed as they are by mongrel dogs and donkey carcasses. . . . It would seem to be more distant from the rest of Peru than London is. A cold egotism is found here, so that no one cares about anybody else's sufferings. . . . We expect to leave in about five or six weeks, and meanwhile we wait patiently [for a ship], battling against mosquitoes."

Before he departed, Humboldt was keen to observe the transit of Mercury across the sun predicted for November 9, 1802. In astronomy, a *transit* is the crossing of a smaller body in front of a larger one. Since Earth is the third planet in the solar system, only the two inner planets, Mercury and Venus, cross the sun from our perspective, appearing as a small, dark dot on the larger, brighter body (Mercury appears about 1/200 the size of the sun; Venus is somewhat larger). On average, Mercury makes a transit every eight years, while it can be more than a century between transits of Venus. The first complete transit of Mercury had been observed in 1677 by the great English astronomer Edmond Halley, and in the intervening years scientists had studied the phenomenon in order to measure the size of the planet. Humboldt, however, was keen to observe this transit because it also provided an excellent means of measuring longitude. Using his chronometer, Humboldt could precisely time the transit at Callao, then later compare his observations to those taken by astronomers in Europe. The difference in time between his measurement and theirs would yield the distance in longitude between the two places.

That was, if the perpetual overcast allowed him to view the transit at all. Situated on the coastal plain (which had been pushed above the Pacific's waters by the same tectonic forces that created the Andes), Lima enjoyed a pleasant, mild climate. However, for eight months of the year, a thick fog, called *garúa*, blanketed the city. In this case, though, Humboldt was lucky. As he wrote in *Aspects of Nature*, "I had, at the critical moment, the rare good fortune of a perfectly clear day, during a very unfavorable season of the year, on the misty coast of Low Peru. I observed the passage of Mercury over the Sun at Callao, an observation which has become of some importance towards the exact determination of the longitude of Lima and of all the south-western part of the New Continent," since the location of those places had been fixed by their relationship to Lima. Humboldt found that the transit began at Callao five hours, eighteen minutes, and eighteen seconds later than at Paris, putting the port at

77 degrees, 6 minutes, and 3 seconds west longitude. Once again he had left his mark on the maps of South America. "Thus," he wrote, "in the intricate relations and graver circumstances of life, there may often be found associated with disappointment [over missing the rendezvous with Baudin], a germ of compensation."

In 1835, Robert FitzRoy, captain of the *Beagle*, confirmed Humboldt's measurements. That ship anchored at Callao for six weeks, but because of political turmoil, Darwin was able to see very little of Lima. As during Humboldt's stay, the sky was constantly overcast. Darwin complained in *The Voyage of the Beagle* that conditions had worsened since Humboldt's visit. "The inhabitants, both here and at Lima, present every imaginable shade of mixture, between European, Negro, and Indian blood. They appear a depraved, drunken set of people. The atmosphere is loaded with foul smells, and that peculiar one, which may be perceived in almost every town within the tropics, was here very strong. . . ."

As if that weren't bad enough, "The city of Lima is now in a wretched state of decay: the streets are nearly unpaved; and heaps of filth are piled up in all directions, where the black gallinazos, tame as poultry, pick up bits of carrion." Still, the city showed evidence of past glory. "The houses have generally an upper story, built on account of the earthquakes, of plastered woodwork; but some of the old ones, which are now used by several families, are immensely large, and would rival in suites of apartments the most magnificent in any place. Lima, the City of the Kings, must formerly have been a splendid town. The extraordinary number of churches gives it, even at the present day, a peculiar and striking character, especially when viewed from a short distance."

It was during his stay in Lima that Humboldt came across guano, a natural fertilizer that would play an important role in European and American agriculture for much of the nineteenth century. During those years, the humble substance—the dried droppings of seabirds—would become one of the most coveted natural resources on earth.

For thousands of years, mankind had been spreading manure on crops to produce more vigorous plants and more bountiful harvests. The practice is still important because, except for carbon, hydrogen, and oxygen, which are supplied by water and air, plants must derive all their nutritional needs from the soil in which they grow. These essential nutrients—numbering four-teen, from major ones such as calcium to trace elements such as molybdenum—may be naturally lacking, or may in time be used up by the plants or carried away by water. Fertilizers, including animal excrement, add nutrients back to the soil—especially nitrogen, phosphorus, and potassium, all of which are needed by plants in substantial quantities. Just as Europeans had long spread manure on their crops, pre-Columbian Indians had for centuries fertilized their fields with guano, which, being more concentrated, is much higher in nutritional value.

The rocky islands off the coast of Peru, where bird droppings were piled up to 150 feet deep, produced a particularly potent form of guano, since the arid climate allowed it to dry without be-ing leached of its nutrients. Recognizing the substance's im-mense potential as a commercial fertilizer, Humboldt introduced guano to Europe, where the French chemists Louis-Nicolas Vau-quelin and Antoine-François, comte de Fourcroy, confirmed his assessment. Later, German scientist Justus von Liebig, a protégé of Humboldt and the founder of organic and agricultural chem-istry, became an advocate of the fertilizer as well. In 1841, when British chemist John C. Nesbit announced that guano contained thirty-three times the nutrient value of manure, the guano boom was born.

Over the next four decades, from 1840 to 1880, it's estimated that twenty million tons of guano were exported from Peru (espe-cially from the Chinchas Islands) for use in the United States and Europe. In 1856, believing that American farmers were be-ing price gouged, the U.S. Congress even authorized individual citizens to seize guano-rich islands anywhere in the world that were not already claimed by a foreign government. An island-hopping land rush ensued to claim the precious guano, and in the coming decades dozens of atolls and cays were annexed to the United States, including Midway and Christmas islands in the Pa-cific. But by the end of the nineteenth century, the boom was run-ning out, due in part to the depletion of the sources and in part to the invention of less expensive synthetic fertilizers.

ON CHRISTMAS EVE 1802, Humboldt, Bonpland, and Carlos Montúfar sailed from Callao, bound for Guayaquil (in present-day Ecuador), some seven hundred miles to the northwest.

During the long voyage on the small ship the *Causino*, Humboldt followed his usual routine of measuring water temperature and speed. Local fishermen had known for centuries that a powerful, cold current flowed from the tip of Chile to northern Peru, from just offshore to about six hundred miles out to sea. However, no one had ever studied the flow, which is now known to produce a number of important climatological and economic effects along the western coast of South America.

The current originates in the Antarctic, but its coldness is due to the fact that it wells up from deep in the Pacific. The prevailing winds push the warm surface waters away from land, and they are replaced by much cooler, plankton-rich water from far below. By holding warm, moist air offshore, the current is responsible for the South American coastal desert. It also is the reason that Antarctic animals like penguins and fur seals are found in the Galápagos, west of Ecuador. More to the point, the plankton attracts an abundance of fish and birds, making the current the most productive marine ecosystem on the planet, accounting for twenty percent of the world's fish catch and positioning Peru as one of the leading fishing nations on earth.

But powerful though it is, the current is not immune to ecological threat. The life-giving flow is disrupted by El Niño ("the Child"), the periodic wintertime warming of the Pacific off the coast of Ecuador that is named after the Infant Jesus because the effect occurs near Christmas. During a persistent El Niño pattern, which can last more than a year, warm, nutrient-poor waters devastate the fisheries. Heavy rains and flooding also occur in South America in these years, while Southeast Asia and Africa are struck by drought. In North America, El Niño produces heavy rains across the Southeast, unusually warm weather along the Pacific Coast, and a relatively quiet hurricane season in the Atlantic. Thousands of deaths and tens of billions of dollars in property damage can result worldwide. Though there is no consensus on the reason for these extended El Niños, many scientists suspect that it is caused by air pollution trapping atmospheric warmth through the "greenhouse effect."

Though Humboldt did not discover this great upwelling of cold water—"I may claim the merit of having been the first to measure its temperature and rate of flow," he was still objecting in 1840—it was named after him anyway. Ironically, the Humboldt Current (now also known as the Peru Current) has become the principal means by which the baron's name has survived into our own time. Even today it is the phenomenon that is most likely to ignite a spark of recognition on the part of those not otherwise familiar with Humboldt's manifold achievements.

JUST AS HIS BREAKTHROUGHS in the hard sciences would transform oceanography, volcanology, plant geography, magnetism, and other fields, Humboldt's investigations of the native peoples of South America would revolutionize the study of anthropology. As the first prominent European to appreciate the great indigenous cultures of the New World and to suggest that pre-Hispanic peoples were more than brutal savages in need of "civilizing," Humboldt greatly enhanced the world's interest in, and recognition of, America's ancient inhabitants. Indeed, it was largely through their reading of Humboldt that American writers such as John Lloyd Stephens, whose 1841 book *Incidents of Travel in Central America, Chiapas, and Yucatan* drew attention to the Maya culture, and the previously mentioned William Hickling Prescott were inspired to undertake their landmark studies of these great civilizations of the past.

Humboldt was also among the first to suggest that the earliest Americans had immigrated from Asia, citing similarities in calendars, legends, and religious symbols. "I regard the existence of a former intercourse between the people of western America and those of eastern Asia as more than probable," he wrote, "though it is impossible at the present time to say by what route and with which of the tribes of Asia this influence was established. . . . We know that adventurers navigated the vaster Chinese seas. . . . May not accident have led to similar expeditions to Alaska and California?" Today, this Asian immigration has been proven by genetic testing, but the question of how and when it occurred is still in question. Though not unanimous, the consensus view has long been that the first Americans migrated out of Asia over a land bridge between Siberia and Alaska, sometime between twenty thousand and fourteen thousand years ago. Recent radiocarbon-dating evidence, however, suggests that the settlement in Siberia thought to be the immigrants' departure point is actually too recent to have been used for that purpose, fueling speculation that the early settlers of North America may have arrived by boat.

Moreover, Humboldt's writings also began to chip away at the assumption of European racial superiority. If the native peoples of South America were not simple heathens but actually the inheritors of a great indigenous culture, by what right did Spain, Portugal, and other powers seize their lands and enslave the Indians under the guise of rescuing them from savagery? "A darker shade of skin color is not a badge of inferiority," Humboldt argued. "The barbarism of nations is the direct consequence of oppression by internal despotism or foreign conquest. It is always accompanied by progressive impoverishment and a diminution of public fortune. Free and powerful institutions remove such dangers." That is, any cultural debasement was not the result of an innate failing but rather the consequence of centuries of oppression, first by their own dictatorial rulers and later by paternalistic and self-serving Europeans. The would-be architects of South American independence were Creoles (whites of European descent), not Native Americans, and they certainly didn't consider the Africans or the Indians their peers. Nevertheless, for Humboldt, the self-proclaimed child of the French Revolution, there was reason to hope that, with national independence, the nonwhite races of South America would also attain the benefits of freedom that were their rightful patrimony.

ON FEBRUARY 15, 1803, Humboldt and Bonpland sailed from Guayaquil for Mexico. As their small ship pulled away from shore, the travelers could hear Cotopaxi booming, as if in a final salute, some 150 miles inland. Humboldt didn't record his thoughts on departing South America. Yet leave-takings and ocean voyages typically put him in a melancholy mood, and from the trail of wistful comments he left on departing other locales, we can assume that this one was also tinged with regret. It had been an incredible two-and-a-half-year odyssey, through dense rain forests, down great rivers, across barren páramos, and over snowcapped mountains, to places where few if any men had ever gone before. Everywhere, Humboldt had opened himself totally to the environment, registering impressions, making connections, and formulating new ideas that would push our think-ing in unprecedented directions—in the sciences, history, and even the fine arts. The most important articles he would bring back from the New Continent would be ideas, he had said. And: "Whether in the Amazonian forests or on the ridge of the high Andes, I was ever aware that *one* breath, from pole to pole, breathes *one* single life into rocks, plants and animals, and into the swelling breast of man." Thanks to his unique, unifying vi-sion of the world, no one would ever look at South America, or nature, or mankind, in the same way again.

But his journey was not yet over. There would be more than another year of discovery before Humboldt left America for good. Before him lay Mexico.

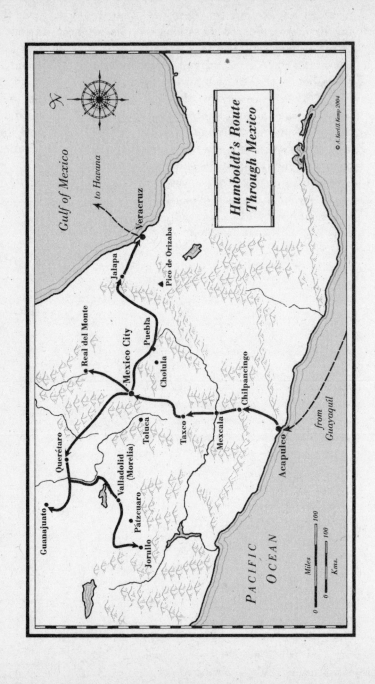

Humboldt's Route Through Mexico

© A.Karl/J.Kemp 2004

Gulf of Mexico

to Havana

Veracruz

Jalapa

Pico de Orizaba

Real del Monte

Mexico City

Puebla

Cholula

Chilpancingo

Querétaro

Toluca

Taxco

Mexcala

Acapulco

from Guayaquil

Valladolid
(Morelia)

Guanajuato

Pátzcuaro

Jorullo

PACIFIC OCEAN

Miles

Kms.

ELEVEN

New Spain

AS THE *ATLANTE* APPROACHED NEW SPAIN, ITS PASSEN-gers and hands scoured the eastern horizon. It had been thirty-three miserable, squall-filled days since the corvette had left Guayaquil, and everyone on board was eager to sight dry land. But as the captain held his course, expecting to intercept the coast at any time, Humboldt—taking his own sightings, as always—realized that the ship's charts had misplaced the port of Acapulco. It was an important error, since not only was the city the hub of Spain's trade with the Far East, but the coordinates of many other places had been reckoned from there; if Acapulco had been mischarted, then so had much of Mexico's Pacific Coast.

At last, the shout went up from the lookout, and on March 23, 1803, the ship dropped anchor. Humboldt found the port of Aca-pulco ("Place of the Reeds" in the Aztecs' Náhuatl language) "one of the finest natural harbors in the known world . . . , forming an immense basin cut in granite rocks open towards the south-southwest, and possessing from east to west more than six thou-sand meters in breadth [about three and a half miles]." Whereas Havana was cheerful and welcoming, on first sight Acapulco was brooding and forbidding, owing to the great cliffs guarding the western side of the bay. "I have seen few situations in either hemisphere of a more savage aspect," Humboldt ventured. "I would say at the same time more dismal and more romantic. This rocky coast is so steep that a vessel of the line [the largest class of warship] may almost touch it without running the small-est danger, because there is everywhere from ten to twelve fath-oms [about sixty to seventy feet] of water."

The place had been occupied by Native Americans for thousands of years. The Spanish had arrived in 1512, and Hernán Cortés had established a port there in 1523. But by the turn of the nineteenth century, Acapulco had become a forlorn backwater, with a population of only about four thousand and a reputation for sheltering escaped convicts and slaves. At the time of Humboldt's visit, the city was receiving scarcely ten ships a year—mainly coming up the coast from Guayaquil and Lima with commodities such as copper, oil, wine, quinine, sugar, and cocoa, and loading up for the return passage with woolens, a little cochineal (dried insects used to made a red dye), and perhaps some contraband goods from Asia.

By far the most important of these arrivals—the highlight of the year in Acapulco—was the annual *nao*, or galleon, from Manila. For 250 years, Acapulco had been the only Spanish city in the Western Hemisphere permitted by the Crown to trade with the Orient. Sometime from the middle of July to early August, the *nao* would depart the Philippines, laden with textiles and spices, and arrive in New Spain in November or December. "Whenever the news arrives at Mexico that the galleon has been seen off the coast," Humboldt wrote, "the roads are covered with travelers, and every merchant hastens to be the first to treat with the supercargos. . . ." Most of these dealers would be disappointed, since the lion's share of the merchandise had already been sold by previous arrangement to the great mercantile houses in Mexico City. From the capital, the goods would be distributed throughout Mexico, and some would even find their way to Spain, via Veracruz on Mexico's Gulf Coast. Then in February or March, the galleon would begin its six-thousand-mile return trip to the Philippines, loaded mainly with Mexican silver but also with cochineal, cocoa, wine, oil, and wool. In addition to trade goods, the ship would be crowded with passengers in each direction. In 1804, the year of Humboldt's visit, the galleon took on a contingent of seventy-five monks bound for the Far East.

As soon as the *Atlante* dropped anchor, even before their baggage was unloaded, Humboldt rushed ashore with his astronomical instruments to begin the measurements that would fix the port's location. "On my trip to Acapulco," he explains, "I was constantly engaged in improving the points of reckoning by observations of the sun and moon. Enormous errors in longitude caused by strong currents render navigation in the latitudes equally long and expensive." He and Bonpland climbed up "naked rocks of strange appearance. They were scarcely sixty meters [about two hundred feet] above sea level, and appeared to be torn by the prolonged effects of the earthquakes so frequent on this coast." (The city had been severely damaged by an earthquake in 1776, necessitating the rebuilding of el Fuerte de San Diego, the pentagonal stone fort guarding the harbor.) After several days of sightings, Humboldt confirmed his suspicion: Acapulco lay as much as five miles to the west of where some current charts had placed it. As a result, the maps of New Spain would be redrawn, to accommodate not just the port but all the other towns whose positions were fixed with relation to it.

Yet Humboldt had not come to Mexico intent on physical exploration. In 1800, the Spanish had ceded to France a huge tract of what had been New Spain, extending from the Rocky Mountains in the west to the Gulf of Mexico in the south. But a vast territory remained, including present-day Mexico and a great swath of land reaching from the current U.S. state of Texas all the way to northern California. At the time of Humboldt's arrival, New Spain was a well-settled, prosperous colony, far different from the trackless Orinoco and Casiquiare. Though there was still some cartographic fine-tuning to be done in Mexico— such as fixing the exact location of Acapulco—the viceroyalty was already well mapped.

Humboldt had come to New Spain to secure passage to the Philippines, the next port of call on his projected round-the-world voyage. But he considered the viceroyalty much more than a way station en route to the Far East. Mexico had long been the crown jewel of Spain's New World empire. By 1800, its population had swelled to nearly six million (compared to 7.25 million in the United States that year). With extensive deposits of silver and gold, as well as fertile agricultural lands, Mexico was contributing about six million pesos to the Spanish treasury annually—an incredible one fifth of the mother country's yearly budget. And that figure didn't even include millions of pesos in trade conducted between New Spain and Old.

To have spent all this time in America and not to have investigated Spain's premier holding would have been unthinkable to Humboldt, whose curiosity ranged, beyond the natural sciences, to subjects as diverse as history, art, anthropology, and economics. Years before, when as a university student he had crossed Germany, Belgium, France, and England with Georg Forster, he had studied those countries' architectural treasures, manufactories, and mines. Now he had an unprecedented opportunity to compare the development of New Spain to that of Europe, as well as the Spanish colonies to the south.

Arriving in Mexico, Humboldt "could not avoid being struck with the contrasts between the civilization of New Spain and the scanty cultivation of those parts of South America which had fallen under [his] notice." In historic, prosperous Mexico, thanks to his royal passport, Humboldt was granted the access and cooperation needed to study the country's economic and political organization with a thoroughness never before possible. With his passion for collecting and correlating data, he relished the opportunity to advance the world's understanding of this important region. "This contrast [between New Spain and South America]," he continued, "excited me to a particular study of the statistics of Mexico, and to an investigation of the causes which have had the greatest influence on the progress of the population and national industry." Whereas in South America he had focused primarily on the manifold connections within the natural world, here in Mexico Humboldt would concentrate on the complex relationships between the land and its people.

His geodesic measurements made, Humboldt didn't linger in Acapulco. At the beginning of April, he, Bonpland, and Carlos Montúfar set out toward Mexico City, nearly two hundred miles to the north. After crossing the narrow coastal plain, the travelers plunged into the Sierra Madre del Sur, the rugged, geologically convoluted range that dominates the current Mexican states of Guerrero and Oaxaca. One of the most doggedly mountainous regions in all Mexico, the sierra is a complex jumble of temperate, forested peaks (some reaching over twelve thousand feet) and exceptionally hot and humid valleys, where Humboldt recorded temperatures as high as 104 degrees. Yet as the party pressed on, neither the demanding terrain nor the unbearable heat prevented him from taking exacting readings of geological features, latitude, and longitude.

Continuing his painstaking measurements all the way to Mexico City, Humboldt would incorporate this data into a unique visual cross-section of the territory that showed the locations of landforms as well as their height, thickness, and other distinguishing characteristics. Though a few such maps had been attempted in the past, none had ever been based on instrument readings, and Humboldt's cross-section would be widely imitated in years to come. Even today, the technique is considered an indispensable tool of geological research, for its ability to synthesize a wealth of disparate data into an easily readable graphic presentation.

Not content to limit this geologic survey to the area between Acapulco and the capital, Humboldt would later draw on statistical archives in Mexico City and extend his cross-section all the way to 42 degrees north latitude, taking in the current U.S. states of Utah, Arizona, New Mexico, and Colorado. For decades to come, this expanded map would form the basis for exploration in the southwestern United States, and it would be utilized by railroad surveyors as late as the 1850s. In fact, it was largely in recognition of this wealth of geographical data that early mapmakers named so many places in the United States—rivers, counties, marshes, and cities—in Humboldt's honor, including Humboldt County, California, and Humboldt Peak in Colorado.

As the travelers climbed up the coastal mountains toward Mexico City, they came to the town of Chilpancingo ("Place of the Wasps"), located on the Río Huacapa about fifty miles northeast of Acapulco, where they encountered pine and oak forests studded with lichens, ferns, and orchids, including dozens of species that grew nowhere else in the world. They also finally got a much-appreciated breath of cooler mountain air. About eighty miles north of Chilpancingo lay the celebrated town of Taxco, once site of the richest silver mine in the world. Built on a plunging hillside and traced by picturesque, twisting streets, Taxco had been founded in 1529 over an Aztec city known as Tlachco ("the Place of the Ball Game"). In 1531, the first Spanish mine in all of North America had been dug there—for tin. But silver had been unearthed instead, and by the end of the sixteenth century Taxco had been producing more of the precious metal than anyplace else in the New World. Gradually, the richest and most accessible veins had played out, the miners had moved on, and the city had entered a two-century decline—until 1716, when a Spaniard named José de la Borda had literally stumbled on another rich vein (according to legend, his horse exposed the gleaming ore when it kicked a rock). Tremendous wealth had once again flowed into Taxco, and the fortunate Borda had constructed houses, schools, roads, and a magnificent, gilded baroque church called the Templo de Santa Prisca.

At the time of Humboldt's arrival, Taxco was no longer the world's most productive silver vein (that honor had shifted to La Valenciana, outside Guanajuato), but the pits were still booming. Eager to get his first look at the famous Mexican mines, Humboldt settled into a nearby house along with Bonpland and Montúfar. At the time, Mexico was producing nearly two million pounds of silver a year—a staggering ten times as much as all the silver mines of Europe combined. There was no ignoring the importance of the precious metal, either to Mexico, where three quarters of the silver remained, or to Spain, whose economy had come to rely on the regular infusion of New World revenue. Yet Humboldt was surprised to discover that every hundred pounds of ore from the Mexican mines yielded only three or four ounces of silver. "It is not then, as has been too long believed, from the intrinsic wealth of the minerals," he concluded, "but rather from the great abundance in which they are to be found in the bowels of the earth and the facility with which they can be wrought that the mines of America are to be distinguished from those of Europe."

In his career as a mining inspector in Germany, Humboldt had introduced ingenious innovations for the safety and education of his men, and he noted with satisfaction that the Mexican miner was treated relatively well. While Indians had previously been enslaved and set to work in the pits under horrible conditions, by 1803, such work, whether by Indians or mestizos, was voluntary. Moreover, not only was the Mexican miner much better compensated than the average farm worker in New Spain, he was actually the best-paid miner in the world. Even so, not all workers were able to resist the temptation offered by the precious ores they handled. "Honesty is by no means so common among the Mexican as among the German or Swedish miners," Humboldt noted, "and they make use of a thousand tricks to steal very rich minerals. As they are almost naked, and are searched on leaving the mine in the most indecent manner, they conceal small morsels of silver in their hair, under their armpits, and in their mouths, and they even lodge in their anus cylinders of clay which contain the metal. These cylinders are sometimes of the length of thirteen centimeters [five inches]. It is a most shocking spectacle," he found, "to see hundreds of workmen, among whom there are a great number of very respectable men, searched on leaving the pit or gallery. A register is kept of the minerals found in the hair, in the mouth, or other parts of the miners' bodies. In the mine of Valenciana the value of these stolen minerals amounted between 1774 and 1787 to the sum of 900,000 francs."

Unlike Peru's silver mines, those in New Spain did not seem to be in decline. Still, Humboldt saw numerous possibilities for improvement in their operation. "In taking a general view of the mineral wealth of New Spain," he wrote, "far from being struck with the value of the actual produce, we are astonished that it is not much more considerable." In fact, the mines' productivity could be increased by as much as threefold, he believed, through a variety of technical improvements such as more judicious use of gunpowder in blasting, narrower shafts, improved communication between adjoining mines, and a more efficient means of drawing off groundwater. "But we must repeat here," he added, "that changes can only take place very slowly among a people who are not fond of innovations, and in a country where the government possesses so little influence on the works which are generally the property of individuals, and not of shareholders. It is a prejudice to imagine that on account of their wealth the mines of New Spain do not require the same intelligence and economy which are necessary to the preservation of the mines of Saxony."

Leaving Taxco, the travelers continued northeast toward Mexico City, till they came to the town of Cuernavaca. Once known by the Indian name Cuauhnáhuac (Place at the Edge of the Forest), the area had been an important agricultural center for centuries before the arrival of the Spanish. After his conquest of Mexico, Hernán Cortés had received Cuernavaca as part of his huge land grant from the Spanish Crown, and taking up residence, he had built Mexico's first sugar mill there about 1530. Over the ensuing years, the town, with its pleasant, springlike climate, had become a popular retreat for the wealthy, including José de la Borda, the silver baron of Taxco, and later the emperor Maximilian. To this day Cuernavaca remains a fashionable though overcrowded retreat from nearby Mexico City.

From Cuernavaca, Humboldt and the others rode up a forested ridge, where they had their first glimpse of the green, cultivated Valle de Mexico, some seventy-five miles long and forty wide. After his first view of the valley, Cortés had written King Charles I calling it "a fairyland such as cannot be seen in Spain." Three centuries later, the impression was scarcely muted. In the distance, Humboldt could make out Mexico City, now the capital of New Spain and before that the heart of the Aztec Empire. Just as he had been moved by the plight of the Inca in Peru, here in New Spain Humboldt would become fascinated by the history and culture of the Aztecs, who had been overrun and enslaved by the Spanish.

IN THE DAYS OF THE AZTECS, the teeming city in the valley was known as Tenochtitlán, or "the Place of the Cactus Fruit." Located at more than seven thousand feet above sea level, nearly equidistant between the Pacific Ocean and the Gulf of Mexico, Tenochtitlán had been constructed on a 2,500-acre man-made island in the great lake of Texcoco. By A.D. 1500, not only was the city the greatest metropolis in the Western Hemisphere, with a quarter-million inhabitants, it was one of the largest cities in the world—twice as populous as London at the time and twenty times larger than Madrid. As capital of the Aztec civilization, the great city in the lake presided over a magnificent empire encompassing 125,000 square miles in central Mexico and nearly four hundred subject cities.

Like the rise of the Inca to the south, the Aztecs' ascendancy had been relatively recent. About A.D. 1250, the nomadic Mexica people had migrated southward from northern Mexico. By 1325 they had founded their city on the lake where, according to legend, an eagle had been seen perching on a cactus. Then, around the middle of the fifteenth century, they had expanded rapidly, through a combination of diplomacy and conquest, and had established the vast empire today known as Aztec—a collective name for the Mexicas and numerous subject peoples linked by language, trade, religion, and other commonalities.

In 1502, Montezuma II ("He Who Angers Himself"), the son and great-grandson of previous emperors, had succeeded his uncle Ahuitzotl on the throne. Like the inca, the Aztec emperor embodied both secular and religious authority. However, the Aztecs did not worship their king as a literal deity thought to be descended from the gods. The Aztec line of succession was not even strictly hereditary; the heir to the throne was chosen from within the royal family by a council of nobles, and was generally a brother, cousin, or some other close male relation of the deceased monarch.

Though the emperor had important religious duties, principal responsibility for Aztec worship fell to a class of priests, who wielded enormous political influence as well. Indeed, religion was central to the lives of all Aztecs, and it was no coincidence that the Great Temple at Tenochtitlán was constructed in the geographical middle of the city. Out of the pantheon of perhaps sixteen hundred Aztec gods (including two hundred major ones), the chief deity during the time of Montezuma II was Huitzilopochtli, god of the sun. According to the Aztecs' apocalyptic mythology, they were living in the last of five eras of mankind's existence on earth. Just as each previous epoch had concluded in cataclysm—wrought by wild animals, wind, fire, and floods—their own age was destined to end in devastating earthquakes, when horrible creatures would stalk the land. An Aztec song captures this sense of borrowed time and looming disaster:

> Ponder this, eagle and jaguar knights,
> Though you are carved in jade, you will break;
> Though you are made of gold, you will crack;
> Even though you are a quetzal feather, you will wither,
> We are not forever on this earth;
> Only for a time are we here.

Though they believed there would be no escaping this devastation, it was thought to be forestalled by Huitzilopochtli's daily progress through the heavens. But to sustain him on his unceasing, life-giving rounds, the sun god demanded the nourishment of human blood. Emperors and priests regularly bled themselves from the ears, tongue, extremities, and penis, but these offerings alone were not sufficient. Huitzilopochtli's great hunger could be sated only by human sacrifice on a huge scale, and great numbers of prisoners and slaves were ritually butchered for this purpose. When the new temple to Huitzilopochtli was consecrated in 1487, the lines of victims, estimated by various historians to number from twenty thousand to eighty thousand, were said to lead off toward all four compass points as far as the eye could see.

The conquistador Bernal Díaz del Castillo wrote an eyewitness account of the conquest of Mexico from the vantage of the common soldier. While the work shouldn't be taken as the last word on the Conquest, it is a rousing tale of adventure. According to Díaz's description of a typical Aztec sacrifice, the victim was held down on a stone block, and the chief priest would "strike open the wretched Indian's chest with flint knives and hastily tear out the palpitating heart which, with the blood, they present to the idols in whose name they have performed the sacrifice. Then they cut off the arms, thighs, and head, eating the arms and thighs at their ceremonial banquets. The head they hang up on a beam, and the body of the sacrificed man is not eaten but given to the beasts of prey."

In addition to religious observance, all other aspects of Aztec society were also regulated to their most minute detail. The duties of each class were carefully enumerated, and laws were stringent. Punishment was uncompromising; as with the Inca, public execution was the standard penalty for theft. Women were responsible for domestic chores such as weaving, while the men raised crops such as corn, beans, sweet potatoes, and chiles, and skilled craftsmen produced fine sculptures, paintings, metalwork, and ceramics.

Also like the Inca, the Aztecs had no wheeled vehicles, no metal tools, and no arch. Yet they constructed great temples and palaces, and they were accomplished astronomers. To record the passage of time, they invented two calendars—one of 365 days to mark the solar year and a more important cycle of 260 days (20 weeks of 13 days each) to regulate religious observances. Because the two calendars resynchronized every fifty-two years, that period served as the Aztec "century." Though the Aztecs had no alphabet, their scribes used glyphs to record histories, genealogies, laws, court records, land deeds, financial accounts, tax rolls, and the myriad other details necessary for the administration of a vast empire extending across central Mexico. In fact, much of our knowledge of Aztec culture is derived from the great books, or codices, that escaped destruction by the Spanish.

Warfare was also central to Aztec society, the essential means through which the empire was preserved and extended. "War is your task," each newborn male was traditionally reminded by the midwife, and every boy was schooled in the military arts. The tactics of Aztec warriors have been described as unsophisticated, little more than large-scale one-on-one combat using weapons such as slings, spears, bows, clubs, and obsidian swords. But the empire's army was large, sometimes reaching as many as two hundred thousand men, and it proved an intimidating, effective instrument against neighboring cities. However, in battle, it was not the death of the enemy that was the tactical aim, but his capture, so that he could be sacrificed at Tenochtitlán.

Despite their military supremacy, by the reign of Montezuma II the Aztecs' world had developed troubling fissures. With further expansion frustrated by the Tarascan and the Chichimec peoples in the north and the Maya in the south, the empire had reached the limits of its growth. In addition, various vassal cities, such as Tlaxcala and Cholula, were chafing over the Mexicas' ever-escalating demands for prisoners and tribute, and rifts had even developed with some longtime allies, such as the city of Texcoco, located just across the lake from Tenochtitlán. Then, beginning about 1502, a series of omens had been observed—a comet, a flood, a fire in the Great Temple—all of which, according to Mexica holy men, warned of looming disaster. Soon afterward, mysterious, white-skinned strangers had been reported in the Caribbean, and a trunk had washed ashore from the Gulf of Mexico containing bizarre clothing and other strange artifacts. Increasingly apprehensive, Montezuma considered constructing a huge new temple to Huitzilopochtl, in the hope of appeasing the insatiable sun god.

Then, in 1518, these portents seemed to be borne out when sails (or as the words of the Aztec laborer who spotted them were later recorded, "mountains . . . floating in the sea") were sighted off the eastern coast, near Cape Rojo, in the current Mexican state of Veracruz. Montezuma dispatched emissaries, who met with the strangers, Spanish explorers who had sailed from Cuba under the command of Juan de Grijalva. Establishing the pattern of appeasement that would characterize his dealings with the Spaniards, Montezuma dispatched nobles with lavish gifts of fine cloaks and gold, for which he received some glass beads and hardtack in return. The strangers left, all who had seen them were commanded to silence on pain of death, and the crisis passed. But the following year, 1519, the white men came again. Cortés had landed.

Hernán Cortés was born around 1485 in the Castilian city of Medellín, the only child of a soldier and a notary's daughter. The nephew of the conquistador Diego de Velázquez, Cortés traveled first to Hispañola in 1504, then in 1511 joined his uncle in Cuba, where he assisted in the conquest of the island and most likely witnessed the execution of the cacique Hatuey the following year. Though their relationship had turned stormy, in 1518 Velázquez commissioned Cortés to lead an expedition to Mexico. When Velázquez rescinded the orders at the last minute, Cortés sailed anyway, leaving Havana on February 18, 1519. According to Bernal Díaz del Castillo, who accompanied him, Cortés had "five hundred and eight [Europeans], not counting the ships' captains, pilots, and sailors, who amounted to a hundred . . . [and] sixteen horses or mares. There were eleven ships, large and small, and one that was a sort of a launch. . . . There were thirty-two bowmen and thirteen musketeers, ten brass guns and four falconets [light cannon], and much powder and shot." Whatever the complement of crossbows, these few hundred adventurers were an unlikely match for one of the greatest empires the world had ever seen.

After making friendly contact with Indians on the island of Cozumel, off the Yucatán Peninsula, Cortés cruised the eastern coast, where he encountered resistance at Champotón and Cintla. At the city of Cempoala, he persuaded the Totonac people to rebel against Tenochtitlán and added two thousand Indian warriors to his ranks. Reboarding their ships, the Spaniards continued up the coast until Holy Thursday 1519, when they landed near the present-day city of Veracruz. On Easter Sunday, an emissary of Montezuma arrived. "He took out of a *petaca*—which is a sort of chest," Díaz wrote, "—many golden objects beautifully and richly worked, and then sent for ten bales of white cloth made of cotton and feathers—a marvelous sight. There were . . . quantities of food—fowls, fruit, and baked fish. Cortés received all this with gracious smiles, and gave them in return some beads of twisted glass . . . , begging them to send to their towns and summon the people to trade with us, since he had plenty of these beads to exchange for gold."

In Tenochtitlán, hearing descriptions of the strangers and their outlandish clothes, food, animals, and weapons, Montezuma panicked. His first impulse was to flee the capital, but cooler-headed advisers managed to dissuade him. Most unnerving was the uncertainty concerning the interlopers' identity. Were they ambassadors of a powerful and distant king, as they claimed, or invaders bent on destruction? Were they previously unknown deities, or exiled gods returning to reclaim their birthrights? If the latter, their leader might well be Quetzalcoatl ("Feathered Serpent"), the generally beneficent god who had introduced agriculture, science, the arts, and other hallmarks of civilization. According to one legend, Quetzalcoatl had been expelled from Tula, the capital of the Aztecs' precursors the Toltecs, after which he had sailed from the eastern coast on a raft of serpents, vowing to return.

Recent scholarship has cast into doubt the traditional explanation that the Aztecs considered Cortés a god. However, besides the place of their landing, several other coincidences could have suggested that the strangers' leader was indeed the Feathered Serpent. For one thing, Quetzalcoatl was generally depicted in vaguely human form with pale skin and was sometimes shown wearing a beard. Also, the newcomers' leader, on hearing of the Aztecs' predilection for human sacrifice, had spoken out against the practice, which Quetzalcoatl was known to staunchly oppose. In recent years, the Aztecs had increasingly given themselves over to murderous offerings to Huitzilopochtli. Could Quetzalcoatl have returned to put an end to the sacrifices and to seek his due? Paralyzed by indecision, Montezuma plied the newcomers with gifts in an effort to appease them, just as he had hoped to satisfy Huitzilopochtli with ever-escalating orgies of bloodletting.

At Veracruz, Cortés became aware of grumbling among some of his men who were loyal to Velásquez. With their provisions gone, thirty-five men already wounded in skirmishes with local tribes, untold thousands of Indian warriors before them, and not even any legal authority to undertake the expedition, the dissenters, in Díaz's words, "were sighing to go home." Faced with the threat of mutiny, Cortés made his famous, audacious gambit of dismantling his ships on the beach (the vessels weren't burned, as is often claimed). Afterward, according to Díaz, the commander "made a speech to the effect that we now understood what work lay before us, and with the help of our lord Jesus Christ must conquer in all battles and engagements. We must be properly prepared, he said, for each one of them, because if we were at any time defeated, which God forbid, we should not be able to raise our heads again, being so few. He added that we could look for no help or assistance except from God, for now we had no ships in which to return to Cuba. Therefore we must rely on our own good swords and stout hearts."

In mid-August, the Spaniards began their march to Tenochtitlan, swelling their ranks with Indian allies, especially the Tlascans. Ever more agitated, Montezuma several times sent ambassadors to meet the advancing host, plying the white men with rich gifts and offers of friendship, while at the same time discouraging them from entering the capital. But the Spanish were not to be deterred. As they approached the great lake of Texcoco, the invaders were awed by the Indian cities through which they passed. "And when we saw all those cities and villages built in the water, and other great towns on dry land, and that straight and level causeway leading to Mexico, we were astounded," Díaz wrote. "These great towns and *cues* and buildings rising from the water, all made of stone, seemed like an enchanted vision. . . . Indeed, some of our soldiers asked whether it was not all a dream. . . . It was all so wonderful that I do not know how to describe this first glimpse of things never heard of, seen, or dreamed of before."

On the outskirts of the capital, Montezuma came out to greet them, and Díaz recorded the Europeans' first glimpse of the Aztec emperor. "The great Montezuma descended from his litter, and these other great Caciques supported him beneath a marvelously rich canopy of green feathers, decorated with gold work, silver, pearls. . . . It was a marvelous sight. The great Montezuma was magnificently clad, in their fashion, and wore sandals . . . the soles of which were of gold and the upper parts ornamented with precious stones. . . ." In addition, "There were . . . many more lords who walked before the great Montezuma, sweeping the ground on which he was to tread, and laying down cloaks so that his feet should not touch the earth. Not one of these chieftains dared to look him in the face."

Cortés dismounted, and the two leaders bowed deeply to each other. The Spaniard presented Montezuma with a necklace of colored glass beads strung with gold, which he placed around the emperor's neck. Through interpreters, the two men exchanged friendly speeches, and the Spaniards joined the procession into the city. "Who could now count the multitude of men, women, and boys in the streets, on the rooftops, and in canoes on the waterways, who had come out to see us?" Díaz asked. "It was a wonderful sight and, as I write, it all comes before my eyes as if it had happened only yesterday. . . . So, with luck on our side, we boldly entered the city of Tenochtitlán or Mexico on 8 November in the year of our Lord 1519."

Though the Spanish were treated as honored guests, they soon began to fear that they had stumbled into a trap. Surrounded by thousands of Montezuma's soldiers, cut off from their Indian allies, they realized that their continued existence depended totally on the mercy of their hosts. Rather than risk their safety to what they supposed to be the Aztecs' fickleness, Cortés and his lieutenants—as Pizarro later would do in Peru—elected to make a bold play and seize the emperor prisoner. Listing the slights supposedly suffered at the hands of the Mexicas, Cortés told Montezuma, "I have no desire to start a war on this account, or to destroy this city. Everything will be forgiven, provided you will now come quietly with us to our quarters, and make no protest. You will be as well served and attended there as in your own palace. But if you cry out, or raise any commotion, you will immediately be killed. . . ." Thus was the lord of the Aztecs made a captive in his own city.

With Montezuma in the Spaniards' power, the Aztecs hesitated to take any action that might put the emperor in jeopardy. But eventually, unable to tolerate the stalemate any longer, the Indians besieged the Spanish in their compound. When Montezuma climbed onto the roof and ordered his warriors to quit the attack, Díaz reports, he was told by one of his own captains, "O lord, our great lord, we are indeed sorry for your misfortune and the disaster that has overtaken you and your family. But we must tell you that we have chosen a kinsman of yours as our new lord [Cuitlahuac]." They said moreover "that the war must be carried on, and that they had promised their idols not to give up until we were all dead. They said they prayed every day . . . to keep him free and safe from our power, and that if things ended as they hoped, they would undoubtedly hold him in greater regard as their lord than they had done before. And they begged for his forgiveness."

No sooner was this reply given than the attack resumed in a shower of stones and spears. The Spanish soldiers guarding Montezuma were taken unawares, and the emperor fell, mortally wounded in the head, arm, and leg. He died soon after, surrounded by Cortés and the other Spaniards, who, Díaz said, wept for him "as though he were our father." Learning of Montezuma's death, the Aztecs pressed their assault with renewed fury, and Cortés concluded that the only hope of survival was to break out of the city. This was achieved on la Noche Triste, "the Sad Night," of June 30 to July 1, 1520, when the Spanish suffered heavy casualties in savage fighting over the city's causeways.

Retreating around the north shore of Lake Texcoco, the remnants of Cortés's army sought refuge in the friendly city of Tlaxcala. Nearly a year later, having received both Spanish and Indian reinforcements, Cortés marched again on Tenochtitlán and, after a siege of ninety-three days, finally succeeded in capturing the city, and with it the Aztec Empire, on August 15, 1521. By then, the once-great capital was in ruins. According to Díaz, "All the houses and stockades in the lake were full of heads and corpses. . . . It was the same in the streets and courts. . . . We could not walk without treading on the bodies and heads of dead Indians. I have read about the destruction of Jerusalem, but I do not think the mortality was greater there than here in Mexico, where most of the warriors who had crowded in from all the provinces and subject towns had died. . . . The dry land and the stockades were piled with corpses. Indeed, the stench was so bad that no one could endure it. . . ." Historians have variously estimated the number of Aztec dead from 40,000 to 250,000. Of 1,800 Spaniards landing in Mexico between 1519 and 1521, it is thought that 1,000 never returned.

Another Mexica song commemorated the Aztec subjugation:

> *It was called the jaguar sun.*
> *Then it happened*
> *That the sky was crushed.*
> *The sun did not follow its course.*
> *When the sun arrived at noon,*
> *Immediately it was dark;*
> *And when it became dark*

Jaguars ate the people. . . .
The giants greeted each other thus:
"Do not fall down, for whoever falls
Falls forever."

Eschewing Tenochtitlán after the slaughter, Cortés established his government at nearby Cuauhnáhuac (present-day Cuernavaca), and the ruined capital became virtually deserted. Though the fighting had ceased, the surviving Aztecs, suffering from famine and smallpox, continued to die long afterward. For his part, Cortés found it "more difficult to contend against my own countrymen than against the Aztecs." At the urging of his uncle Velásquez, a commissioner was named to investigate Cortés for malfeasance, cruelty, and nepotism, but he was acquitted of any wrongdoing and named captain-general and governor of New Spain. Then in 1526, another commissioner was appointed to investigate him, and Cortés was stripped of the governorship, though he was named a marqués and retained his title of captain-general. In 1540, he returned to Spain to reclaim what he considered his just reward from the Crown, but having fallen out of royal favor, the conqueror of Mexico was studiously ignored for seven years. He died of dysentery in Seville in 1547, as he was preparing to return to Mexico.

IT WAS CORTÉS who had named the conquered land New Spain. As he had written his king in the first flush of discovery, "From what I have seen and understood concerning the similarity between this country and Spain, in its fertility, its size, its climate, and in many other features of it, it seemed to me the most suitable name for this country would be New Spain of the Ocean Sea, and this in the name of Your Majesty I have christened it." But by 1800, the earlier name for the country, Mexico, was enjoying a new wave of popularity as the powerful Creoles became increasingly disenchanted with the government of Old Spain.

The Spaniards had occupied Mexico City since 1521, when Cortés had founded it as his new capital. Like Tenochtitlán before it, Mexico City was by the time of Humboldt's arrival one of the largest cities in the world, with a population of perhaps 250,000 at a time when New York had only 60,000 and

Philadelphia, the largest city in the United States, could boast just 70,000. Mexico City was also one of the most opulent metropolises in the world, with broad streets, impressive public buildings, fine parks and plazas, and lavish mansions. Joel Poinsett, first U.S. ambassador to Mexico, from 1825 to 1829 (and namesake of the Poinsettia flower, which he introduced to the United States), wrote, "The streets . . . are all well-paved and have sidewalks of flat stones. The public squares are spacious and surrounded by buildings of hewed stone, and of very good architecture. The public edifices are vast and splendid . . . and have an air of solidity and even magnificence. . . . There is an air of grandeur in the aspect of this place. . . ."

To Humboldt, the capital was "the City of Palaces." At its heart was the vast central square, the Zócalo, that had once been the Aztec ceremonial center. Along the west side, the viceroy's palace occupied the former site of Montezuma II's residence, and on the northern side, the baroque cathedral had been built over the site of a rack where thousands of sacrificial skulls were once displayed. Thus in the Spanish epoch, even as in Montezuma's day, the square, like the capital around it, symbolized the marriage of ecclesiastical and political authority. Though the Aztec pantheon had been supplanted by the Trinity, religion still played a predominant role in Mexican life and government. (To this day Mexico remains one of the most devoutly Catholic countries in the world.) At the time, Mexico City boasted more than a hundred churches, twenty-three monasteries, and fifteen convents, all reflecting the tremendous wealth and power of the Church, which controlled huge tracts of land, collected tremendous revenues, and operated outside the scope of secular law. To Humboldt, the difference between the old religion and the new was largely superficial, at least as far as the Indians were concerned. "It is not a dogma which has replaced another," he observed, "but one set of ceremonies which has been substituted for another. The natives know nothing of religion but the outward forms of worship. . . . The feast days of the Church, the fireworks that go with them, the processions interspersed with dances and oddly costumed marchers are for the poorer Indians a rich source of entertainment."

Not that secular institutions had been neglected in the capital. Thanks to the flood of Mexican silver, the city also boasted a university, a public library, a botanical garden, an academy of fine arts, and, of particular interest to Humboldt, a state-of-the-art school of mining directed by Andres del Rio, a classmate from Freiberg. The mint, which Humboldt found "the richest and largest in the whole world," had from 1726 to 1780 coined gold and silver worth half again as much as all the mints of France combined. In fact, Humboldt calculated that the silver extracted from every mine in Europe over the course of a full year would keep Mexico City's mint operating for a scant two weeks.

The travelers settled into a comfortable house near the library, where Humboldt spent many hours engaged in research. He also pored over the mint archives and myriad other public records, as he amassed the statistics for his *Political Essay on the Kingdom of New Spain.* The Aztec ruins just off the main square were being excavated during this time, and the famous round calendar stone now in the National Anthropological Museum had just been unearthed. Fascinated by the country's ancient inhabitants and impressed by their obvious erudition,

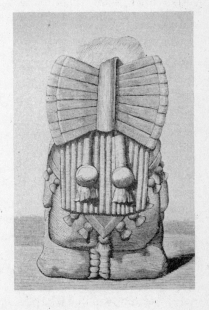

Statue of an Aztec priestess.
From Humboldt's *Views of the Cordilleras and Monuments of the Indigenous Peoples of America.*

Humboldt sketched some of the recently discovered sculptures, which he later reproduced in his *Researches Concerning the Institutions and Monuments of the Ancient Inhabitants of America*, the work, published in 1810, that introduced Europe to the culture of America's pre-Hispanic peoples.

Taking time out from their stay in the capital, the visitors also journeyed to the immense, pyramid-studded ruins at Teotihuacán ("Place Where Gods Are Made"), about thirty miles to the north. Examining the extensive complex, Humboldt concluded that the structures appeared even more ancient than those of the Toltecs, predecessors of the Aztecs who had come to dominate the Valley of Mexico about A.D. 900. Later archeologists have proven him right: We now know that the culture of Teotihuacán flourished from about A.D. 300 to 900, when it was the largest city in the Americas and the capital of the most extensive pre-Hispanic empire in the New World, stretching from north-central Mexico as far as present-day Guatemala.

In Mexico City, the course of Humboldt's journey took yet another unexpected turn: Instead of continuing on to the Philippines and hence around the world, he decided to extend his stay in Mexico, then to return to Europe via Cuba and the United States. He gave several reasons for this change of plans. To his friend the French astronomer Jean-Baptiste-Joseph Delambre, he explained, "As for the Philippines, I have given them up temporarily, for . . . I am eager first to publish the results of this expedition." To the secretary of the Institut National Humboldt also wrote, "I can think of nothing but of preserving and publishing my manuscripts." But in that same letter, he cited several other considerations as well: "The damaged state of our instruments, the futility of our efforts to replace them, the impossibility of meeting Captain Baudin, the lack of a ship that could bring us to the enchanted islands of the South Pacific, but, above all, the urgent need to keep pace with the rapid advancement of science which must have taken place during our absence, these are the motives for the abandonment of our project of returning via the Philippines. . . ." Perhaps other, unnamed factors were at work as well. Was he, like many other travelers before and after, enchanted by the beauty and the history of Mexico and eager to see more of the country? Or, after four strenuous years, were he and Bonpland simply worn out, having reached the end of even their amazing reserves? They were certainly homesick: "How I long to be in Paris!" Humboldt exclaimed in his letter to the Institut.

And so it was decided: Humboldt and Bonpland, still accompanied by Carlos Montúfar, would extend their stay in Mexico, eventually wending their way to the eastern coast, where they would find a ship to Cuba, then to the United States, and finally to Europe. Leaving Mexico City, they traveled north and west for the months of August and September 1803.

Their first destination was the thirteen-mile-long canal-and-tunnel complex at Nochistongo, constructed to control the floods to which Mexico City, built on Lake Texcoco, was susceptible. The most extensive water-control project of the colonial era and one of the great public works projects of the time, the canal had been designed and begun by Enrique Martínez, a Spanish engineer of French extraction. Martínez's design called for a canal and a five-mile-long tunnel through the relative lowlands of Nochistongo. Begun in 1607 and employing more than fifteen thousand Indian workers, the project had been completed in an astonishing ten months. But the tunnel, eleven feet wide and fourteen feet high, had proved inadequate, and various additions and extensions had been made over the ensuing decades. Finally, in 1767, the huge project had been brought to fruition, after enormous expense and the loss of an estimated seventy thousand workers' lives.

In early August, eager to explore Mexico's mineral wealth in more detail, Humboldt and the others set out for what at the time was the richest silver mine in the world, La Valenciana, located outside lovely Guanajuato, some 170 miles northwest of Mexico City. Dramatically perched on a series of steep ravines at the foot of the Sierra de Santa Rosa, Guanajuato was a major city, with a population of about seventy thousand. "One is astonished," Humboldt found, "to see in this wild spot large and beautiful edifices in the midst of miserable Indian huts. The house of Colonel Don Diego Rul, who is one of the proprietors of the mine of Valenciana, would be an ornament to the finest streets of Paris and Naples." Though the city had been founded in 1559, after the discovery of gold and silver, Guanajuato's current boom traced back to only 1768, when a partnership led by Antonio de Obregón y Alcocer had unearthed the richest vein of silver in the known world. Obregón, elevated to the title conde de la Valenciana, became one of the richest men in Mexico and, in thanks for his incredible good fortune, built the spectacular, gold-and-silver-filled Iglesia de San Cayetano on a hill overlooking the city, hard by the entrance to the mine itself. At the time of Humboldt's visit, La Valenciana employed more than twenty thousand and produced an astonishing one fifth of the world's annual silver extraction.

The former mining inspector was determined to learn everything he could. "I climbed all mountains using my barometer," he wrote. "In Valenciana I descended three times to the bottom of the mine, two times in Rayas, in Mellado, in Fraustros, in Animas, and in San Bruno. I visited the mine of Villalpando, spent two days in Santa Rosa and in Los Álamos. . . . I had a dangerous fall on my back in Fraustros, and experienced extreme pain for fourteen days due to a sprain of the base of my spine!" In his diary he referred to his stay in Guanajuato as one of the most exhausting periods of his life, quite a statement considering some of the other trials to which he had subjected himself.

With its extensive deposits of gold and silver, the area around Guanajuato had long been one of the richest regions of New Spain. A few years after Humboldt's departure, the region also became the cradle of Mexican independence. For decades, living conditions had been deteriorating among the largely Indian workers on the large estates, due to factors ranging from drought to population growth, to a shift from maize to cash crops such as wheat and fruit. By September 1810, the poverty and outrage had escalated to the point where Miguel Hidalgo, parish priest in the town of Dolores, not far from Guanajuato, was moved to issue his famous *Grito de Independencia*, calling for revolution, racial equality, and social justice. Dolores was the first town to fall to the rebels, followed by nearby San Miguel (home of Ignacio Allende, another leader of the revolt). Guanajuato itself was captured soon afterward, when its Alhondiga de Granaditas, a grain storehouse converted into a Spanish fortress, fell on September 28. But the uprising did not win the support of either the wealthiest Creoles in the area, or the more autonomous Indian workers in other parts of the country. In January of the following year, government forces retook Guanajuato and crushed the rebellion, displaying the severed heads of the revolutionary ringleaders—Hidalgo, Allende, Juan Aldama, and Mariano Jiménez—in iron cages hung from the four corners of the Alhondiga. A guerrilla war dragged on, and it was not until 1821, eleven years later, after liberal changes in Spain caused the conservative Creole elite to join the revolution, that Mexico finally won its independence. And even afterward, the country's history would be rife with political turbulence, revolution, and, in the mid-nineteenth century, a disastrous war with the United States.

Though all these events were still in the future, Humboldt sensed the political forces taking shape in Mexico, just as he had in the other Spanish possessions in South America. Whatever the future government of Mexico, he argued, the country's huge untapped potential must be developed to the advantage of all its citizens, including the Indians who had suffered the most under the three centuries of Spanish rule. It was to this mistreated and neglected segment of the population that he turned in the closing words of the *Political Essay on the Kingdom of New Spain*: "May this labor begun in the capital of New Spain," he hoped, "be of utility to those called to watch over public prosperity! And may it in an especial manner impress upon them this important truth, that the prosperity of the whites is intimately connected with that of the copper-colored race, and that there can be no durable prosperity for the two Americas till this unfortunate race, humiliated but not degraded by long oppression, shall participate in all the advantages resulting from the progress of civilization and the improvement of social order!"

On September 9, the travelers left Guanajuato and headed south, stopping at the gracious city of Valladolid (since renamed Morelia in honor of another hero of the War of Independence, José María Morelos y Pavón, who was born there in 1765). On September 19, about 150 miles west of Mexico City, they reached the volcano of Jorullo (thought to be an Indian word meaning "Paradise," a reference to the lush tropical surroundings). Humboldt was particularly eager to see Jorullo, as it was one of the few volcanoes formed in historical times.

On September 29, 1759, following a period of earthquakes and subterranean explosions, the mountain had risen from the surrounding fields, spewing hot ash and triggering mud flows. After six weeks, the cone had reached 820 feet, and by the time it had ceased erupting in 1774, it towered 4,400 feet above the surrounding plain and formed a crater over a mile wide. From eyewitness accounts of Jorullo's creation and from its bubblelike shape, Humboldt took the volcano as evidence for Leopold von Buch's "crater of elevation" theory, the enormously influential but ultimately disproved idea that volcanoes were formed by the upwelling of subterranean gases. Humboldt believed the fact that horses' hooves produced a peculiarly hollow sound on the earth nearby was a corroboration of that view.

The Indians living in the area had a different theory of Jorullo's origin. "In their opinion," Humboldt wrote, "the volcanic eruption was the work of the monks, the greatest, no doubt, which they have ever produced in this hemisphere!" Apparently, a group of Capuchin missionaries had stayed at a nearby plantation and, "not having had a favorable reception . . . they poured out the most insulting imprecations against the place. . . . They prophesied that the farm would be consumed by fire, and that soon afterward the air would cool off so that all mountains would be covered with snow and ice. The first of these maledictions having come true, the Indians regard the gradual cooling of the volcano as the sinister foreboding of perpetual winter. It is in this manner that the Church preys on the credulity of the natives so as to render their ignorant minds the more submissive."

From Jorullo, Humboldt's party continued to the Indian town of Pátzcuaro, located on the shore of a large and lovely lake. The travelers then trekked eastward to Nevado de Toluca, an extinct, fifteen-thousand-foot-high volcano with a lake in its crater. Humboldt became the first man to scale the volcano, taking barometric readings to determine the upper limit of vegetation and the lower limit of the snowcap. "As it is an honorable object for the exertions of scientific societies to trace out perseveringly the cosmical variations of temperature, atmospheric pressure, and magnetic direction and intensity," he wrote in *Aspects of Nature*, "so it is the duty of the geological traveler, in determining the inequalities of the earth's surface, to attend more particularly to the variable height of volcanoes." Toward this end, as he ranged over the Mexican isthmus, he measured, in addition to Jorullo and Toluca, the heights of the snowcapped, still-active Popocatépetl and the extinct volcanoes Cofre de Perote and Orizaba.

Moreover, Humboldt discovered that, whereas the Andean volcanoes ran in a north-south direction, those in Mexico formed a generally east-west line, today called the Trans-Mexican Volcanic Belt. Here was yet more evidence that, rather than being independent phenomena, as Werner and the neptunists believed, volcanoes shared a subterranean link and tended to congregate along fissures in the earth's crust. Thus the suspicions concerning the origin of volcanoes that Humboldt had reached in South America were confirmed by his study of the Mexican mountains.

Returning to Mexico City, Humboldt and Bonpland began to pack their extensive collections—now including Aztec codices and sculptures in addition to new geological and botanical specimens—for the long journey back to Europe. On January 20, 1804, they and Carlos Montúfar set out eastward from the capital, toward Puebla, known as "the City of Tiles" for the beautiful hand-painted ceramics decorating many of its buildings. Founded in 1535, on the road between Mexico City and Veracruz, Puebla had grown by the time of Humboldt's visit to a city of nearly fifty thousand, boasting one of the country's finest cathedrals and the first theater in all of Mexico. From Puebla, the travelers continued east to Cholula, which had been a major city and important religious center for nearly two millennia. With a population of about one hundred thousand at the time of Cortés's arrival, the city had chosen to resist the Spaniards, for which they had suffered an estimated six thousand dead and a sacking by Cortés's allies the Tlascans.

From Cholula, Humboldt's party passed the old Aztec town of Jalapa, then began their descent from the highlands toward the coast. On February 19, they arrived at Veracruz, ending their Mexican adventure at the place where Hernán Cortés had begun his. Despite the bad anchorage and numerous sandbanks, Veracruz was ideally located for trade with Spain and the Caribbean, and for centuries the city had been Mexico's principal eastern port, servicing some five hundred ships a year at the time of Humboldt's visit. Yet, with its tropical heat, it was also considered Mexico's "principal seat of yellow fever (*vómito prieto*). Thousands of Europeans landing in Mexico at the period of the great heats fall victims to this cruel epidemic," Humboldt explained. "Some vessels prefer landing at Vera Cruz in the beginning of winter, when the tempests begin to rage, to exposing themselves in summer to the loss of the greater part of their crew from the effects of the *vómito*, and to a long quarantine on their return to Europe." So severe were the outbreaks that they impeded the economy of the entire country, which received necessities from Europe through Veracruz and shipped goods to market via the port. Fortunately, there was no outbreak during Humboldt's stay. The packet for Havana was due to depart in three weeks, leaving him ample time to verify the port's location and explore the environs.

IN HIS YEAR in New Spain, Humboldt had collected an enormous amount of information about the land and its people. Taking detailed measurements wherever he went, he had improved the country's maps and had introduced an original and widely copied cartographic technique depicting a cross-section of the country's landforms. Exploring Mexico's volcanoes, he had confirmed his earlier ideas about mountain formation and had moved even further into the vulcanist camp. Studying the Aztec and other indigenous cultures, he had made important realizations concerning their Asian origin and their fantastic history, which he was about to reveal to Europe.

Moreover, he was leaving a considerable legacy within Mexico's borders. For Humboldt recognized Mexico's unique place among the countries of the New World—populous, thriving, but nevertheless harboring tremendous untapped potential. Spain had been so singly focused on the viceroyalty's obvious mineral wealth that it had neglected the country's many other resources— agricultural, manufacturing, commercial, human—which in the long run, Humboldt realized, would have made a much more lasting contribution to the colonial economy. Mexico's annual agricultural production, he pointed out, actually had a significantly greater monetary value than the vaunted gold- and silver-mining operations. "If we consider the people of New Spain and their commercial connections with Europe, it cannot be denied that in the present state of things the abundance of the precious metals has a powerful influence on the national prosperity. It is from this abundance that America is enabled to pay in specie for the produce of foreign industry, and to share in the enjoyments of the most civilized nations of the old continent. Notwithstanding this real advantage," he warned, "it is to be sincerely wished

that the Mexicans, enlightened as to their true interest, may rec-
ollect that the only capital of which the value increases with time
consists in the produce of agriculture, and that nominal wealth
becomes illusory whenever a nation does not possess those raw
materials which serve for the subsistence of man or as employment for his industry."

In 1811, Humboldt would publish his *Political Essay on the
Kingdom of New Spain*. Like the later *Political Essay on the Island of Cuba*, the book would draw together diverse data regarding population growth, agriculture, mines, manufactories,
commerce, and the distribution of wealth, all collected from personal observation or government archives. A masterful synthesis
of all these disparate elements, *New Spain* provided the first systematic, detailed examination of conditions in the colony and the
problems it faced as it entered the nineteenth century. Though
stuffed with statistics, the work was also intended as a plea—
especially to Charles IV of Spain, to whom the book is dedicated—
to recognize Mexico's enormous nonmineral wealth and to harness
it in a way that would benefit all its citizens.

The indefatigable, perspicacious Humboldt made such a favorable impression on the colonial authorities during his stay
in New Spain that they offered him a ministerial position. Conversely, a few years later, the leaders of the independence movement, taking inspiration from his republican ideals and his faith
in Mexico's future, would adopt him as a hero of their own cause.
Afterward, Benito Júarez, the Mexican president revered for his
moral courage and enlightened government, conferred on Humboldt the title "Benefactor of the Nation." And in the middle of
the nineteenth century, fifty years after the Prussian's departure, President Ignacio Comonfort bestowed on him the Grand
Cross of the Order of Guadelupe and proposed christening a
town in his honor. In several Mexican cities, streets are still
named for him and houses where he stayed have been converted
into museums. Though in Mexico, as in the world at large, Humboldt's name has dimmed over the ensuing centuries, his spirit
lives on there, in the reverence for cultural heritage and the continuing struggle for social progress.

ON MARCH 7, 1804, Humboldt, Bonpland, and Carlos Montúfar
sailed for Havana. There they reclaimed the crates of geological,

botanical, and zoological specimens they had stored for safekeeping. Then on April 29, 1804, they sailed from Cuba on the Spanish frigate *Concepción*. On their earlier visit, some three years before, Humboldt had found Havana "one of the gayest and most picturesque" ports in the Western Hemisphere. But now, as he watched the quays and fortresses slip by, we can imagine him in a more pensive mood, his ever-active mind ranging over all he had seen and accomplished in the New World.

After five years' traveling in uncertain conditions, he must have been eager to find himself again in familiar, comfortable surroundings. And surely he was looking forward to his reunion with Wilhelm, Caroline, and their growing family. Yet that anticipation would have been tinged with the poignant realization that the greatest adventure of his life was drawing to a close. Earlier, in the letter to his friend Delambre posted from Mexico City, Humboldt confessed, perhaps melodramatically, his ambivalence over returning to staid, safe Paris. "It is every man's duty to seek that position in life in which he can best serve his generation," he had written. "I almost think that to fulfill my destiny I ought to perish in a crater or be drowned on the high seas. This at least is my present opinion . . . though I can readily believe that with advancing age and the enjoyments of European life I may yet live to change my views."

Despite his sensitivity to melancholy, Humboldt wasn't one to brood over the past for very long. Bursting with restless energy and insatiable curiosity, he was constantly embarking on new projects and making new plans. So as the coast of Cuba receded from view, he must have considered his homecoming with increasing excitement. He would have to get Delambre and von Buch and the others to bring him up to date on all the scientific advances made while he was away. And what stories he would have to tell—he could imagine their expressions when he described the near-capsize in the Orinoco, or showed them rocks from Chimborazo. And what new information he had to share— from the first detailed investigations of American volcanoes, to the location of the earth's magnetic equator, to the discovery of thousands of previously unknown plants, to his researches on the incredible, neglected history of the continent's pre-Hispanic inhabitants. What a privilege to be entrusted with the first in-depth scientific exploration of the New World, and with the dis-

semination of all that new information to professionals and lay-
men alike. Over the five years of his journey, he had collected
enough material to fill dozens of volumes (in the end, there
would be thirty based on his American travels). Indeed, though
he didn't know it at the time, he was about to spend the remainder of his fortune and much of the rest of his life acting as the
principal emissary from the New Continent to the Old, and re-
vealing to the rest of the world the full scope of America's natural
and man-made wonders.

But Humboldt's American adventure wasn't quite over. For
the *Concepción* was bound for Philadelphia, not Spain. He must
also have been looking forward to this visit to the United States,
the infant republic for which he held such affection and harbored
such high hopes. But first there were more than fifteen hundred
waterborne miles to travel between Havana and Philadelphia.
After a brief hiatus following the 1802 Treaty of Amiens, the war
in Europe had resumed, with Spain allied with France against
England, Austria, and Russia. As a result, the *Concepción*, like
any Spanish vessel, would be subject to seizure by enemy warships. There was also the ever-present threat posed by natural
hazards such as storms. The frigate was buffeted by a hor-
rendous, week-long gale in the Bahama Straits, during which
Humboldt worried obsessively over the safety of their scientific
specimens—especially since he still had no idea what had happened to the two other parts of the collection shipped to Europe
three years before. (If he had known that the portion entrusted
to Juan González had been lost off the coast of Africa, Humboldt's anxiety would have been all the greater.)

But the *Concepción* survived the gale, and twenty-four days
out of Havana, the frigate entered the Delaware River and docked
safely at the port of Philadelphia. Humboldt's Latin American
odyssey had come to an end. The dissemination of his discoveries
and the building of his legacy were about to begin.

TWELVE

Washington, Paris, and Berlin

AS THE *CONCEPCIÓN* TACKED UP THE DELAWARE RIVER, the spires of Philadelphia hove into view along the western bank. With its cobbled, poplar-lined streets and handsome brick buildings, the city seemed more familiar, more European, than the capitals of South America that Humboldt had come to know over the past five years. Though the national government had relocated to Washington four years before, Philadelphia remained the cultural hub of the young republic, and Humboldt was eager to call at the prestigious American Philosophical Society and catch up on the scientific advances of the past five years.

He was also hoping to arrange a meeting with Thomas Jefferson, who remained as president of the Philosophical Society even after having been elected president of the nation. Jefferson was by all accounts a man of rare accomplishment, sensitivity, and intellect, and Humboldt was a longtime admirer; in fact, the possibility of paying his respects was a primary purpose of this visit. The year before, the president had acquired from the French a huge swath of territory, extending from St. Louis to New Orleans, that had once been part of New Spain—the Louisiana Purchase. Humboldt flattered himself that, as Jefferson dispatched his own corps of discovery to explore this new territory, he would be interested in what his visitor had gleaned about the land during all those days spent in the government archives of Mexico City.

Soon after his arrival, Humboldt wrote a letter in French to the president, who was located in the new capital of Washington, D.C., emphasizing their common political and scientific interests:

Mr. Jefferson:

Arrived from Mexico on the blessed ground of this republic . . .
I feel it my pleasant duty to present my respects and express
my high admiration for your writings, your actions, and the
liberalism of your ideas, which have inspired me from my
earliest youth. . . .

For moral reasons I could not resist seeing the United
States and enjoying the consoling aspects of a people who
understand the precious gift of Liberty. I hope to be able to
present my personal respects. . . . I am quite unaware whether
you know of me already through my work on galvanism and
my publications in the annals of the Institut National in Paris.
As a friend of science, you will excuse the indulgence of my
admiration. I would love to talk to you about a subject that you
have treated so ingeniously in your work on Virginia, the teeth
of mammoth which we too discovered in the Andes. . . .

While he waited for a response, Humboldt sought out the
city's scientific elite, including physician Caspar Wistar, botanist
Benjamin Smith Barton, and physician (and signer of the Decla-
ration of Independence) Benjamin Rush. Combing through the
public library, he was delighted to come across an item in a scientific journal announcing the "arrival of M. de Humboldt's manuscripts at his brother's house in Paris, by way of Spain." It was
his first indication that any of the precious specimens had safely
reached their destination. Humboldt also discovered that some of
his European correspondents had forwarded his letters to the
French journals, which had been reporting on the progress of his
expedition. But he was taken aback to learn that his death had
been announced in the press on three separate occasions—most
recently in Acapulco, where he had supposedly succumbed to a
tropical disease. Humboldt's arrival in Philadelphia was also
duly noted in the city's newspapers. *Relf's Philadelphia Gazette
and Daily Advertiser* reported, "Baron de Hombott [sic] arrived
in this city on Wednesday night."

Within days, Humboldt had a cordial reply from the president. Having just dispatched Lewis and Clark on their great expedition of discovery (they started up the Missouri River in May
1804, the same month that Humboldt arrived in Philadelphia),

Jefferson was eager to hear more about the formerly Spanish territories. "A lively desire will be felt generally to receive the information you will be able to give," the president wrote. "No one will feel it more strongly than myself, because no one perhaps views this New World with more partial hopes of its exhibiting an ameliorated state of the human condition. In the new position in which the seat of our government is fixed, we have nothing curious to attract the observations of a traveler, and can only substitute in its place the welcome with which we should receive your visit, should you find it convenient to add so much to your journey."

Soon after, Humboldt, Bonpland, Montúfar, and Charles Wilson Peale (painter, amateur scientist, and friend of Jefferson), set out for Washington. Peale described the long coach ride in his diary: "The Baron spoke English very well, in the German dialect. . . . It was amusing to hear him speak English, French, and Spanish Languages, mixing them together in rapid Speach. He is very communicative and possesses a surprising fund of knowledge, in botany, mineralogy astronomy Philosophy and Natural History . . . he has been travelling ever since he was 11 years of age and never lived in any one place more than 6 months together, as he informed us." (This latter version of his life wasn't the first or last time that Humboldt would exaggerate his adventures for effect.)

Washington City was still very much under construction, but Humboldt was warmly received by the capital's luminaries, who were greatly impressed by his erudition and energy. Secretary of the Treasury Albert Gallatin wrote his wife, "We all consider him as a very extraordinary man, and his travels, which he intends publishing on his return to Europe will, I think, rank above any other productions of the kind. . . . He speaks twice as fast any anybody I know, German, French, Spanish, and English altogether. But I was really delighted, and swallowed more information of various kinds in less than two hours than I had for two years past in all I had read or heard. . . . The extent of his reading and scientific knowledge is astonishing. . . ."

Dolley Madison, wife of Secretary of State (and future president) James Madison, was also taken with the visitor. As she wrote her sister, "We have lately had a great treat in the company of a charming Prussian Baron von Humboldt. All the ladies say they are in love with him, notwithstanding his want of personal charms. He is the most polite, modest, well-informed, and interesting traveler we have ever met, and is much pleased with America. . . . He had with him a train of philosophers who, though clever and entertaining, did not compare with the Baron."

President Jefferson hosted Humboldt and his companions at a luncheon at the White House and later invited them to Monticello, where they had long discussions ranging over meteorology, agriculture, astronomy, and other shared scientific interests. Jefferson's secretary William Burwell, who was present at these meetings, wrote that Humboldt was "of small figure, well made, agreeable looks, simple unaffected manners, remarkably sprightly, vehement in conversation and sometimes eloquent. Jefferson welcomed him with greatest cordiality and listened eagerly to the treasure of information."

As well he should have. Though Humboldt's expedition through New Spain had been sponsored by the Spanish government, and the information he'd collected there was of a decidedly sensitive nature (including the most accurate maps ever created of the country and the most extensive data ever compiled on its mines), Humboldt now suffered no qualms of conscience over sharing this intelligence with a neighboring foreign power. The president was particularly eager for any information that might bear on the new Louisiana Territory: "Can the Baron inform me what population may be between those lines of white, red or black people?" he asked in a note to Humboldt. "And whether any & what mines are within them? The information will be thankfully received." As the nineteenth century progressed and the United States expanded inexorably westward, Humboldt's maps would prove an invaluable resource for explorers, generals, and engineers for decades to come.

As time neared for his departure, Humboldt sent Jefferson a note of warm thanks and friendship: "I have had the honor to see the First Magistrate of this great republic living with the simplicity of a philosopher, receiving me with such great kindness as I shall always remember. . . . I take leave in the consolation that the people of this continent march with great strides toward the perfection of a social state, while Europe presents an immoral and melancholy spectacle. . . . I sympathize with you in the hope ... that humanity can achieve great benefit from the new order of things to be found here. ... "

The two men would correspond for many years, and Jefferson's high regard for Humboldt is obvious in his letters. In one dated December 6, 1813, he addresses the other as "my friend" and assures him of his "constant attachment" as well as his "affectionate esteem and high respect and consideration." The friendship, rooted in their shared political philosophy and common love of science, would endure for more than twenty years, until Jefferson's death in 1826.

On July 9, 1804, Humboldt and his companions departed the New Continent aboard the French frigate *La Favorite*. Though he had made an enduring impression on the United States, Humboldt's reception there was a faint glimmer of what was awaiting him in Europe.

THE TRANSATLANTIC VOYAGE, for once, was swift and smooth and free of British warships, and *La Favorite* dropped anchor in Bordeaux on August 1, following a passage of only twenty-three days. While Bonpland visited his brother in La Rochelle, Humboldt and Montúfar took the coach to Paris. After five years of constant travel, Humboldt must have been relieved to be home at last—or at least what passed for home to the undomesticated, peripatetic bachelor. But in a letter to a friend, he also admitted his uncertainty over being back in Europe. "I confess that it was with a heavy heart that I parted from the bright glories of tropical lands," he wrote. "I quite dread this first winter. Everything will be so strange, and it will take me some time to settle down again."

The French capital had changed tremendously during his absence. When Humboldt had left for Madrid in 1798, the Directoire had been tottering, the national treasury had been empty, and Napoleon had been leading the French army in Egypt. But by 1804, Bonaparte had been installed as first consul for life, financial and political stability had been restored, and France had emerged as the most powerful nation in Europe. In fact, having ruled as a virtual dictator for years, Napoleon was about to be crowned emperor, with powers more sweeping than those of any Bourbon monarch.

Humboldt detested absolutism in any form, but more than

ever, Paris was the cultural, scientific, and publishing capital of Europe, and he believed that only there could he realize the ambitious series of books based on his American travels. Thus the republican Humboldt made the first in a series of political compromises with European royalty, in this case establishing himself in the capital of an expansionist imperial power—and one that would shortly be at war once again with his native Prussia. Despite his genuine and oft-voiced convictions, he felt compelled to set aside his principles and establish himself in the place he felt would be most conducive to his research and writing.

Like Napoleon before him, Humboldt was received in Paris as a conquering hero. Caroline, who was in the city for her brother-in-law's arrival, reported to Wilhelm the sensation that Alexander was making. "It has rarely fallen to the lot of any private individual to create so much excitement by his presence or to give rise to such widespread interest," she wrote. And in another letter: "Alexander continues to make the greatest impression. His collections are immense, and to work them over, to compare and develop his ideas, will require from five to six years. . . . He is not in the least aged . . . his face is decidedly fuller and his liveliness of speech and manner has increased—as far as that is possible. . . ." In fact, she was afraid all the attention was going to his head. "Alexander lets himself be swept off his feet by French charm," she added with a whiff of disapproval. But in the end she decided, he "is always the same. Impossible to describe him. He is such an incredible mixture of charm, vanity, soft feelings, cold, and warmth as I have ever seen in anyone else."

Partly it was the force of his personality—the tremendous energy, the penetrating intellect, the enormous erudition—that struck all who came in contact with the thirty-five-year-old explorer. Everyone seemed to recognize in him a larger-than-life figure who comes along once in a generation, if then. But it was not only Humboldt's huge intellectual capacity that captured the public imagination. No one could deny his amazing physical achievements as well—penetrating rain forests where no European had ever ventured, climbing higher than anyone had thought possible, cheating death on more than one occasion. Humboldt had laid bare a wild, unknown continent, returning with fabulous stories, bizarre plants and animals, and new insights into nature itself. Though later in life he would also be revered for his humanitarianism and his generosity, it was the captivating blend of mental brilliance and physical daring that generated this first, galvanizing rush of celebrity—as if, in a later century, Albert Einstein had been the first to conquer Mount Everest.

In typical fashion, Humboldt responded to this new celebrity with a relentless commitment to his work. "He was awfully busy in the first eight days," Caroline reported. "He is with Kohlrausch [the Prussian ambassador] quite a bit, meets him as early as six in the morning, works and talks all the time." Humboldt himself wrote of this period, "I am working with greater effort than ever, and trust that my publications will turn out to be less immature than my last. My health is better than it has ever been," he added, despite a partial paralysis in his right arm, which he attributed to rheumatism contracted in the rain forest; the condition would trouble him for the rest of his life. To a friend Humboldt admitted feeling overwhelmed by the extent of what he had taken on. "I've really bitten off more than I can chew," he confessed.

In mid-October, Humboldt presented the first reports of his journey at a crowded, highly anticipated meeting of the Institut National des Sciences et Arts, the organization created by the French government in 1795 to promote science, the beaux arts, and literature. Also in October, the first exhibition of Humboldt's botanical collection opened at the Jardin des Plantes, where it proved a sensation. Meanwhile, the Bureau of Longitude Studies and the Observatory were reviewing his copious barometric and astronomical measurements, and artists had been hired to begin copying his sketches of plants and ancient Indian monuments. As in Washington, everyone was amazed by Humboldt's intellectual scope. "This man combines an entire academy in him," marveled great French chemist Claude-Louis Berthollet.

However, one contemporary who didn't share the general excitement over Humboldt's return was the first consul himself. Suspicious of Humboldt's Prussian nationality, the emperor-to-be was also aware of his liberal politics. Perhaps he was even jealous of his fame, which was said at the time to be second only to Bonaparte's own. Though Napoleon was a member of the Institut National, he declined to attend Humboldt's lecture there. And when the two met at a gala in the Tuileries following the emperor's coronation on December 2, 1804, Napoleon remarked coolly, "I understand you collect plants, monsieur." Humboldt confirmed the supposition, whereupon the emperor replied, "So does my wife," and turned his back.

"The Emperor Napoleon," Humboldt worried, "behaved with icy coldness to Bonpland and seemed full of hatred towards me." But the antipathy between the two men (who had been born in the same year) was mutual, and Humboldt made a point of avoiding the emperor's weekly receptions. In fact, the abrupt meeting at the coronation gala was their only face-to-face encounter. Convinced that the other was an enemy spy, Napoleon had him followed by the secret police and once even ordered him expelled from France. When Humboldt persuaded a friend to intercede, the emperor asked, "But doesn't he devote his time to politics?" "No," was the reply, "I've never heard him talk about anything but science." Bonaparte reluctantly rescinded his order.

It was at Napoleon's coronation that Humboldt also met Simón Bolívar. Bolívar had been a boy of sixteen when Humboldt and Bonpland had arrived in his native city of Caracas, Venezuela, in 1799. The two hadn't met at the time, because young Simón, son of one of the city's wealthiest Creole families, had been sent to Europe to give his education a final polish. A child of the Enlightenment and a student of Rousseau, Voltaire, and Jefferson, Bolívar (like Humboldt) deplored Napoleon as a perversion of the principles of the French Revolution. But at the same time he couldn't help admiring the Corsican's martial acumen and personal charisma—qualities he would later strive to emulate—and it was perhaps for this reason that he chose to attend the installation of the new emperor.

When Humboldt and Bonpland were introduced to Bolívar at the gala, the conversation quickly turned to politics. Humboldt suggested that, following Spain's revocation of its liberalized trade policy two years before, Venezuela was ripe for revolution—if only the right leader would step forward to focus the opposition. Bonpland was also encouraging, expressing the view that "revolutions themselves bring forth great men who are worthy of carrying them out." In August 1805, Humboldt and Bolívar met again in Rome, and again discussed South American independence. Then, on August 15, Bolívar climbed Rome's Aventine Hill and made his famous vow to dedicate himself to freedom.

The following year, Francisco Miranda, another native of Caracas, led a British-supported revolt in Venezuela. But most of his countrymen remained loyal, the coup faltered, and Miranda fled to England. By 1807, Bolívar was back in Venezuela, still bent on revolution. Three years later, the republicans' moment finally arrived, after France had invaded Spain and taken King Ferdinand VII prisoner. When Spain's final defeat seemed imminent, Miranda returned to Venezuela, deposed the Spanish viceroy, and installed himself as dictator. On July 5, 1811, the cabildo, or legislature, declared the country's independence. But in 1812 a devastating earthquake struck Caracas, and in its aftermath the city was reoccupied by loyalists. Miranda was imprisoned, and his aide, Simón Bolívar, took his place at the head of the movement.

Over the next decade, Bolívar and his troops persevered through several seemingly fatal reversals, finally ousting the Spanish for good in 1822. That year, he was named president of the newly formed Gran Colombia, a federation of present-day Colombia, Panama, Ecuador, and Venezuela. But the distances proved too vast and the political and social differences too great, and even El Libertador was unable to hold the federation together. After Gran Colombia was disbanded in 1830, Caracas became the capital of independent Venezuela. Bolívar died of tuberculosis on December 17 of that year, a virtual outcast in the nation he had liberated. As he commented bitterly, "Independence is the only benefit we have gained, at the cost of everything else." But to the end Bolívar counted himself a friend and admirer of Humboldt, whom he lauded as "the true discoverer of South America" for exploring more of the continent's six million square miles than anyone before, for captivating the world with his depictions of the region's aesthetic and scientific wonders, and, most important, for inspiring South Americans to recognize the vast resources of their own continent and to utilize them for social and political progress.

PREOCCUPIED THOUGH HE WAS with his publications after his return to Paris, Humboldt traveled to Switzerland to conduct a series of experiments on atmospheric physics with his new friend French chemist Joseph-Louis Gay-Lussac. Afterward, he went on to Italy for a delayed reunion with his brother, Wilhelm. Then, grudgingly, he journeyed to Berlin to pay his respects to

King Friedrich Wilhelm III. Thus it was 1807 before Humboldt's hugely ambitious publication program finally began to see the light of day.

The first to appear was *Aspects of Nature*, which, though based only in part on his travels on the New Continent, would prove one of the most popular of all his books, as well as his personal favorite. Consisting of eight essays, ranging from the physiognomy of plants to the perils of the Llanos, from the structure of volcanoes to the majesty of the Orinoco, the work was meant to capture the direct experience of nature, as though the reader himself were enjoying the magnificent view from atop a tall mountain or hearing the rain forest come alive at night. The book was intended to provide the wider public with the same deep pleasure that Humboldt had always found in the aesthetic and intellectual enjoyment of nature's wonders. As he wrote in the preface, "To minds oppressed with the cares or the sorrows of life, the soothing influence of the contemplation of Nature is peculiarly precious. . . . May they, 'escaping from the stormy waves of life,' follow me in spirit with willing steps to the recesses of the primeval forests, over the boundless surface of the Steppe, and to the higher ridges of the Andes."

Also in 1807, Humboldt released the first of his American publications. Originally planning only eleven works, he confidently wrote a friend, "Considering the remarkable energy of my disposition, I expect to see the whole thing out of my hands in a couple of years, or at most in two and a half years, as I'm now impatient to discharge my cargo in order to embark on something new." In reality, the monumental set, published under the collective title *Voyage to the Equinoctial Regions of the New Continent, Made in 1799–1804*, would run to thirty volumes and wouldn't be completed for more than three decades. Prepared with the aid of various collaborators and sporadic help from Bonpland (who was listed as coauthor), the books were by far the most ambitious publication program of the period.

The first volume to appear was the *Essay on the Geography of Plants*, the seminal work that Humboldt had started immediately after his descent of Chimborazo. Dedicated to his old friend Goethe, the *Essay* would prove one of Humboldt's most enduring contributions to natural history, correlating plant growth with physical factors such as soil type, exposure to sun and wind, temperature, and especially height above sea level. Just as vegetation changes predictably with latitude, Humboldt demonstrated, plant distribution also varies systematically with altitude: From the ocean to 3,000 feet above sea level (on the equator) is found the zone of palms, followed by those of ferns (to 4,900 feet), oaks (to 9,200 feet), evergreen shrubs (to 10,150 feet), herbs (to 12,600 feet), and finally grasses and lichens (to 14,200 feet). By codifying how geographic features influence plant growth, Humboldt created a whole new branch of science, still known by the name he gave it, plant geography, which has had a huge impact not only on our understanding of botanical processes, but on horticulture and commercial agriculture as well.

In 1810 appeared the two-volume, oversized, lavishly illustrated *Researches Concerning the Institutions and Monuments of the Ancient Inhabitants of America*, the work that established Native American cultural studies—including art, architecture, language, astronomy, and religion. Though the book would prove tremendously influential, not everyone in Europe was ready to receive its message. As the reviewer for the conservative British *Quarterly Review* put it, "We cannot admit with our author that a nation so barbarous as the Mexicans had any knowledge of the *cause* of eclipses. . . . A picture language or such rude representation of the objects of sense as village children chalk on walls and barn-doors, are the first and rudest efforts to record ideas, . . . and with both of these even the wild Hotentots, . . . the very lowest perhaps of the human race, appear to be acquainted. . . . The Mexicans may have advanced but, we believe, not a great way, beyond the village children. . . ." The reviewer continued, "We have dwelt but little, and that little will perhaps be thought too much on those cycles and calendars, those chronologies and cosmogonies extracted out of the—to us, at least—unintelligible daubings designated under the name of the 'Codices Mexicani.' To M. de Humboldt, however, they would appear to be of first-rate importance. . . ." Fortunately, not everyone was so dismissive of American indigenous culture.

Next came the three-volume *Political Essay on the Kingdom of New Spain*, the first comprehensive study of Mexico, which, published in 1811, the year after the colony's unsuccessful revolt against Spain, proved particularly timely. Incorporating Humboldt's original research as well as a wealth of information culled from government archives, the book became the prototype work of political geography, exploring the intricate connections between the physical and man-made worlds.

In 1814 began publication of the long-awaited series of books intended for a general audience, the *Personal Narrative of Travels to the Equinoctial Regions of the New Continent*. Ultimately comprising three volumes, this was the chronicle that would thrill the young Charles Darwin. However, the work didn't prove to be the breezy adventure story that the public was anticipating. Packed with scientific data and technical digressions, the books were more a physical description of South America and an account of the social and political conditions that Humboldt had found there. With the prepublication destruction of Volume Four, dealing with the Andes, the work was also incomplete.

The *Quarterly Review* praised Humboldt's *Personal Narrative*, for "seeing every thing, and leaving nothing unsaid of what he sees;—not a rock nor a thicket, a pool or a rivulet,—nay, not a plant nor an insect, from the lofty palm and the ferocious alligator, to the humble lichen and half-animated polypus, escapes his scrutinizing eye, and they all find a place in his book. . . . he is so deeply versed in the study of nature, and possessed of such facility in bringing to bear, that we may say of him . . . that he never quits a subject until he has exhausted it." The review went on, in a less laudatory vein, "But this very facility . . . becomes a fault in the personal narrative of voyages or travels; at least the bulk of readers will be very apt to lay down the book on finding the thread of the story perpetually interrupted by a learned disquisition of a dozen pages on the geognostical constitution of a chain of mountains, or the lines of isothermal temperature." Humboldt's dear friend the French mathematician and physicist Dominique François Arago was more succinct: "Humboldt, you really don't know how to write a book. You write endlessly, but what comes out of it is not a book, but a portrait without a frame."

Between 1814 and 1834, Humboldt published a three-volume history of the early voyages of discovery to the Americas, called *Critical Examination of the History of Geography of the New Continent and of the Progress of Nautical Astronomy in the 15th and 16th Centuries*. Particularly notable in this work was his solution of the riddle of how the New Continent had come to be named America, after Florentine explorer and mapmaker Amerigo Vespucci. For centuries, Vespucci had been suspected of appropriating the name himself, in what would have been one of history's most blatant acts of self-promotion. But through original bibliographic research, Humboldt proved that the honor had actually been bestowed by sixteenth-century German cartographer Martin Waldseemüller on his 1507 map (which was also the first to depict North and South America as independent continents and not part of Asia). Indeed, it is doubtful that Vespucci, who died in 1512, ever even saw the map that made him famous.

In 1828, Humboldt published the two-volume *Political Essay on the Island of Cuba*, another pioneering work of political geography, which was noteworthy, in addition to being the most comprehensive study ever made of the island, as Humboldt's most extensive, impassioned argument for the abolition of slavery throughout the world.

During these years, Humboldt published works of a more technical nature as well—on geology, zoology, barometric and astronomic measurements, and his new graphic technique of isothermal lines. As Humboldt immersed himself in these various projects, it became clear that his earlier estimate of a few years to complete his American books had been wildly optimistic. It also became apparent that the books would cost far more than his original estimate. Encompassing more than fourteen hundred illustrations, many of them hand colored, the works were in fact the most costly publication program ever undertaken by a private individual, with no expense spared in design, artwork, and printing. Humboldt also advanced generous fees to his various collaborators, specialists who ensured the scientific accuracy of the publications. His own perfectionism added further to the huge cost—when he was dissatisfied with the illustrations, he destroyed them, along with their printing plates, and had them redone at enormous personal expense. Moreover, while some of the works were meant for a general readership, sales of the more specialized volumes were predictably modest. With huge production costs and uneven sales, the massive undertaking gave rise to the pinched financial circumstances in which Humboldt, once independently wealthy, would find himself for the rest of his life. The American journey itself had taken more than a third of his inheritance; the publications it produced would consume the remainder.

Other factors besides financial worries were weighing on Humboldt during this time. In some ways, the years after his return to Europe should have been the most satisfying of his life, as he reveled in worldwide celebrity and reaped the scientific windfall from his expedition to the New World. But this period, coming after his greatest adventure, was also tinged with a sense of anticlimax. And Europe's wars and political turmoil took a toll on him as well. In fact, for the rest of his life, the fervent republican would find himself in treacherous circumstances, making uncomfortable compromises with monarchs for whom he had little respect and with whom he shared little common philosophical ground.

When Humboldt visited Berlin in 1805 to pay his respects to Friedrich Wilhelm III, the king appointed him a chamberlain to the royal court. The republican Humboldt was uncomfortable with the title from the beginning. "Pray do not mention that . . . I was made a chamberlain!" he implored in a letter to a friend. But he didn't feel he could decline the honor, and in any event, he needed the modest stipend that went with it. Worse yet, the retainer wasn't nearly enough to cover his expenses, which, in addition to producing his publications, included maintaining a standard of living appropriate to someone of his class, complete with coach and personal servant. Thus, Humboldt was forced to swallow his convictions and approach the king for periodic additional grants.

Though Friedrich Wilhelm was eager to have Humboldt in Berlin, where his presence would add to the luster of the royal court, Humboldt managed to beg off, on the grounds that only in Paris could he find the technical, artistic, and editorial assistance he needed to complete his books. Then in early 1813, Prussia, having been defeated by Napoleon at Jena and forced to sign the humiliating Treaty of Tilsit, rejoined the coalition against France. Humboldt's brother, Wilhelm, and Wilhelm's son, Theodor, joined the army, as did German patriots of all ages. Wilhelm felt Alexander should do the patriotic thing and rush home to join the volunteers, but even at that time of national crisis, the younger Humboldt remained in the comfort of the enemy capital.

"I must confess quite candidly . . . ," Wilhelm wrote Caroline, "that I cannot endorse Alexander's stay in Paris. It is true that he could not do anything for the war comparable in importance to what he did in Paris. . . . But the honorable thing is not to weigh profits. To value one's own personality so highly and spare oneself is beyond all my estimation of a good character."

But Alexander, though intensely anxious over the safety of his brother and nephew, had no particular patriotic feeling for Prussia, and he felt no compunction about refusing his country's call. "German patriotism is a high-sounding phrase," he later wrote a friend. "In 1813 it served to fire the hearts of German youth. . . . And what has resulted from that infinite waste of blood and treasure? The probable outcome was already evident in 1814, when the crowned heads congregated in Paris" to restore the Bourbon dynasty in France and to stifle democratic reform across the continent. When Wilhelm was awarded the Iron Cross for his wartime service, Alexander quipped, "I would have preferred the Southern Cross." Yet despite his lingering disapproval, Wilhelm, who had been Prussian minister of education, intervened on his brother's behalf and secured a generous grant from the king, which helped to alleviate Alexander's financial crisis and to underwrite the ongoing publication program.

However, by 1827, Charles X was on the throne of France, the ultramonarchists held sway, and the French government had instituted press censorship and other instruments of political repression. Finding this reactionary climate increasingly untenable, Humboldt at last considered a return to Berlin, though he had earlier compared the city to a desert and had confidently written Kunth, his old tutor, "You can be sure that I shall never find it necessary to set eyes on the spires of Berlin again." Eager to secure Humboldt's return, King Friedrich Wilhelm offered him a generous stipend and permission to spend several months a year in Paris, if he would only agree to return to the Prussian capital. In fact, the king's position wasn't as much of an offer as an ultimatum. "My dear Herr von Humboldt!" he wrote. "You must by now have finished the publication of the works you considered could only be finished in Paris. I therefore cannot allow you to extend your stay any further in a country which every true Prussian should hate. I accordingly await your speedy return to the Fatherland. Yours affectionately, Friedrich Wilhelm."

On his return to his native city, Humboldt accepted the title privy councilor and became the king's primary advisor on all matters scientific and artistic. When Friedrich Wilhelm III died in 1840 and was succeeded by Friedrich Wilhelm IV, the new monarch proved even more dependent on Humboldt, appointing him to his state council, having him review all his personal correspondence, and even demanding he spend several evenings a week with the royal family. The irony of a famous and vocal proponent of republican government finding himself an intimate member of a royal court did not go unnoticed, and Humboldt suffered some rebuke on account of it, especially after some of his private correspondence was made public in which he satirized the monarch.

Thus Swiss-American naturalist and Humboldt protégé Louis Agassiz later found it necessary to defend his mentor for this compromise. Humboldt "has been blamed for holding his place at court," Agassiz noted, "while in private he criticized and even satirized severely everything connected with it. . . . We may wish that this great man had been wholly consistent, that no shadow had rested upon the loyalty of his character, that he had not accepted the friendship and affection of a King whose court he did not respect and whose weaknesses he keenly felt. But," he went on, "let us remember that his official station there gave him the means of influencing culture and education in his native country, in a way which he could not otherwise have done, and that in this respect he made that noblest use of his high position. His sympathy with the oppressed in every land was profound."

In fact, Humboldt did use his position at court to further scientific and cultural projects, such as the building of a new observatory in Berlin, and to steer the monarch toward a liberal course whenever possible. Besides persuading the king to modify academic restrictions on Jewish intellectuals in Germany, Humboldt helped to push through a (largely symbolic) law granting automatic freedom to any slave entering Prussia. He also warned the king to implement reforms in order to avoid the popular revolts that rocked Europe in 1848. His counsel was ignored, but during the ensuing unrest Humboldt remained unmolested by the mob, despite his position at court. After the revolt was crushed, he even led the huge funeral march for the fallen revolutionaries—certainly a distinction unique among members of a royal court in those turbulent days. As Humboldt himself wrote, somewhat defensively, "I have always sustained the firm conviction that a fancy uniform must never prevent me from defending the eternal principles of political freedom and constitutional institutions, a faith that I continuously expressed in my writings, speeches, and friendships."

But Humboldt's sense of malaise during these years in Europe ran deeper still. In fact, all the honors, all the soirées and salons, even the compulsive work schedule, were never enough to dispel his essential loneliness. "I live in the past rather than the present," he confided to Wilhelm, "and find it difficult to get used to a life without love and attention." His older brother's early prediction—"Happy he will never be, and never tranquil, because I cannot believe that any real attachment will ever steal his heart"—appeared to be coming to pass. Indeed, after his return to Europe, Alexander began to experience periods of depression, just as he had as a youth. In a letter to Caroline, he confessed to "frequent fits of melancholy and annoying stomach pains." And he expressed the dire opinion "There is no really deep feeling in man which is not painful. Such is our lot."

Humboldt once described his life as "agitated yet not really fulfilled," and there is a real sense that his manic activity—the brutally long hours, the frenetic social schedule, even the journey that made him famous—was intended to mask an underlying anxiety. "Alexander is not only unique in his great knowledge, but of truly good disposition, warm, helpful, and capable of sacrifice," Caroline wrote. "But he lacks the quiet satisfaction of his inner self and of his ideas. . . . Because of that, he does not understand people, though he craves their company." For his part, Wilhelm believed that Alexander never overcame his painful early years with their cold and domineering mother. "Neither he nor I will ever alter fundamentally," he suggested, "considering that both of us were subjected in our youths to an austere and lonesome upbringing, so that the influence of later years can count for nothing. . . ." But if that were true, why, in middle age, was Wilhelm happily surrounded by family while Alexander was still alone? "From his boyhood on," Wilhelm explained to Caroline, "Alexander strove toward outer activities, whereas I chose a life dedicated to the development of the inner man. Believe me, dearest, in this lies the true value of life."

Another explanation is that, as one whose primary attraction was to other men, Humboldt in his time and place simply didn't have the same opportunities for establishing lasting relationships or enjoying the usual pleasures of home and family. Still, after his return to Europe, Humboldt attempted to fill his life with people, just as he tried to fill his hours with work. He formed numerous friendships with gifted, generally younger men, such as historian and future statesman François-Pierre-Guillaume Guizot, who would later lead France's constitutional monarchists.

We should be careful not to read too much into the fact that these scientific friendships were exclusively with men, since at the time that field of endeavor was not open to women. In addition, there were two notable exceptions in his life where Humboldt seems to have been drawn to members of the opposite sex. The first was a harmless adolescent infatuation with Henriette Herz, the beautiful wife of physician Marcus Herz, who had been a mentor to Alexander and Wilhelm in Berlin. The second was in 1808, when judging from a surviving letter, the thirty-nine-year-old Humboldt had a short-lived but intimate relationship with a woman named Pauline Wiesel. "You know my joys and sorrows," he wrote her. "We were very happy here [Paris] for two long weeks, but I could have guessed that it wouldn't last for very long. . . . I embrace you intimately. Everything here is so empty and desolate. I'd walk for twelve hours just to see you. We will be close to each other for ever. You know me. Please write to me soon."

What, if anything, do these attractions tell us about the nature of Humboldt's relationships with men? The consensus view is that, in all likelihood, none of these male friendships was overtly sexual. However, some of his later relationships do exhibit a particular intensity reminiscent of the feverish bonds that Humboldt formed as a schoolboy. His first intimate friendship after his return to Paris was with the French physicist Joseph-Louis Gay-Lussac, to whom he gave the facetious nickname "Potash." Nine years Humboldt's junior, Gay-Lussac traveled with him to Italy and Berlin, and the two shared a one-room apartment in the French capital for several years. Eventually,

Humboldt took up residence near a new friend, the handsome, twenty-three-year-old German painter Carl von Steuben. Visiting Steuben every day, he reveled in the young man's company, which he called his "only joy." Generous with him, as with all his friends, Humboldt advanced Steuben's career by commissioning botanical illustrations and a life-size portrait of himself, arranging commissions from the royal family, and even dedicating a book to him.

In a similar vein, Humboldt later became attached to the young, impecunious German mathematician and self-proclaimed genius Gotthold Eisenstein, whom he supported for a time and on whose behalf he shamelessly pestered the king for grants. "My affection for you is not merely grounded on your remarkable gifts," Humboldt wrote Eisenstein. "My heart is drawn to you by your gentle, amiable character, and by your proneness to melancholy, to which you must not yield, I implore you for heaven's sake." Humboldt's championship eventually won Eisenstein a place in the German Academy of Science, and when the young man contracted consumption soon afterward, Humboldt paid his medical bills, then mourned bitterly when his friend succumbed to the disease.

But the man with whom Humboldt would share his most intimate friendship, beginning in 1809 and stretching over nearly five decades, was the brilliant French mathematician and physicist Dominique-François Arago, who was fifteen years younger than he. It was "an affection," in the words of one biographer, "which, without any sexual overtones, can only be described as the great passion of Humboldt's life."

Born in the small city of Perpignan, near Toulouse in southern France, Arago won admission to the École Polytechnique and was later appointed director of the Paris observatory and permanent secretary of the Institut National. In 1806, he was with Jean Biot making geographical measurements near the Spanish border, when war broke out between the two countries, and he was imprisoned in Spain and Algiers for nearly three years. Incredibly, during his long confinement, Arago managed to conceal his sheaf of notes under his shirt, and when he was finally repatriated in 1809, he triumphantly submitted the tattered pages to the French authorities. Hearing of the young man's pluck, Humboldt sent him a note, and when Arago later came to Paris, he looked up the older man to thank him. Shortly thereafter, the two became inseparable. As Wilhelm wrote in 1814, "You know [Alexander's] passion always to cling to one person who strikes his fancy temporarily. Right now it is the astronomer Arago, from whom he cannot be separated. . . ."

Though Arago shared Humboldt's liberal political views, he wasn't willing to make the same kind of compromise with authority that the other had, and more than once Humboldt found himself in difficult circumstances due to their friendship. On one occasion, when King Friedrich Wilhelm III was visiting Paris, Arago refused on principle to show the monarch around the observatory, despite Humboldt's repeated pleadings. Finally, the king took the extraordinary step of disguising himself as a commoner and, with Humboldt in tow, managed to get his tour. But as he conducted them through the building, Arago made a scathing attack on Prussia for its shabby treatment of France after Napoleon's defeat. Humboldt pulled Arago aside and warned, "Moderate your tone; you are speaking to the king." To which Arago replied, "I thought so; that's why I've been so frank."

After his return to Berlin, Humboldt missed Arago desperately. By now occupied with his wife and young family, his friend didn't have the same emotional need for Humboldt that the older man had for him, and Arago proved an unreliable correspondent. Humboldt hung on his infrequent letters, begging for just a few lines. He also greatly anticipated his periodic journeys to Paris, when he could visit him. "I am longing to see you again . . . ," he wrote in 1841, when Prussia was threatening war with France once more, this time over Egypt; "if my health remains good, I should like to start for Paris early in the spring and stay for some months; I am saddened by the uncertainty that this stay which promises so much pleasure for me . . . could be inconvenient for you and less agreeable than at former times in our lives . . . ," he continued. "Incapable of doing anything that could displease you in the complicated situation, accustomed to look upon your slightest wish as an order, I should be happy if you would have the kindness to send me three lines. . . . Can you refuse me this kindness? It would be the nicest present from you for the new year. . . ."

Arago's reply came three months later: "I cannot, I will not believe that you can ask seriously whether I should find pleasure in your journey to Paris. Do you really feel doubt in my attachment? Any uncertainty on this point would be cruelly wrong. Apart from my family, you are beyond comparison the person I love most in the world." Though on this unequal basis, their friendship endured until Arago's death in 1853.

After his return from South America, Humboldt also experienced a number of personal losses. The first was in 1811, when Carlos Montúfar, who had followed him to Paris, returned to Ecuador to join the rebellion against Spanish rule. Enlisting in a revolutionary army raised by his father, who had been such a kind host to Humboldt and Bonpland in Quito, the young Montúfar was captured by the colonial authorities and shot.

Aimé Bonpland also met a hard fate in South America. After their return to Paris, Humboldt secured an annual pension for Bonpland from the French government. He listed him as coauthor of the thirty-volume *Voyage to the Equinoctial Regions of the New Continent*, and he was relying on him to classify and prepare for publication their huge botanical collection. However, Humboldt was repeatedly frustrated by Bonpland's lack of attention to the work. Finally, in September 1810, he wrote him: "You haven't sent me a line on the subject of botany. I beg and beseech you to persevere until the work is complete, for . . . I have received only half a page of manuscript. . . . It's an object of the greatest importance, not only in the interests of science but for the sake of your own reputation and the fulfillment of the contract you entered into with me in 1798. So do pray send us some manuscript . . . ," he implored. "The public is under the impression that you have lost all interest in science over the last two years. . . . I embrace you affectionately, and in the course of a month I shall know whether you still love me sufficiently to gratify my wishes."

The source of Bonpland's apparent lack of interest isn't clear. It may be that his forte lay in fieldwork rather than desk work. Or perhaps his botanical acumen simply wasn't equal to the daunting prospect of classifying three thousand species new to science. Bonpland had always been the junior partner in the relationship, and some have suggested that he grew resentful after their return to Europe, when Humboldt captured the lion's share of acclaim. Then again, maybe Bonpland was simply occupied with other work, having been placed in charge of the empress

Josephine's extensive gardens at Malmaison. Whatever the reason, he edited only four of the seventeen botanical volumes published from the journey, and Humboldt was forced to retain Karl Sigismund Kunth, the nephew of his and Wilhelm's old tutor, to finish the work.

Over the next few years, Bonpland grew very close to Josephine Bonaparte, and was said to be with her at her deathbed in 1814. At the age of forty, having lost his patron and dear friend, Bonpland married a twenty-four-year-old Frenchwoman of dubious reputation. Two years later, he and his young wife left Paris and returned to South America, where he had been offered a professorship of natural history at Buenos Aires.

However, in December 1821, while on a plant-collecting expedition in the southern Andes, Bonpland passed into the disputed frontier between Argentina and Paraguay. Though he had official permission to enter the area, he and his party were attacked by Paraguayan cavalrymen, who suspected them of spying. Everyone else was killed, and Bonpland, having suffered a sword wound to the head, was taken prisoner and held under house arrest in a remote outpost. His young wife deserted him. In Europe, Humboldt worked desperately to elicit his friend's freedom, marshaling other celebrities to intervene on his behalf and even writing the Argentine president, all without success.

When Bonpland was finally released in 1830, he settled in an out-of-the-way part of Uruguay, where he continued his botanical studies, created a museum, and ultimately received a ranch from the Uruguayan government in recognition of his services. Settling down with an Indian woman, with whom he had several children, he never totally gave up the idea of returning to Paris. But by the time of his death, in 1858 at the age of eighty-five, Bonpland hadn't set foot in Europe for forty-two years.

In March 1829, Humboldt sustained another blow, when his sister-in-law, Caroline, whom he considered "the common bond of our family," died of cancer. Wilhelm himself died not long after, in April 1835. Though he didn't achieve the worldwide celebrity of his younger brother, Wilhelm, hewing closer to the path that his mother had planned for both her sons, became a distinguished linguist and influential government official. As Prussian minister of education, he was responsible for modernizing the nation's educational system and founding Berlin University (now known as Humboldt University). In fact, he managed to exert an important liberalizing influence on the nation before leaving public service in 1819 in response to the government's increasingly reactionary policies.

Though Wilhelm had often played the disapproving older brother, he was Alexander's only close remaining relative. "I did not think that my old eyes could have shed so many tears," the younger Humboldt wrote his publisher after his brother's death. "I am the unhappiest of men." Years later, Alexander commemorated his brother by quoting, at the end of one of his own works, an essay Wilhelm had written on the brotherhood of man: "If we would indicate an idea which, throughout the whole course of history, has ever more and more widely extended its empire . . . it is that of establishing our common humanity—of striving to remove the barriers which prejudice and limited views of every kind have erected among men, and to treat all mankind, without reference to religion, nation, or color, as one fraternity, one great community. . . . Thus deeply in the innermost nature of man, and even enjoined upon him by his highest tendencies, the recognition of the bond of humanity becomes one of the noblest leading principles in the history of mankind."

In his later years, Alexander also proved himself a great friend and mentor to young scientific talent. As one of the most famous men in the world, he was able to use his influence to benefit other researchers, and over the course of his long life, many found encouragement and protection under his aegis. These included Claude-Louis Berthollet, a collaborator of Antoine Lavoisier who discovered the reversibility of chemical reactions, among other contributions to that science; zoologist Achille Valenciennes, assistant to Georges Cuvier and coauthor of the landmark twenty-two-volume work *History of Fishes*; and Justus von Leibig, founder of organic and agricultural chemistry, who called Humboldt "a kind friend and powerful patron for my scientific studies." Mathematician Carl Friedrich Gauss wrote, "One of the most wonderful jewels in Humboldt's crown is the zeal with which he lends his assistance and encouragement to genius."

Another who benefited from Humboldt's early help was the great naturalist Louis Agassiz, to whom Humboldt extended financial support, introductions to prominent scientists, and assistance in emigrating to America. "Be happy in this new undertaking," Humboldt wrote as the young man left Europe, "and preserve for me the first place of friendship in your heart." Agassiz believed there was scarcely an important scientist in the world who wasn't in Humboldt's personal debt, and after becoming a famous professor at Harvard himself, he never forgot the older man's unselfish patronage. Years later, he would deliver an impassioned eulogy to a large and distinguished gathering in Boston, on the occasion of Humboldt's centenary, in which he confessed that he "loved and honored the man whose memory brings us together. . . . No man impressed his century intellectually more powerfully . . . ," he continued. "Nor is it alone because of what he has done for science, or for any one department of research, that we feel grateful to him, but rather because of that breadth and comprehensiveness of knowledge which lifts whole communities to further levels of culture, and impresses itself upon the unlearned as well as upon students and scholars."

Owing to his celebrity, Humboldt received thousands of letters a year and was frequently sought out by those eager to pay their respects. Some Americans even claimed that they had journeyed to Europe expressly to meet the great man. All remarked on the aging Humboldt's hospitality, unaffected modesty, and undimmed intellect. Toward the end of Humboldt's life, the American writer Bayard Taylor interviewed him for the *New York Tribune*. "He came up to me with a heartiness and cordiality which made me feel that I was in the presence of a friend, gave me his hand, and inquired whether we should converse in English or German . . . ," Taylor wrote. "The first impression made by Humboldt's face was that of a broad and genial humanity. . . . A pair of clear blue eyes, almost as bright and steady as a child's, met your own. . . . His wrinkles were few and small, and his skin had a smoothness and delicacy rarely seen in old men. His hair, although snow white, was still abundant, his step slow but firm, and his manner active almost to the point of restlessness . . . ," Taylor found. "He talked rapidly with the greatest apparent ease, never hesitating for a word. . . . He did not remain in his chair more than ten minutes at a time, frequently getting up and walking about the room. . . ."

Charles Darwin was less impressed when he met the hero of his youth for the first and only time. As he recounts in his autobiography, "I once met at breakfast at Sir Roderick Murchison's house, the illustrious Humboldt, who honoured me by expressing a wish to see me. I was a little disappointed with the great man, but my anticipations were probably too high. I can remember nothing distinctly about our interview, except that Humboldt was very cheerful and talked much." It seems an ungenerous assessment from the man who had claimed that he owed every professional achievement to his early reading of Humboldt's works.

Just as Humboldt used his fame to encourage young talent, he also lent his celebrity to some of the very first international scientific collaborations. In 1828, as acting president of the Association of German Naturalists and Physicians, he presided over the organization's annual meeting in Berlin, which drew six hundred of Europe's most prestigious scientists, in addition to the king and assorted government ministers. Despite the organization's title, its purpose was not to disseminate scientific information only throughout Germany, but also to promote the international spread of information. As Humboldt said in his opening address, "The Association . . . appears to us to be one of the most striking effects of the increased facility and desire of communication between different countries." Ironically, it was Humboldt, the great synthesizer, who suggested that the meeting break up into sections based on scientific discipline, in recognition of the increasing specialization of the sciences. It was a procedure that would be increasingly adopted around the world.

"The Association of German Naturalists and Physicians . . . ," Humboldt continued, "well deserves to be imitated in other countries. . . ." And indeed it was, serving as a model for similar organizations in Europe and America, including the British Association for the Advancement of Science (founded in 1831) and the American Association for the Advancement of Science (established in 1848). Thus began the international cooperation that researchers today take for granted, in which scientific information, with few exceptions, is seen as the rightful property of all mankind, just as Humboldt envisioned.

Successful as his chairmanship was, Humboldt minimized his own role at future annual meetings, always claiming the press of work or, as he got older, the infirmities of age. In fact, he had lost his taste for the get-togethers, which he felt had degenerated into "a theatrical spectacle, where in the midst of endless feasting the vanity of learned men finds ready gratification."

However, beginning about 1830, Humboldt was the impetus behind an important early international research project, when he persuaded the British government to undertake a series of magnetic measurements in their colonies around the world—including Canada, Australia, New Zealand, St. Helena, and the Cape of Good Hope—in order to extend the map of the world's geomagnetic field, an area of research he had pioneered that was of enduring interest to him. In recognition of his early fostering of international collaboration in the sciences, the German government has created the Alexander von Humboldt Foundation to administer the country's principal program of research grants for foreign scholars.

THOUGH HE HAD LEFT AMERICA with the hope of returning one day, Humboldt never visited the New Continent again. In fact, his only other major expedition was to Siberia, where he traveled in 1829, at the age of sixty, at the invitation of tsar Nicholas I. Like Charles IV of Spain some three decades before, the tsar was hoping that Humboldt would discover precious minerals—in this case gold, platinum, or diamonds in the Ural Mountains. And indeed, Humboldt did discover the country's first diamond mine, in Siberia. Over a period of eight months, he and his party traveled some 11,500 grueling miles by coach, taking magnetic measurements and making scores of geologic and geographic observations. The journey through Russia, he wrote to Wilhelm, provided "one of the great moments of my life." In 1843, Humboldt published *Central Asia, Researches on the Mountain Chains and Comparative Climate*, a three-volume work that yielded a wealth of new information on Russia and her resources.

Late in life, Humboldt embarked on an incredibly ambitious new multivolume project, which he called *Cosmos: A Sketch of the Physical Description of the Universe.* Its title deriving from the Greek for "world" or "order," the massive work attempted to outline all knowledge about the physical sciences in a way that would reveal to the intelligent lay reader the order underlying the universe's apparent chaos. The last time a single author would undertake such a wide-ranging summary, *Cosmos* proved Humboldt's final, greatest, and most popular publication.

He himself called it "the work of my life," and indeed, the idea

Kosmos.

Entwurf

einer physischen Weltbeschreibung

von

Alexander von Humboldt.

Erster Band.

Naturae vero rerum vis atque majestas
in omnibus momentis fide caret. si quis
modo partes ejus ac non totam complectatur
animo.   **Plin.** H N. lib. 7 c. 1.

Stuttgart und Tübingen.

J. G. Cotta'scher Verlag.

1845.

Frontispiece of *Cosmos*.

had been gestating for five decades. In 1797, before he'd even left for America, he'd written, "I have conceived the idea of a physical description of the world. As I feel the increasing need for it, so also I see how few foundations exist for such an edifice." In 1820, when he'd still been in Paris, he'd toyed with the idea again, but had already been overextended with his other publications. Then in 1834, he'd confessed, "I have the crazy notion to depict the entire material universe, all that we know of the phenomena of universe and earth, from spiral nebulae to the geography of mosses and granite rocks, in one work—and in a vivid language that will stimulate and elicit feeling. . . . But it is not to be taken as a physical description of earth: it comprises heaven and earth, the whole of creation."

The idea had come to the fore again between fall 1827 and spring 1828, when Humboldt had delivered a series of sixty-one lectures at Berlin University encompassing the entire scope of the physical sciences—astronomy, geology, plant geography, geomagnetism, ocean currents, and myriad other topics. Meant to counter the unscientific, romantic speculations of the German "nature philosophers," the university lectures proved so popular that Humboldt was induced to repeat them, as sixteen simplified installments intended for a general audience—including the entire spectrum of Berlin society, from the royal family to impoverished students—who packed a Berlin hall each week. In the words of a contemporary newspaper report, "By the lucid manner in which he grasps the facts discovered by himself and others in various branches of science and arranges them in one comprehensive view, [Humboldt] throws so clear a light upon the boundless region of the study of Nature that he has introduced a new method of treating the history of science." The lectures proved the phenomenon of the season, and a commemorative medal was struck to memorialize the event. Humboldt's publisher offered him a large advance to publish the lectures as they had been delivered, but he insisted on thoroughly reworking the material before setting it in print.

As a result, it was the better part of two decades before the first volume of *Cosmos* appeared, in 1845. The second of five eventual volumes was issued two years later, and both books became immediate best-sellers. Volume One sold out in two months,

and by 1851, when Humboldt estimated eighty thousand copies
had been shipped, it had been translated into virtually every European language. After the release of the second volume, Humboldt's publisher wrote him breathlessly, "In the history of book
publishing, the demand is epoch making. Book parcels destined
for London and St. Petersburg were torn out of our hands by
agents who wanted their orders filled for the bookstores in Vienna and Hamburg. Regular battles were fought over possession
of this edition, and bribes offered. . . ."

Reviews were lavish in their praise, and the king ordered another commemorative medal struck. On the obverse was a profile
of Humboldt; on the reverse, set against a background of constellations, tropical plants, and surveying equipment, were six letters: *ΚΟΣΜΟΣ*—the word that contained an entire universe and
that summarized, more than any other, Humboldt's vision of
the single, unifying force of nature. "Everything is interrelated,"
he had said, and *Cosmos* was the work whereby he sought to
prove it.

Cosmos offered a brilliant capstone to Humboldt's life, and a
premier example of his synthesizing impulse. Besides clearing
away the speculative musings of the nature philosophers, the
books were intended "to increase by a deeper insight into its
essence the enjoyment of Nature. I cannot share the apprehension
caused by a certain narrow-mindedness and clouded sentimen-
tality that Nature loses its magic, the charm of the mysterious
and the sublime, by any study of its forces . . . ," Humboldt wrote.
"We cannot concur with Burke when he says that ignorance of
Nature only is the source of admiration and the feeling of the
sublime."

Though Humboldt wanted to share, through *Cosmos*, his lifelong love of Nature with a wide audience, he also had a more
practical purpose in mind. With his unshakable belief in progress,
Humboldt realized that in an increasingly technical age, mankind's well-being would be determined in large measure by how
well we mastered the transformative power of science. "Those
countries which lag behind in industry," he warned, "in the application of mechanics and technical chemistry, in the careful selection and utilization of natural products, where the respect for
such activities does not permeate all classes of society, will

unfailingly decline in prosperity. They will sink faster when their neighbor states, with an energetic exchange between science and industry, go forward with renewed vitality."

Cosmos was intended to advance the betterment of mankind, by aiding the dissemination of technological know-how and by making the benefits of science available to all nations. In fact, he envisioned science as a uniting factor among peoples, "making common cause against ignorance and prejudice," as a contemporary article in the *American Review* phrased it. "If the world is ever to be harmonized," the piece continued, "it must be through a community of knowledge, for there is no other universal or non-exclusive principle in the nature of man. . . . A careful examination discovers [the first few chapters of *Cosmos*] to be an exposition of the very spirit of liberal culture." Thus, with *Cosmos*, Humboldt sought to promote the humanitarian, democratic causes that he had espoused his entire life.

As he grew older, Humboldt's internationalist outlook helped to spread his reputation throughout the world. His portrait hung on every continent—reportedly even in the palace of the king of Siam—and in 1852 the British Royal Society awarded him the Copley Medal, its highest honor. In the expanding, practical-minded United States, blessed with tremendous natural resources and incredible natural beauty, Humboldt's egalitarian, forward-thinking message made him enormously popular. Ever since his brief visit in 1804, he had shown a strong affection for the young republic and had followed events there with excitement—such as westward expansion and the discovery of gold in California—and apprehension—such as the continuation of slavery and the annihilation of the Native Americans. Americans had returned his affection, and dozens of places on the continent—rivers, mountains, counties—had been named in his honor. *Cosmos* and *Aspects of Nature* found a huge readership in the United States, and to the image of a brilliant scientist and liberal republican, Humboldt added that of much-loved author. With the influx of German immigrants in the nineteenth century, the Prussian's reputation surged even higher, and in 1857, U.S. secretary of war John B. Floyd wrote him, "Never can we forget the services you have rendered not only to us but to all the world. The name of Humboldt is not only a household word throughout our immense country, from the shores of the Atlantic to the waters of the Pacific, but we

Humboldt in 1857.

From *Life of Alexander von Humboldt* by Karl Bruhns (editor), 1872.

have honored ourselves by its use in many parts of our territory, so that posterity will find it everywhere linked with the names of Washington, Jefferson, and Franklin."

In 1850, Humboldt published Volume Three of *Cosmos*. Writing from nine at night to three in the morning, after his official duties were done, he pressed ahead, getting only four or five hours' sleep per night. Then in 1857, as the work dragged on, Humboldt suffered a slight stroke, and he worried that he wouldn't be able to finish his culminating work. Living beyond his contemporaries, he wrote, "I have buried all my race," and he took to quoting Dante that life was "a race to death." In 1858, he published the fourth volume, though he also had a premonition that the following year would be his last. On April 19, 1859, he sent the fifth and final volume of *Cosmos* to his publisher. Two days later, he took to his bed. On May 6, at two-thirty in the afternoon, four months short of his nintieth birthday, he died peacefully, with his niece Gabriele and her brother-in-law at his side. Beside his bed, they found a quotation he had copied from *Genesis*, the last words the prolix Humboldt ever wrote: "Thus the heavens and the earth were finished, and all the host of them."

Humboldt lay in state in his own library, and huge crowds came to pay their respects. On May 10, his body was carried in a solemn funeral procession to the Berlin Cathedral, where the prince regent met the bier on the front steps. As the cortege moved through the crepe-shrouded streets to the music of Chopin's Funeral March, Humboldt's laurel-and-azalea-wreathed hearse was preceded by four royal chamberlains. Twenty students marched alongside the casket, carrying palm fronds, and behind walked Humboldt's family, followed by hundreds more students, professors, government ministers, members of parliament, magistrates, municipal authorities, churchmen, members of the academies of Science and Arts, and, finally, thousands of ordinary citizens. It was said to be the grandest nonmilitary funeral in the city's history, with the exception of one—the tribute to the fallen revolutionaries of 1848, which Humboldt himself had led. The next day the peripatetic scientist was laid to rest at Tegel, in the family graveyard next to Wilhelm and Caroline.

Humboldt's Spirit

AS WORD OF HUMBOLDT'S DEATH FILTERED AROUND the world, there was an outpouring of sadness and reverence befitting a beloved international celebrity. All the New York papers, like those around the globe, carried the story on their front pages. The *Herald* lauded him as "one of the greatest men of his age or of any other age" and suggested, "Perhaps there is not in the annals of mankind the name of another man who had lived to the same age and produced such an amount of intellectual work, and that, too, of the highest order. . . . He had a gigantic intellect, from which nothing in nature or in science appeared to be hid. He could grasp all subjects, and he appeared to know everything. . . . *Cosmos* is his imperishable monument, which will endure as long as the earth which it describes."

The *Tribune* averred, "His fame belonged not only to Europe, but to the world; and in this country especially, probably no man who was known to us only through the medium of his scientific writings was held in equal reverence and admiration. . . . But what will ever distinguish Humboldt from the mass of physical inquirers who had preceded him, is his study of the universe as a harmonious whole, and his search for the laws of order, beauty, and majesty beneath the apparent confusion and contradictions of isolated appearances." The *Evening Post* called him simply "The most eminent scientific man of the nineteenth century" and added, "Nature was to him a marvelous whole, all the parts of which were intimately connected with each other, and conspired to one grand and harmonious result." In addition, the articles lauded him for his practicality, demonstrated friendship for

America, generosity, unwavering belief in progress, unstinting work ethic, and unaffected manners.

Ten years later, on the centennial of his birth, Humboldt's memory was still very much alive. In New York, where he had never even set foot, the *Times'* headline proclaimed, CELEBRATION GENERALLY THROUGHOUT THE COUNTRY. The city "seemed to be in holiday dress," a proclamation was presented by the mayor, a bronze bust was unveiled in Central Park to the cheers of "an immense throng" of at least twenty-five thousand ("two thirds of whom were Germans"), and a banner proclaiming, "1769—HUMBOLDT—1869," was unfurled over City Hall. There was "an imposing" parade through streets hung with American and German flags, an even bigger torchlight procession that evening, and a grand banquet for 250 invited guests.

In Boston, three different events were held, including an organ concert and a two-hour eulogy by Louis Agassiz at the Music Hall, sponsored by the Society of Natural History and attended by such luminaries as Governor Claflin, Henry Wadsworth Longfellow, James Russell Lowell, and Oliver Wendell Holmes. In the evening the city government sponsored a reception at Horticultural Hall, where Julia Ward Howe and Oliver Wendell Holmes recited poems written for the occasion and where a letter from the Quaker poet and abolitionist John Greenleaf Whittier was also read. Other cities across the country had their own Humboldt celebrations, including Washington, Cleveland, Richmond, Chicago, Cincinnati, Memphis, and Baltimore, among many others. In Germany, the day, predictably, was marked by "a national demonstration" and, in Berlin, the laying of the cornerstone for a Humboldt monument.

We may well ask, If Humboldt was so widely celebrated and so beloved during his long life and even a decade afterward, why has he been largely forgotten in our own time (with the exception of university-level geography textbooks)? One factor is actually the same one that made him great—the wide-ranging nature of his intellect. Though Humboldt was a tireless innovator in the sciences, by the end of his long life, he'd become an anachronism. Above all he was a generalist, intent on examining every natural process and shaping the myriad discordant data into a coherent whole, as in *Cosmos*. However, by the mid-nineteenth century, science was progressing so rapidly that it was increasingly

The New-York Times.

VOL. XVIII....NO. 5610.   NEW-YORK, WEDNESDAY, SEPTEMBER 15, 1869.   PRICE FOUR CENTS.

HUMBOLDT.

The One Hundredth Birthday of the Philosopher.

Celebration Generally Throughout the Country.

Unveiling of the Bust at the Central Park.

ORATION BY DR. FRANCIS LIEBER.

Procession, Banquet and Operation in this City.

SIXTEENTH RECEPTION IN BOSTON.

Eulogistic Address by Professor Agassiz.

IN THIS CITY.

The front page of *The New York Times* on the centennial of Humboldt's birth.

becoming the province of specialists, as shown by the trend to re-place university departments of "Natural Philosophy" with the narrower disciplines that we know today. Science was also be-coming the province of mathematics, and though Humboldt was a great advocate of careful quantification, the integration of data into mathematical formulas was never his forte. (For instance, after discovering how the earth's magnetic field varies with lati-tude, he had left it to his friend Carl Gauss to work out the for-mulas predicting the values for any point on earth.) Ironically, the generalizing worldview that Humboldt championed, which in an earlier time had made him the most celebrated naturalist in the world, increasingly seemed static, soon to be overshadowed by the more dynamic model proposed in Darwin's *On the Origin of Species*, published in the year of Humboldt's death.

Then, too, Humboldt never made a single great, paradigm-changing discovery with which his name would be forever linked, as Darwin had with natural selection. Many of Humboldt's dis-coveries were so fundamental that they came to be taken for granted, and their source was eventually forgotten. His strength lay rather in opening avenues of research that could then be de-veloped by others. In fact, Humboldt was so successful in culti-vating a new generation of scientists that they soon outstripped him, though none would approach him in sheer breadth of know-ledge and expertise. As a result Humboldt assumed, in the words of a biographer, "the anonymity of a great teacher," remembered with gratitude by his students but remaining in the background.

Changes in America itself also contributed to Humboldt's eclipse. As German immigrants became assimilated into the larger society, their heroes tended to be those who made their mark in North America, not in the hemispheres to the south or east. And as the tremendous national dramas of civil war and westward expansion played themselves out, the United States produced no dearth of homegrown heroes. With the country's attention shift-ing inward, Humboldt, though memorialized in numerous place names around the globe, including the Humboldt Current, was gradually nudged aside.

Owing to accident of birth, personal inclination, and sheer longevity, Humboldt had either the good fortune or the bad luck to inhabit the cusp between the Enlightenment and the Roman-tic Period. Instead of standing firmly for either Old or New, he straddled both camps, melding a cool rationalism with emotional warmth and aesthetic awareness. Similarly, he combined a passion for scientific generalization with a compulsion for quantifcation, and he even abetted early efforts at scientific specialization that would soon make his sweeping *physique générale* seem outmoded. Is it possible that this very individuality, this defiance of easy classification, is yet another reason for the gradual fading of his memory? In any event, there is no doubt that by nurturing talented newcomers with perspectives different from his own, Humboldt helped to usher in the era in which he himself seemed out of place.

Still, Humboldt was one of the creators of this modern world that we take for granted. Making the first in-depth scientific expedition through the Amazon Basin and the Andes, he opened up an entire continent to scientific and geographic study, redrawing maps, discovering thousands of new species, and canvassing South America so thoroughly that his contemporaries complained that he had left nothing for them. His studies of New World volcanoes reshaped geology by helping to replace Werner's neptunist paradigm, first with Hutton's vulcanist model and then with Lyell's synthesis of the two competing camps. He also greatly increased our understanding of volcanic processes, including the discovery that volcanoes tend to be found along fissures in the earth's crust.

He created the science of plant geography, and his painstaking data collection laid the basis of modern physical and political geography, oceanography, and climatology. Besides confirming that the earth's magnetic field varies predictably with latitude and discovering the planet's magnetic equator, he was the first to observe magnetic storms. He was instrumental in focusing scientists' attention on the need for accurate, systematic data collection, and he created such fundamental techniques of data presentation as isotherms and geological profiles. He was an unstinting supporter of scientific talent and an important early advocate of international scientific collaboration. A lifelong liberal and humanist, he pioneered the study of indigenous American cultures, including the Inca and the Aztecs, and he did everything in his power to bring science within the grasp of the layperson.

Humboldt was a firm believer in the leveling power of science, its ability to improve the life of every member of society regardless of social class. One could argue that in our own times science has been used to advance totalitarian causes as often as democratic ones. However, Humboldt would respond that "scientific literacy" (as we now call it) is more crucial than ever in today's technologically driven republics, where citizens are expected to cast informed ballots on such issues as cloning, global warming, and environmental conservation. In any event, Humboldt's creed has been proven beyond all doubt—science has improved the lot of mankind, spectacularly if not evenhandedly, despite the problems of technological growth. And Humboldt himself played a key role in that process.

Late in life, Humboldt summed up his own contribution with typical understatement: "I have never been able to hoodwink myself as I have always been surrounded by people who were superior to me. My life has been useful to science less through the little I have contributed myself than through my efforts to let others profit of the advantages of my position. I have always had a just appreciation of the merits of others, I have even shown some acumen in the discovery of new talent. I like to think that, while I was at fault to tackle from intellectual curiosity too great a variety of scientific interests, I have left on my route some trace of my passing."

By opening up a new continent to scientific inquiry, by laying the foundation of so many branches of modern science, by revolutionizing research methods through careful observation and measurement in the field—and especially by urging the scientific enterprise toward the search for unifying principles—Alexander von Humboldt played a crucial part in creating science as we know it today. Inheriting the broad scientific vision of those who had gone before, such as Thales and Newton, he bequeathed that philosophy to those who would come later, such as Darwin and Hubble. Today, as science continues to strive for the Big Picture with such all-encompassing theories as chaos and superstrings, Humboldt's quest to grasp the unity of nature, far from being a dusty anachronism, is as vital as ever. For throughout history, the synthesizing impulse has proved a powerful, even world-changing, tool for understanding the universe, capable of penetrating the intricate, contradictory web of surface phenomena to reveal the universal, unified *cosmos* beneath—that fundamental, unchanging phenomenon we call *truth*.

BIBLIOGRAPHY

HUMBOLDT'S WORKS

Throughout this book, I have quoted and paraphrased liberally from Humboldt's own works, especially the following. For additional Humboldt publications, see Appendix I.

Humboldt, Alexander von. *Aspects of Nature in Different Lands and Climates; with Scientific Elucidations.* Translated by Mrs. Sabine. Philadelphia: Lean and Blanchard, 1850.

————. *Cosmos: A Sketch of the Physical Description of the Universe.* Translated by E. C. Otté. Introduction by Nicolas A. Rupke. Volumes 1 and 2. Baltimore: The Johns Hopkins University Press, 1997.

————. *Political Essay on the Kingdom of New Spain.* Translated by John Black. Edited and with an Introduction by Mary Maples Dunn. New York: Alfred A. Knopf, 1972.

————, and Aimé Bonpland. *Personal Narrative of Travels to the Equinoctial Regions of America During the Years 1799–1804.* [Includes *Political Essay on the Island of Cuba.*] Translated by Thomasina Ross. Volumes 1–3. London: George Routledge and Sons, Limited, 1895.

————. *Researches, Concerning the Institutions and Monuments of the Ancient Inhabitants of America, with Descriptions & Views of Some of the Most Striking Scenes of the Cordilleras.* Translated by Helen Maria Williams. Volumes 1 and 2. London: Longman, 1814.

HUMBOLDT BIOGRAPHIES

After the works of Humboldt himself, I have drawn most heavily from the following biographies. While I have differed with these authors at points, their work has been invaluable:

Botting, Douglas. *Humboldt and the Cosmos.* New York: Harper & Row Publishers, 1973.

De Terre, Helmut. *Humboldt: The Life and Times of Alexander von Humboldt, 1769–1859.* New York: Alfred A. Knopf, 1955.

Kellner, L. *Alexander von Humboldt.* New York: Oxford University Press, 1963.

OTHER SOURCES

Anna, Timothy A. *The Fall of the Royal Government in Mexico City.* Lincoln: University of Nebraska Press, 1978.

Bernier, Olivier. *The World in 1800.* New York: John Wiley & Sons, Inc., 2000.

Boorstin, Daniel J. *The Discoverers.* New York: Random House, 1983.

Collier, Richard. *The River That God Forgot: The Story of the Amazon Rubber Boom.* New York: E. P. Dutton & Co., Inc., 1968.

Collier, Simon, Harold Blakemore, and Thomas E. Skidmore, general editors. *Cambridge Encyclopedia of Latin America and the Caribbean.* New York: Cambridge University Press, 1985.

Cvancara, Alan M. *A Field Manual for the Amateur Geologist.* Revised Edition. New York: John Wiley & Sons, Inc., 1995.

Darwin, Charles. *On the Origin of Species by Means of Natural Selection.* New York: Bantam Books, 1999.

———. *The Voyage of the Beagle.* New York: Modern Library, 2001.

Díaz, Bernal. *The Conquest of New Spain.* Translated and with an Introduction by J. M. Cohen. New York: Penguin Books, Ltd., 1963.

Eliot, Charles William. *Voyages and Travels: Ancient and Modern.* The Harvard Classics, Volume XXXIII. New York: P. F. Collier & Sons, 1909–14.

Fagg, John Edwin. *Cuba, Haiti, and the Dominican Republic.* Englewood Cliffs, N.J.: Prentice-Hall, Inc., 1965.

Faul, Henry, and Carol Faul. *It Began with a Stone: A History of Geology from the Stone Age to the Age of Plate Tectonics.* New York: John Wiley & Sons, Inc., 1983.

Fison-Roche, Roger. *A History of Mountain Climbing.* Paris, New York: Flammarion, 1996.

Furneaux, Robin. *The Amazon: The Story of a Great River.* New York: G. P. Putnam's Sons, 1969.

Gould, Stephen Jay. *I Have Landed: The End of a Beginning in Natural History.* New York: Harmony Books, 2002.

Greene, Mott T. *Geology in the Nineteenth Century: Changing Views of a Changing World.* Ithaca, N.Y.: Cornell University Press, 1982.

Hemming, John. *The Conquest of the Incas.* New York: Harcourt Brace Jovanovich, Inc., 1970.

Herndon, William Lewis. *Exploration of the Valley of the Amazon.* New York: Grove Press, 2000.

Jones, Steve. *Darwin's Ghost: The Origin of Species Updated.* New York: Random House, 2000.

Keay, John, general editor. *History of World Exploration.* London: Paul Hamlyn Publishing, 1991.

Kendall, Ann. *Everyday Life of the Incas.* New York: Dorset Press, 1973.

Laudan, Rachel. *From Mineralogy to Geology: The Foundations of a Science, 1650–1830.* Chicago: University of Chicago Press, 1987.

Lewis, William. *Exploration of the Valley of the Amazon.* Edited and with an Introduction by Hamilton Basso. New York: McGraw-Hill Book Company, Inc., 1952.

Lyell, Charles. *Principles of Geology.* Volumes 1–3. London: John Murray, 1830–33.

Mason, Stephen F. *A History of the Sciences.* New Revised Edition. New York: Macmillan General Reference, 1962.

Prescott, William H. *The Conquest of Mexico.* New York: Julian Messner, Inc., 1948.

———. *The Conquest of Peru.* New York: The Heritage Press, 1957.

Radin, Paul. *Indians of South America.* Garden City, N.Y.: Doubleday, Doran & Company, Inc., 1942

Revkin, Andrew. *The Burning Season: The Murder of Chico Mendes and the Fight for the Amazon Rain Forest.* Boston: Houghton Mifflin Company, 1990.

Roosevelt, Theodore. *Through the Brazilian Wilderness.* New York: C. Scribner's Sons, 1914.

Schurz, William Lytle. *This New World.* New York: E. P. Dutton & Co., Inc., 1954.

Simons, Geoff. *Cuba: From Conquistador to Castro.* New York: St. Martin's Press, 1996.

Smith, Anthony. *Explorers of the Amazon.* New York: Viking Penguin, Inc., 1990.

Sobel, Dava. *Longitude: The True Story of a Lone Genius Who Solved the Greatest Scientific Problem of His Time.* New York: Walker & Co., 1995.

Soustille, Jacques. *The Daily Life of the Aztecs on the Eve of the Spanish Conquest.* New York: The Macmillan Company, 1962.

Sterling, Tom. *The Amazon.* Time-Life Books, 1975.

Thomas, Hugh. *Conquest: Montezuma, Cortés, and the Fall of Old Mexico.* New York: Simon & Schuster, 1993.

Tutino, John. *From Insurrection to Revolution in Mexico: Social Bases of Agrarian Violence, 1750–1940.* Princeton: Princeton University Press, 1986.

Vega, Garcilaso de la. *Royal Commentaries of the Incas and General History of Peru.* Austin: University of Texas Press, 1966.

Wilson, Edward O. *Consilience: The Unity of Knowledge.* New York: Alfred A. Knopf, 1998.

Wilson, Jason. Introduction to abridged edition of Humboldt's *Personal Narrative of a Journey to the Equinoctial Regions of the New Continent.* New York: Penguin Books, 1995.

Worcester, Donald E. *Simón Bolívar.* Boston: Little, Brown, 1977.

SOURCES

Principal sources are given below; other resources may have been consulted for each chapter as well. See Bibliography for complete bibliographic information on each title. Sources under each subheading are listed in order of importance to my research.

All Humboldt quotations and details of his journey are taken from his *Personal Narrative*, except as noted below.

PREFACE: *Humboldt's Ghost*

Humboldt quotation regarding *la physique générale*: J. Wilson
Goethe quotation: De Terre
Emerson quotation: De Terre
Bolívar quotation: Wendy Moonan, "Jungle Fever Strikes a Collector," *New York Times,* March 30, 2001
Gould quotation: Gould
Tribune quotation: Issue of May 19, 1859
Herald quotation: Issue of May 19, 1859
Jones quotation: Jones
Darwin quotation on the course of his life: Botting
Humboldt quotation about longing for the distant and unknown: Humboldt, *Cosmos*

ONE: *Tegel*

Second Great Age of Discovery: Keay, Boorstin, J. Wilson
Humboldt biographical material: De Terre
Trends in geology: Greene, Laudan
Spain and its New World colonies: Schurz, Bernier
La Condamine: Smith

Humboldt's instruments: J. Wilson, Botting
Humboldt's predecessors and his philosophy: J. Wilson, E. O. Wilson,
　Mason

TWO: *Tenerife*

Prince Henry the Navigator: Boorstin, Keay

THREE: *Cumaná*

Humboldt's letter to his brother from Cumaná: De Terre
Darwin passage: Darwin, *The Voyage of the Beagle*
Slavery statistics: S. Collier
La Condamine: Smith
Trends in botany: Boorstin
Columbus quotation: Schurz
Catholic Church and its missions: Schurz

FOUR: *Caracas*

Basil Hall on earthquakes: Schurz
Creoles and Gachupines: Bernier
Venezuelan independence movement: Worcester

FIVE: *The Llanos*

Description of the Llanos: Humboldt, *Personal Narrative* and *Aspects of
　Nature*
Galvani and Volta: Mason
Humboldt's letter lamenting his failure regarding galvanism: Kellner
Description of hunt for electric eels: Humboldt, *Aspects of Nature*
Introduction to the rain forest: Sterling, Revkin, Furneax
Bates passage on jungle quiet: Sterling
Description of jungle noises: Humboldt, *Aspects of Nature*
Description of piranha: Roosevelt

SIX: *The Orinoco*

Early explorations of the Orinoco: Smith
Sir Walter Raleigh quotation: Eliot

Description of the cataracts: Humboldt, *Aspects of Nature*
White rivers and black rivers: Sterling
Linnaeus quotation: Humboldt, *Personal Narrative*
Rubber: Collier, Revkin
Humboldt quotation on the future of the Amazon Basin: Collier

SEVEN: *The Amazon*

Introduction to the Amazon: Sterling, Smith, Furneaux
Cristóbal Acuña quotation: Radin
Early expeditions on the Amazon: Smith
Description of the Río Negro: Herndon
Article in Brazilian newspaper: De Terre

EIGHT: *Cuba*

Humboldt's letter to his brother regarding Bonpland's illness: De Terre
Humboldt's letter to his brother about continuing his journey: De Terre
Description of the Llanos during the rainy season: Humboldt, *Aspects of Nature*
Humboldt's letter to Willdenow: De Terre
History of Cuba and Havana: Simons, Fagg
Hatuey episode: Simons
Longitude: Sobel
Humboldt's letter to Baudin: De Terre
Columbus quotation on Jardines y Jardinillos: Humboldt, *Personal Narrative*

NINE: *Chimborazo*

Details of the journey and Humboldt quotations: De Terre and Botting, except as noted
Introduction to the Andes: Collier
José Celestino Mutis: Botting
Darwin and mastodon bones: Darwin, *The Voyage of the Beagle*
Speculations on volcanoes: Humboldt, *Aspects of Nature*
Doña Rosa's description of Humboldt: Botting
Humboldt's letter to his brother regarding climb up Pichincha: Kellner
Nonview of the Pacific from the top of Pichincha: Humboldt, *Aspects of Nature*
Ascent of Chimborazo: Humboldt, *Researches Concerning the Institutions and Monuments of the Ancient Inhabitants of America*

History of mountain climbing: Fison-Roche
Lyell's theories and influence: Lyell, Greene, Laudan
Lyell's letter regarding meeting with Humboldt: De Terre
Darwin quotation regarding Lyell: J. Wilson

TEN: *Cajamarca*

Details of the journey: Humboldt, *Aspects of Nature;* De Terre; Botting
Humboldt quotations: Humboldt, *Aspects of Nature,* except as noted
Humboldt quotation on Cotopaxi: De Terre
Humboldt's letter to his brother regarding Indian manuscripts: De Terre
Geomagnetism: Kellner
Quotation regarding decrease of magnetic force from the pole: Humboldt,
    Cosmos
Introduction to the Inca: Kendall, Radin
Description of the inca's walled garden and the Temple of the Sun: Vega
Pizarro quotation regarding Inca roads: Schurz
Spanish conquest: Hemming; Prescott, *The Conquest of Peru*
Conquistador's comments on sighting Cajamarca: Hemming
Introduction to Lima: Bernier
Humboldt's letter with impressions of Lima: De Terre
Guano: "Guano Trade" case study, Trade and Environment Database,
    Mandala Projects Internet site, hosted by American University (www.
    american.edu/TED/)
The Humboldt Current and El Niño: "Humboldt Current," National
    Oceanographic and Atmospheric Administration website (www.na.
    ne fsc.noaa.gov/lme/text/lem13.htm)
Humboldt quotation regarding credit for the current named after him:
    De Terre
Argument concerning Native Americans: Humboldt, *Personal Narrative*
Controversy over Siberian land bridge: Nicholas Wade and John Noble
    Wilford, "New World Ancestors Lose 5,000 Years," *The New York Times,*
    July 25, 2003
Humboldt quotation on the unity of nature: Botting

ELEVEN: *New Spain*

Details of the journey: Humboldt, *Political Essay on the Kingdom of New
    Spain;* De Terre; Botting
Humboldt quotations: Humboldt, *Political Essay on the Kingdom of
    New Spain,* except as noted
Introduction to New Spain: Bernier
Humboldt's cross-section of Mexico: Friedrich, Anke M., "Alexander
    von Humboldt as a Pioneer of Western North American Tectonics."

Paper presented at the Annual Meeting of the Geological Society of America, 2002

Introduction to the Aztecs: Thomas, Soustille, Díaz

Spanish conquest: Thomas; Prescott, *The Conquest of Mexico;* Díaz

Aztec song: Thomas

Doubts that Aztecs considered Cortés a god: Camilla Townsend, "Burying the White Gods," *American Historical Review,* June 2003.

Cortés's threat to Montezuma: Díaz

Mexica song: Thomas

Cortés's comment on his countrymen: Prescott

Cortés's comment on similarity of Mexico to Spain: Schurz

Introduction to Mexico City: Bernier; Humboldt, *Political Essay on the Kingdom of New Spain*

Poinsett quotation: Bernier

Humboldt's letter to Delambre: De Terre

Humboldt's letter to the Institut National: De Terre

Nochistongo: "Enrique Martinez" in "Virtual American Biographies," Virtuology Internet site (www.virtualology.com)

Mexican War of Independence: Bernier, Anna, Tutino

Humboldt's comment on Indians' account of the creation of Jorullo: De Terre

TWELVE: *Washington, Paris, and Berlin*

Details of Humboldt's journey in the United States and all quotations from that period: De Terre, Botting

Humboldt's and Bonpland's activities from arrival in Europe onward and all quotations from that period, except as noted: De Terre, Botting, Kellner

Jefferson's later letters to Humboldt: Positive Atheism Internet site (www.positiveatheism.org)

Changes in Paris: Bernier

Simón Bolívar and the Venezuelan independence movement: Worcester

Agassiz's defense of Humboldt's position in the Prussian Court: From his speech on Humboldt's centenary, as quoted in the *New York Times,* September 15, 1869

Humboldt's quotation of his brother's passage on the rights of man: Humboldt, *Cosmos*

Humboldt's address to the German Association of Naturalists and Physicians, made September 18, 1828: Cornell University's Making of America Internet site (www. library5.library.cornell.edu/moa)

EPILOGUE: *Humboldt's Spirit*

Article on Humboldt's death: *New York Herald,* May 19, 1859

Article on Humboldt's death: *New York Tribune,* May 19, 1859

Article on Humboldt's death: *New York Evening Post,* May 19, 1859
Article on Humboldt's centenary: *New York Times,* September 15, 1869
Some possible reasons for the decline of interest in Humboldt: J. Wilson, Kellner, and the suggestions of John Edwards
Final Humboldt quotation: Kellner

APPENDIX I

Other Works of
Alexander von Humboldt

Below are important works of Humboldt in addition to those listed in the Bibliography. The title appears in the language in which the work was first published.

Florae Fribergensis specimen plantas cryptogamicus praesertim subterraneas exhibens, 1793.

Humboldt's first significant publication, including observations and experiments made on underground plants while he was a mining inspector in Germany; Humboldt argues that plants should not be studied in isolation but as an integral part of the environment in which they are found.

Versuche über die gereizte Muskel—und Nervenfaser nebst Versuchen über den chemischen Prozess des Lebens in der Thier—und Pflanzenwelt (2 volumes), 1797.

A report of Humboldt's experiments in galvanism and nerve conductivity; the work created an international sensation when it was released.

Essai sur la géographie des plantes, 1807.

The seminal work in the science of plant geography.

Des lignes isothermes et de la distribution de la châleur sur le globe, 1817.

The work that introduced isothermal lines; a cornerstone of climatology.

Voyage aux régions équinoctiales du Nouveau Continent, fait en 1799–1804 (With A. Bonpland; 30 volumes), 1807–39.

The most ambitious and expensive publication of its time, including all the scientific findings of Humboldt's American journey.

Examen critique de l'histoire de la géographie du nouveau continent et des progrès de l'astronomie nautique aux quinzième et siezème siècles, 1836–39.

A critical study of the early explorations of the Western Hemisphere; the work in which Humboldt explained how America got its name.

APPENDIX II

Places Named After
Alexander von Humboldt

CITIES, TOWNS, VILLAGES

Guevea de Humboldt, Oaxaca, Mexico
Humboldt, Arizona
Humboldt, Saskatchewan, Canada
Humboldt, Illinois
Humboldt, Indiana
Humboldt, Iowa
Humboldt, Kansas
Humboldt, Minnesota
Humboldt, Nebraska
Humboldt, South Dakota
Humboldt, Tennessee
Humboldt Hill, California

UNITED STATES COUNTIES

Humboldt County, California
Humboldt County, Iowa
Humboldt County, Nevada

BODIES OF WATER

Humboldt Bay, California
Humboldt Bay, New Guinea
Humboldt Current, Pacific Ocean
Humboldt Harbour, California
Humboldt Reservoir, Nevada
Humboldt River, Nevada

Humboldt Salt Marsh, Nevada
Humboldt Sink, Nevada
Little Humboldt River, Nevada

MOUNTAINS AND GLACIERS

Humboldt Glacier, Greenland
Humboldt Mountains, China
Humboldt Mountain Range, Antarctica
Humboldt Mountain Range, New Zealand
Humboldt Peak, Colorado
Humboldt Peak, Venezuela
Humboldt Range, Nevada

PARKS AND FORESTS

Alejandro de Humboldt National Park, Cuba
Humboldt Lagoons State Park, California
Humboldt Redwoods State Park, California
Humboldt-Toiyabe National Forest, Nevada

EXTRATERRESTRIAL FEATURES

Mare Humboldtianum, the Moon

INDEX

ACKNOWLEDGMENTS

FIRST, I'D LIKE TO THANK MY AGENT, DEIRDRE MULLANE at the Spieler Agency, who is a gifted editor and valued friend as well as a talented representative.

Thank you to my editor at Gotham Books, Brendan Cahill, for his confidence, enthusiasm, and deft editorial touch.

Thanks, too, to Patricia Gift and Kay McCauley for their early encouragement of this project, and to Andrés and Nelson de la Torre, for reacquainting me with Alexander von Humboldt.

I am indebted to the New York Public Library, the Ricks Memorial Library of Yazoo City, Mississippi, and the Biblioteca Pública de San Miguel de Allende, Mexico, of whose collections I made extensive use. Thank you as well to the many authors whose works I consulted in the course of my research; this book would not have been possible without theirs.

Thanks to Hugh Lacey of Swarthmore College for his thoughtful comments on portions of the manuscript relating to the history of science; to Frank Baron of the University of Kansas for reviewing passages on the life of Humboldt; to Timothy Hawkins of Indiana State University for his helpful critique of the sections on Latin American history; and to especially John Edwards of the University of Washington for his thoughtful reading of the entire manuscript. Any remaining errors are entirely my own.

I'd also like to thank my friends and family members, who sustained me during this project, including my mother, Marion Rehn, my brother, Bill Helferich, my sister, Marlene Bergendahl, and my mother-in-law, Florence Nicholas.

Finally, my deepest thanks to my beloved wife and first reader, Teresa Avila Nicholas, without whose faith and support this book would never have been begun, no less completed.